# FACING TO THE FUTURE
## —ARCHITECTURE TREND

# 面向未来
## ——建筑趋势2015

### 上册

龙志伟 编著
Edited by: Long Zhiwei

**创意新"空"**——建筑结构新模式
Innovated and new "space"—new model of architecture structure

**以"理"服人**——建筑理念新传承
Convinced "concept"—new heritage of architecture idea

广西师范大学出版社
· 桂林 ·

图书在版编目(CIP)数据

面向未来：建筑趋势 2015 / 龙志伟 编著. —桂林：广西师范大学出版社,2015.2
ISBN 978-7-5495-5679-3

Ⅰ. ①面… Ⅱ. ①龙… Ⅲ. ①建筑设计 Ⅳ. ①TU2

中国版本图书馆 CIP 数据核字(2014)第 152903 号

出 品 人：刘广汉
策　　划：君誉文化
责任编辑：肖　莉
装帧设计：龙志杰
广西师范大学出版社出版发行
（广西桂林市中华路 22 号　　邮政编码：541001）
（网址：http://www.bbtpress.com）
出版人：何林夏
全国新华书店经销
销售热线：021-31260822-882/883
上海锦良印刷厂印刷
（上海市普陀区真南路 2548 号 6 号楼　邮政编码：200331）
开本：646mm×960mm　　1/8
印张：68　　　　　　字数：30 千字
2015 年 2 月第 1 版　　2015 年 2 月第 1 次印刷
定价：568.00 元（上、下册）

---

如发现印装质量问题，影响阅读，请与印刷单位联系调换。
（电话：021-56519605）

# 序 Preface

未来的建筑会是什么样子?我们将会生活在怎样的环境之中?

如果你错过《建筑趋势2013》,那么你没有理由再错过本书——《面向未来——建筑趋势2015》。翻开此书,相信你将会在一系列真实详尽的建筑案例中,找到你想要的答案。

建筑行业经过历史的变迁发展,正逐渐以崭新的面貌焕然一新地呈现在我们面前。科技在进步,生活水平在提高,遵循着这一永恒不变的主题,人们对"未来建筑"也都有自己的看法。智能建筑、绿色建筑、仿生建筑相继应运而生,各家建筑设计公司也纷纷在追求效益的基础上,越来越重视建筑与环境共生、建筑与健康等问题。不论是可持续化、生态化、智能化等设计理念,还是从建筑材料、结构、外形等入手的设计方法,都旨在为人们创造一处真正的自然舒适的居住空间。

作为《建筑趋势2013》的续编,本书系统地网罗了来自NL Architects、Office for Metropolitan Architecture、C. F. Møller Architects、Philippe SAMYN and PARTNERS、 architects & engineers、AART Architects、加拿大CPC建筑设计顾问有限公司等全球范围内数十家一流设计单位的最新案例作品60个。不仅案例更丰富、资料更详尽,在案例分类上,我们也从建筑外形、材料、空间结构、技术、理念5个方面作了更科学、更切实的调整和追求。本书不仅可以带你一睹最具视觉标志性的建筑形态,感受最具独特创意的建筑结构空间,还能在3D建模技术、参数化设计等技术领域中领略最具魅力的建筑风采,在回归自然、绿色环保的建筑理念中触摸来自"雪松木"、"混凝土"的纯朴质感,回味来自"耐候钢"的沉淀余韵,被"幕墙玻璃"、各类"金属板"所折射出来的时尚现代光辉所折服。这不仅是一幅描绘未来建筑的蓝图,更是一本在细节处精雕细琢的图书珍品。

未来的建筑会是什么样子?当你翻开第一页起,我相信,就是你所看到的样子。

What architecture looks like in the future? What environment would we live in?

If you have missed *Architecture Trend 2013*, you have no reason to miss this one – *Facing to the future-Architecture Trend 2015*. Once you open this book, you will find the answers you want with a series of real and elaborate architectural cases.

After years of transition and development, the construction industry is taking on a brand new look. Science is in progression and people's living standard is improving. Following this eternal topic, people have their own views on "future architecture". "Smart architecture", "green architecture" and "bionical architecture" are born one after another. On the basis of pursuing profit, he architectural design companies are attaching more and more importance to the existence of architecture and environment, as well as relationship between architecture and health. Whether the concepts of sustainability, ecologicalization and intelligence, or the design solutions of building material, structure and shape are all aimed at creating a truly natural and comfortable living space for people.

As the continuation of *Architecture Trend 2013*, this book systematically features the latest 60 works by tens of first-class design units home and abroad including NL Architects, C. F. Møller Architects, Philippe SAMYN and PARTNERS, architects & engineers, AART Architects. Besides the more abundant cases and elaborate statistic, we also adjust the classification more scientifically and faithfully from the aspects of building shape, material, spatial structure, technique and concept. This book not only provides you with the most iconic building form, the architectural structure of the most innovative concept, but also make you have a taste of the most charming building style by the 3D modeling technique, parametric design, feel the simplicity of "cedar" and "concrete" in the concept of returning to nature and environmental protection, chew the fantasy of "Cor-Ten steel" and admire the modern fashion of "glass curtain wall" and "metal plates". It is not only a blue print of future architecture, more a carefully-carved and detailed book treasure.

What architecture looks like in the future? Once you open the first page of this book, you will be informed.

# 目录 Contents

## 8 创意新"空"
## ——建筑结构新模式
## Innovated and new "space"
## —new model of architecture structure

| | | |
|---|---|---|
| 10 | 美国纽约HL23公寓大厦 | |
| | HL23 | |
| 18 | 翡翠公寓 | |
| | Jade Apartments | |
| 24 | Zoey | |
| | Zoey | |
| 30 | 英国伯明翰图书馆 | |
| | Library of Birmingham, UK | |
| 42 | 德国耶拿Göpel电子有限公司研发大楼 | |
| | Göpel GmbH, New Construction of a Research and Development Building | |
| 48 | 里约热内卢MAR——艺术博物馆 | |
| | MAR—Art Museum of Rio de Janeiro | |
| 60 | 奥地利格拉茨NIK办公大楼 | |
| | NIK Office Building | |
| 66 | 荷兰阿姆斯特丹Furore公寓大楼 A栋 | |
| | Furore – Block A | |
| 74 | 瑞士苏黎世Cocoon | |
| | Cocoon | |
| 84 | 德国斯图加特奔驰博物馆 | |
| | Mercedes-Benz Museum | |

| | | |
|---|---|---|
| 90 | 澳大利亚悉尼Blakehurst俱乐部<br>The Blakehurst Club | |
| 98 | 昂赞图书馆<br>Library in Anzin | |
| 106 | 俄罗斯莫斯科电视秀亭台<br>Gazebo for TV Show | |
| 112 | 德国纽伦堡Sipos Aktorik GmbH生产车间<br>A Production Hall of a Sipos Aktorik GmbH | |
| 122 | 巴西里约热内卢艺术城<br>Cidade das Artes | |
| 132 | 巴西圣保罗Alphaville俱乐部<br>Alphaville Club | |
| 138 | 荷兰阿尔梅勒羊厩<br>Sheep Stable Almere | |
| 146 | 法国巴黎Asnieres-Sur-Seine学校体育馆<br>Gymnase Scolaire, Asnieres-Sur-Seine | |
| 152 | 荷兰阿姆斯特丹A8高速路公园<br>A8ernA | |
| 162 | 2015年世博会奥地利馆<br>Austrian Pavilion in Expo, 2015 | |
| 166 | 广州杨家声设计顾问有限公司办公室<br>Ben Yeung Office | |

# 174 以"理"服人
## ——建筑理念新传承
### Convinced "concept"
### —new heritage of architecture idea

**176** 韩国昌原国立庆尚大学医院
Changwon Gyeongsang National University Hospital

**186** 绿谷
Green Valley

**192** 韩国大邱图书馆
Daegu Gosan Public Library

**198** 因斯布鲁克办公楼
Office Building in Innsbruck

**210** 苏州中吴红玺
Suzhou Zhongwu Hongxi

**216** 意大利曼图亚多功能综合体
Multi-purpose Complex DUNANT

**222** 北京中国国家美术馆
National Art Museum of China

**228** 内蒙自治区鄂尔多斯基督教堂
Ordos Protestant Church

**236** 德国慕尼黑宝马办公大楼
BMW Office Building

**242**　美国加利福尼亚州湖畔工作室
　　　　Lakeside Studio

**246**　荷兰费嫩达尔Panorama社区中心
　　　　Panorama

**254**　英国Birnbeck岛"村庄"
　　　　Birnbeck Island Village

**262**　Maat-Pita Gaudham寺庙综合楼
　　　　Maat-Pita Gaudham Temple Complex

# 创意新"空"
## ——建筑结构新模

# Innovated and new "space"
## —new model of architecture structure

如果说把建筑外形比作一件"外衣"的话，那么建筑结构则是构成整个建筑的"骨骼"。再美的外形，没有一个坚实、可靠的骨架作支撑，也无法展现其美好的姿态。总体来说，近代建筑的主体结构大体上经历了三个发展阶段：砖（石）木混合结构、砖（石）钢筋（钢骨）混凝土混合结构以及钢和钢筋混凝土框架结构，而便于施工、安全、经济、适用和美观，则是支撑建筑结构不断更新、不断发展的五大原则。

在现代建筑学中，建筑结构设计作为建筑整体设计的一个基础，在某种程度上决定着建筑设计能否实现。因此，如何把握建筑结构的模式，变得尤为重要。

首先，高层建筑结构是中外所有建筑发展的大趋势。一个地区建筑的兴建与当地经济的发展有着密切的关系，而高层建筑往往是社会需求、经济发展的产物。这是因为，高层建筑不但能更有效地利用土地资源，提高空间的利用率，而且还具有体量大、视觉冲击力强、标志性强等特点。本章节中，英国伯明翰图书馆采用"建筑体快叠加"的方式提升建筑高度，德国耶拿电子有限公司研发大楼借助"楼层的交错式排列"拔高建筑，而瑞士苏黎世 Cocoon 则采用螺旋式上升结构，这些措施都是为了最大化地利用上升空间，进而增加建筑使用面积。

其次，随着物质水平的提高，民众的审美要求也在不断提升，因此，建筑结构在美学方面的利用与涉及也在逐步加深。追求美好的事物是人的天性，作为影响建筑结构发展的一个隐性因素，美学将会在未来的建筑设计中占据越来越重的分量。

此外，在建筑结构的发展过程中，越来越多的设计师开始注重对建筑屋顶的利用。现代建筑的屋顶不再像过去那样，利用砖瓦进行覆盖，大多采用平顶的设计方式。设计师们注重对屋顶空间的利用，一方面体现了可持续发展的理念，将建筑可利用的空间面积最大化，另一方面，也充分体现了现代建筑结构的美学以及地区文化魅力。本章节中的澳大利亚悉尼 Blakehurst 俱乐部利用 3D 数字建模软件打造几何体平坦屋顶，通过抬升屋顶，在最大程度上发挥建筑张力，进而形成了一个与户外相连的明亮又通风的空间格局；巴西圣保罗 Alphaville 俱乐部是一个大型综合俱乐部，活动频繁，来往人员较多，其中，俱乐部的网球场坐落在与该建筑相连的较低的地块上，因此，该屋顶被设计成观看网球比赛的观众席，居高临下的设计，便于网球爱好者及时捕捉到运动员在球场上的每一个动感的瞬间；而德国纽伦堡 Sipos Aktorik GmbH 生产车间则因其工业背景，采用屋顶天窗的设计，整个屋顶造型就像一个酒吧，除了能使车间得到丰富的自然采光外，还有利于缓解工人们的工作压力，激发工作热情。

If the architecture appearance is compared to a "coat", the architecture structure will be the "skeleton" constituting the whole building. No matter how beautiful shape, without a solid and reliable framework as support, can not show its wonderful posture. Overall, the main structure of modern architecture generally has experienced three development stages: brick (stone) and wood hybrid structure, brick (stone) and rebar (steel ribs) concrete composite structure, and steel and reinforced concrete frame structure. Moreover, the five principals for the building structure to be constantly updated and developed are the convenient construction, safety, economy, application and beauty.

In modern architecture, the architecture structure design, as a basis for the overall design of the building, determines that whether the design can be achieved in a way. Therefore, it is particularly important that how to grasp the architecture structure model.

First, the high-rise building structure is a general trend both in the domestic and foreign building development. The build of region architecture is closely with the local economic development, and the high-rise buildings are often the product of social needs and economic development. This is because the high-rise buildings are not only more efficient to use of land resources, and improve the utilization of space, but also have a large volume, strong visual impact, prominent signs and other characteristics. In this chapter, like Birmingham library uses the method of "stacking building volumes" to enhance the building height; R & D building in Germany Jena Electronics Co., Ltd. is with the "floor staggered arrangement" to overstate the building; while Zurich Cocoon adopts spiral structure, these measures are designed to maximize the use of rising space so as to increase the usable building area.

Second, with the improvement of the material level, people's aesthetic requirements are also rising, so the building structure is gradually deepened to use and involve in the aesthetic aspects. To pursue the good things is human nature, so the aesthetics will occupy an increasingly important component in the future architectural design as an invisible factor affecting the development of building structure.

In addition, in the process of the development of the architecture structure, more and more designers begin to focus on the building roof utilization. Modern building roof is no longer as the past to use bricks for covering, most of the roofs use flat-top design. Designers also focus on the roof space utilization, which on the one hand reflects the concept of sustainable development to maximize the available space area of the building, on the other hand, also fully embodies the aesthetics of modern architecture as well as and the charms of regional culture. Taking this chapter for example, Blakehurst Club in Sydney, Australia is lifted the roof with 3D digital modeling software to create a flat geometry roof and plays building tension to maximum extent, so there forms a bright and airy space pattern connecting with the outdoor; Sao Paulo, Alphaville Club is a large comprehensive club with frequent activities and many people, and is located on the lower plot which is connected with the building, so the roof is designed to be the spectator seats for watching a tennis match. The condescending design is convenient for tennis enthusiasts to timely capture every dynamic moment from the athletes on the field; meanwhile, the workshops of Sipos Aktorik GmbH in Nuremberg, Germany are adopted the design of the roof skylight due to its industrial background. The entire roof shape is like a bar, which besides makes the workshops get rich natural lighting, and helps people ease the work pressure and stimulate the enthusiasm for work.

# 美国纽约 HL23 公寓大厦

## HL23

设计单位：Neil M. Denari Architects
项目地址：美国纽约

Designed by: Neil M. Denari Architects
Location: New York, U.S.

**项目概况**

HL23是一栋14层的公寓大厦，其独特的狭窄场地与纽约West Chelsea艺术区第23街High Line高线花园相邻。大厦充满现代感和未来感的14层空间，重新定义着这一街区的形象。

**规划设计**

从狭窄的场地条件来看，采用灵活的几何体才能在紧挨着高线公园的场地上形成相对宽敞的建筑。这样的要求使得建筑每层设计一个单元，主居住区由9个单层单元和2个双层单元构成。这些拥有三个独特、连贯立面的单元，正是曼哈顿砌体结构中的珍品。

**建筑设计**

建筑朝向高线公园的西立面拥有雕刻般的表面，小型的窗户设计既保证了住户隐私，又可观看到曼哈顿的优美景色。玻璃幕墙和巨大的不锈钢板覆盖在复杂的悬臂式钢架上，形成了一个极具标志性的建筑立面。

建筑的机械美学无处不在，钢骨交叉支架在玻璃窗上纵横交织着，只需按个钮就能控制机械让玻璃墙面往外滑开。而那些由专门制造卡车车身的钣金冲压厂打造出来的钢板，使整座建筑在阳光下绚烂夺目的同时，还宛如拥有发达肌肉般的漂亮线条，像极了一辆刚毅劲酷的意大利跑车。

Sculptural Façade
Mechanical Aesthetics
Abundant Inner Spaces

雕刻般立面
机械美学
丰富内部空间

**Site Plan**
总平面图

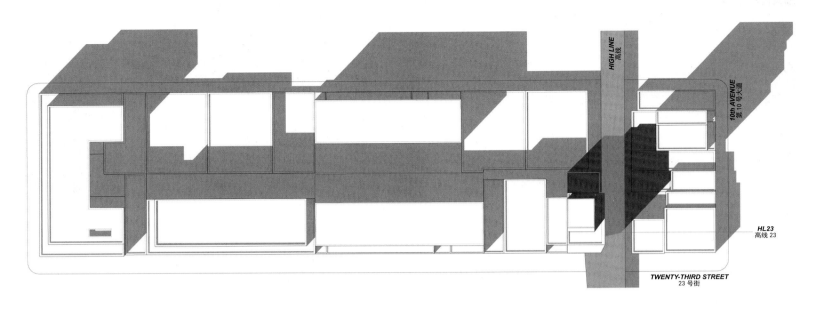

01. **LIVING/DINNING ROOM**
01. 客厅 / 餐厅
02. **KITCHEN**
02. 厨房
03. **BEDROOM**
03. 卧室
04. **POWDER ROOM**
04. 卫生间
05. **BATHROOM**
05. 浴室

01. **ENTRY HALL**
01. 门厅
02. **LIVING / DINNING ROOM**
02. 客厅 / 餐厅
03. **KITCHEN**
03. 厨房
04. **BEDROOM**
04. 卧室
05. **POWDER ROOM**
05. 卫生间
06. **BATHROOM**
06. 卧室

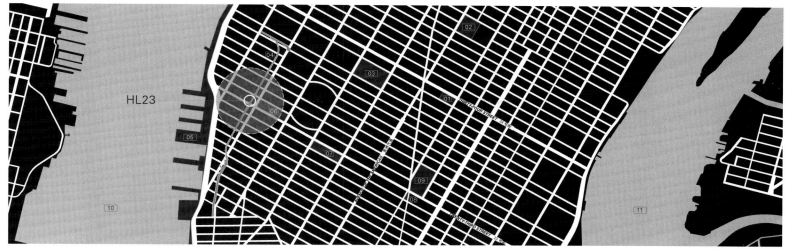

**01. EMPIRE STATE BUILDING**
01. 帝国大厦
**02. BRYANT PARK**
02. 布莱恩特公园
**03. MADISON SQUARE GARDEN**
03. 麦迪逊广场花园
**04. HIGH LINE**
04. 高线
**05. CHELSEA PIERS**
05. 切尔西码头
**06. LONDON TERRACE**
06. 伦敦露台
**07. HOTEL CHELSEA**
07. 切尔西酒店
**08. FLATIRON BUILDING**
08. 熨斗大厦
**09. MADISON SQUARE PARK**
09. 麦迪逊广场公园
**10. HUDSON RIVER**
10. 哈德逊河
**11. EAST RIVER**
11. 东河

*High Line Plan*
高线平面图

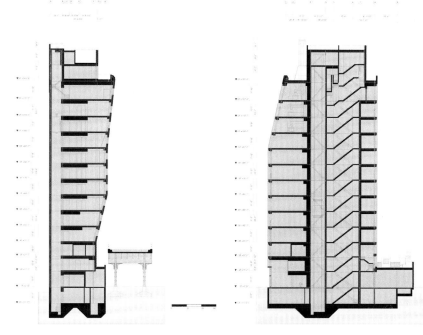

### Profile
HL23 is a 14-floor condominium tower that responds to a unique and challenging site directly adjacent to the High Line at 23rd Street in New York's West Chelsea Arts district.

### Planning Design
For the narrow site, a supple geometry must be found to allow a larger building to stand in very close proximity to the elevated park of the High Line. With one unit per floor and three distinct yet coherent façades, the building consists of nine single floor units and two duplex units, a rarity in Manhattan's block structure.

### Architectural Design
The east façade facing the High Line is formed as a sculptural surface with smaller windows allowing privacy and framed views across Manhattan. Glass curtain wall and stainless steel mega-panels hang on a complex cantilevered steel frame, creating an iconic building façade.
Mechanical aesthetics is reflected throughout the building. Steel cross brackets interweave on the glazed window; only by pushing a button, one can control the machinery to make the glass wall slide outwards. The steel plates made by stamping factory of truck make the whole building shine brightly under the sun, as if it has muscle-like lines. The whole building looks like a firm and cool Italian sports car.

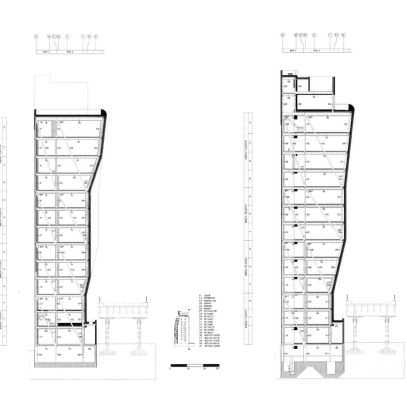

◀ 内部空间——这座建筑奇特的外形也创造了饶有趣味的内部空间变化。较低楼层可以透过像是巨型挡风玻璃的落地窗欣赏到高线花园的各式风景，而从面向西南方的四楼起居室中，可以直接俯瞰室外美景，拥有完全不被阻挡的绝佳视野。在随着建筑物高度增加而扩展的各个楼层里，人们的心情也会随之而改变。

Inner space—the peculiar shape also creates interesting and abundant inner space variation. Through the large French window of windshield, the lower floors own the view of High Line. The living room of fourth floor facing south directly overlooks outdoor interrupted scenery. In the various expanded levels with increase of heights, people's mood would get better.

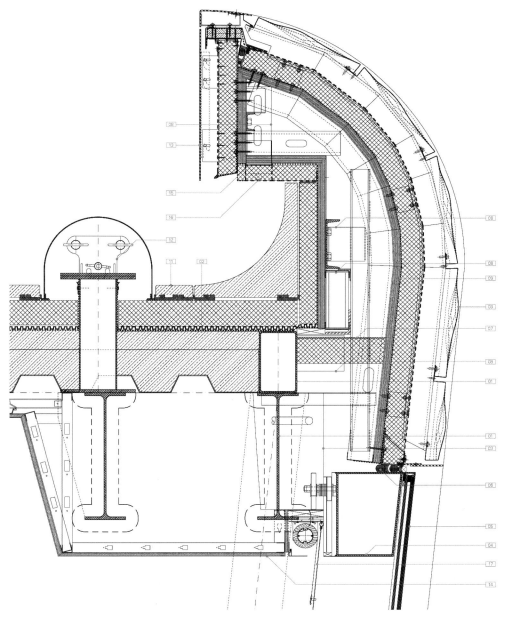

01. **PRIMARY STRUCTURAL STEEL (HHS220×12×5/8) WITH CEMENTITIOUS FIREPROOFING**
01. 主钢结构（HHS220×12×5/8）与胶结防火材料
02. **3"×18 GA. COMPOSITE METAL DECK WITH 4" NORMAL WEIGHT CONCRETE; WITH CONCRETE TOPPING SLOPED TO DRAIN**
02. 3*18 GA. 采用4" 普通混凝土复合金属甲板；用混凝土倾斜的顶部排水
03. **CURTAIN WALL MEGAPANEL SUPPORT BRACKET AND GRAVITY ATTACHMENT ANCHOR**
03. 幕墙拼图方块的支撑架和重力附件锚
04. **STEEL SPREADER BEAM**
04. 钢梁分布
05. **INSULATED GLASS UNIT: 8 HS LOW IRON + SILK SCREEN + 1.52 PVB + 6 HS CLEAR + PLANITHERM + DARK GREY TECHNOFORM WITH AIR + 10 FT LOW IRON**
05. 绝缘玻璃装置：8HS 低铁＋丝网＋1.52 聚乙烯醇缩丁醛＋6HS 空隙＋ PLANITHERM 玻璃＋深灰色泰诺风与空气＋3.048m 低铁
06. **BAKER ROD AND SEALANT**
06. 贝克杆和密胶
07. **4" MINERAL WOOL FIRESTOP AND SMOKE SEALANT**
07. 4" 矿物棉防火和排烟密封胶
08. **FIELD FRAMING**
08. 领域框架
09. **PARAPET MEGAPANEL WITH 4" 16 GA STUD (16" O.C.), 2×5/8" LAYER OF EXTERIOR GRADE GYPSUM BOARD, NON-PERMABLE AIR/VAPOR/RAIN BARRIER MEMBRANE AND 3 1/2" THERMAL INSULATION HOLD BY 4" ALUMINUM STICK PIN (5 PINS PER 24"×48" PIECE OF INSULATION)**
09. 拼图方块护墙采用 4" 16 GA 螺栓（16" O.C.），2×5/8" 外墙等级石膏板，非渗透空气／蒸汽／雨屏隔膜和通过 4" 铝条形襻针（绝缘每 24"×48" 块 5 个襻针）的 3 1/2" 持续保温
10. **RAIN SCREEN SYSTEM WITH 12'LONG STAMPED STAINLESS STEEL PANELS, 3" AIR SPACE, PANELS SUPPORTED ON PURLIN AND HAT CHANNEL**
10. 12' 长的冲压不锈钢面板的雨屏幕系统，3" 领空，桁条和帽子通道的木板支撑
11. **ROOFING SYSTEM. 2-1/2" WHITE CONCRETE PAVER, FILTER FABRIC, 2 LAYERS OF 2" DRAINAGE MAT AND HOT RUBBERIZED ASPHALT ROOF MEMBRANE. CUSTOM CURVED CONCRETE PAVER**
11. 屋面系统。2-1/2" 白色混凝土摊铺机、过滤布、2 层 2" 排水垫与热橡胶沥青顶薄膜。自定义弧形混凝土摊铺机
12. **PERMANENT DAVIT PEDESTAL WITH CUSTOM STAINLESS STEEL COVER**
12. 永久吊基座与定制的不锈钢盖
13. **INNER PARAPET WITH 4MM CLEAR ANODIZED ALUMINUM CLADDING, 3" AIR SPACE, BREATHABLE MEMBRANE, 3" THERMAL INSULATION, SELADHESIVE AIR/VAPOR/RAIN BARRIER MEMBRANE, 2 LAYERS OF 5/8 GLASROC, ALUMINUM PANEL SUPPORT 3" VERT. 16 GA HAT CHANNEL AT EACH PANEL SUPPORT**
13. 内护墙采用 4 毫米的表面阳极处理铝合金覆层，3" 领空、透气、3" 保温、SELADHESIVE 气体／蒸汽／雨屏隔膜、5/8 GLASROC 两层、3" 铝面板支撑的 16 GA 帽子通道
14. **MOLDERSISTANT DRYWALL CEILING**
14. 石膏天花板
15. **OPERABLE AIR VENT 4"× 8", 2 AT EACH PARAPET MEGAPANEL**
15. 可操作的通气孔 4" × 8"，每个拼图方块栏杆 2 个
16. **PROVISION FOR LED STRIP-LIGHT WITH ALUMINUM MOUNTING CHANNEL**
16. 用于 LED 灯光，灯带用铝安装通道
17. **MOTORIZED SOLAR SHADE SYSTEM WITH STANDARD 3" WHITE CLOSURE PLATE; SPRING LOADED TENSION CABLE AND HEMBAR SEALED INSIDE SHADECLOTH**
17. 带标准 3" 白色封板的机动太阳能遮阳系统；弹簧张力电缆和杂封的内密封

# 翡翠公寓
## Jade Apartments

项目地址：悉尼

**Location: Sydney**

非公式化
设计
玻璃幕墙
建筑

Non-formula
Design
Glass Curtain Wall
Construction

**项目概况**

翡翠公寓大楼坐落在悉尼中央商务区，建筑楼高15层，包含27个一、二、三居室的公寓，以及一个拱形的玻璃天花板的双层阁楼。建筑师们通过使用非公式化的设计方法，以丰富生活体验。他们认为，生活环境应该涉及居民的情绪，或者至少提供能够吸引各种生活方式的多样化环境，而不是他们通常给予的平淡无奇、公司化的环境。

**建筑设计**

该建筑是朝北向的，而且它有一个全釉面外观，大型金属穿孔百叶叶片和穿孔纱网的设计让屋主能够自主地控制阳光、美景和隐私。该玻璃幕墙建筑拥有一个转瞬即逝的特性，随着不断通向天空，变得越来越透明，体量不断解构，最终以一个拱形的透明屋顶结束该玻璃幕墙建筑拥有一个转瞬即逝的特性，随着不断通向天空，变得越来越透明，体量不断解构，最终以一个拱形的透明屋顶结束。

该建筑坐落在城市历史的一部分，被4层高的砖石前仓库包围。该塔浮在一个拱形砖石裙楼之上，在视觉上也与毗邻的砖瓦建筑完美地结合在一起，并且呼应这个城市一部分的字符。它的大厅在同一个空间包含了住宅的门厅和咖啡厅/小酒馆，其中在入口有一个拱形天花板。翡翠公寓将外部引入，这个开放性的设计理念延伸到所有的阳台和生活区，由全高的玻璃作进一步的诠释。翡翠公寓的内部设计与结构化的线性元素和弧形玻璃幕墙概念相互影响，与公寓反映了城市生活方式的流动性。

**01** East Elevation
东立面
Sussex Street  Scale: 1:100
莎瑟街  比例: 1:100

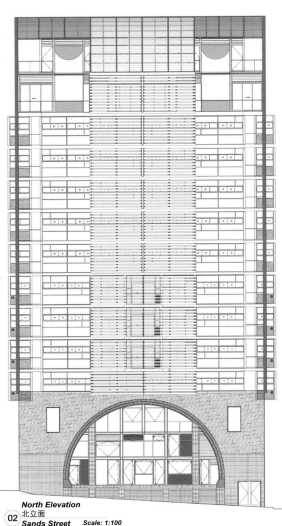

**02** North Elevation
北立面
Sands Street  Scale: 1:100
金沙街  比例: 1:100

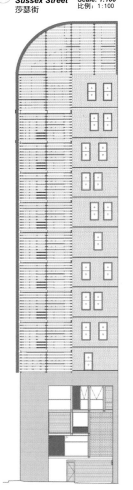

**01** West Elevation
西立面
Sands Street  Scale: 1:100
金沙街  比例: 1:100

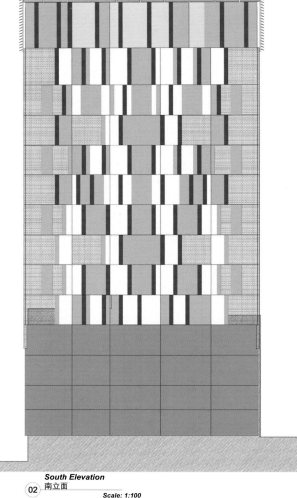

**02** South Elevation
南立面
Scale: 1:100
比例: 1:100

## Project Overview

The Jade Apartment tower is located in the center of Sydney's CBD. The 15-storey tower consists of 27 one, two and three-bedroom apartments, plus double-level penthouses with a vaulted glass ceiling. Architects are using a non-formula design approach to enrich the lifestyle experience, not sterilize it. Living environments should be able to relate to a resident's moods, or at least offer a diverse range of environments that appeal to the various lifestyle journeys, not the bland, corporatized environments they are often given.

## Building Design

The building is north facing yet it has a fully glazed façade. Large perforated metal louver blades and perforated mesh screens allow occupants to control sunlight, views and privacy. The result is a façade that becomes increasingly translucent towards the sky. The glass-walled building has an ephemeral quality which de-materializes as it rises, culminating in the vaulted see-through roof.

The building is located in a historic part of the city surrounded by 4 storey masonry former warehouses. The tower floats over an arched masonry podium which visually complements the adjoining brick buildings and echoes the character of this part of town. Its lobby contains the residential foyer and a cafe/bistro in the same space, which has a vaulted ceiling over the entry. Jade that brings the outside in. This design philosophy of openness extends to all balconies and living areas, which are full-height glass. The interior design of Jade interplays the structured linear elements and the curved glass façade concept, with apartments reflecting the flow of the city lifestyle.

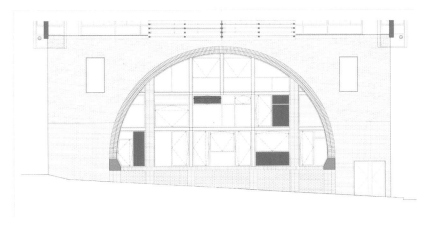

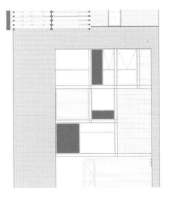

**Detailed Elevation**
细节立面图
*Sands Street North*
金沙街北

# Zoey
Zoey

**建筑师**: WTA Architecture + Design Studio
**开发商**: 甲米地理想国际建设开发总公司
**项目地址**: Noveleta
**建筑面积**: 45 101m²

*Architects: WTA Architecture + Design Studio*
*Client: Cavite Ideal International Construction and Development Corporation*
*Location: Noveleta*
*Area: 45,101m²*

## 项目概况

该项目位于Noveleta，该小镇拥有常住人口41 678人，距离甲米地经济特区1.5km，超过200个产业定位器。但是，时至今日，Noveleta依旧缺乏完善的城市中心，该地区的各种住宅小区零散地分布在城市边缘的各个角落，限制了该地区的经济发展。Zoey项目的建设，将为Noveleta打造一个密集的、多功能型的城市中心，它将为该地区的商业发展提供充足的活动空间，融合更多的资金，吸引更多的人流量。

## 建筑设计

Zoey包括水平地面的商铺、2层楼的排屋、模块化的公寓单位以及行政套房，这些房屋类型将以横纹肌理的方式设置，从而在最大化空间布局的同时，构建出一种理想化的情景模式。该项目还将涉及一个面积达10 000m²的庭院，这个庭院将被划成不同的公共空间，届时允许不同的居住人口和参观者之间的相互作用。在庭院周围，还会有一个盛大的草坪活动区、一个游泳池和设施区、运动区以及一个可容300人的顶级会所。除此之外，Zoey将有一个独特的多层轮廓包裹着这个庭院，其允许更好的空气流通，并从内外庭院有一个引人注目的景致。此衔接会形成峰值和骤降，为住宅创造更独特的空间。

Zoey庭院公寓创建了一个独特的住房结构，将为甲米地的常见城镇呈现城市生活的不同愿景。它是现代庭院的发展，将为Noveleta晋升为直辖市提供坚实的保证。

## Project Overview

The site is located in Noveleta a town of 41,678 people composed of largely residential subdivisions. It is 1.5km from the Cavite Special Economic Zone with over 200 industrial locators. Noveleta up to today lacks a proper town center and the various residential estates, restricting the economic development of the region. It is to be able form a denser more variated urban center for Noveleta. It will provide ample public space for the development of commercial spaces to attract ample foot traffic.

## Building Design

Zoey consists of walk-up ground-level shop houses, 2-storey row houses, modular apartment units, and executive suites. These housing types will be arranged in a striated manner allowing each type to be situated in its ideal level and maximize their location. The units will wrap around a vast pedestrian only courtyard of one hectare. This courtyard will be zoned into distinct public spaces that shall then allow for varying interactions between the different residential demographics and visitors. It will have a grand lawn activity zone, a pool and amenities zone, a sports zone, and a 300-capacity clubhouse. Wrapping around this courtyard, Zoey will have a distinct multi-storey profile which shall allow for better air flow and dramatic views from both inside the courtyard and outside. This articulation will form peaks and dips creating more unique spaces for the residences.

Zoey Courtyard Residence creates a unique housing typology that shall present a different vision of urban life for the common towns of Cavite. It is a modern courtyard development replete with incredible opportunities to further develop Noveleta as an up and coming municipality.

**New City Center
Striated Management**

新城市中心
横纹肌理

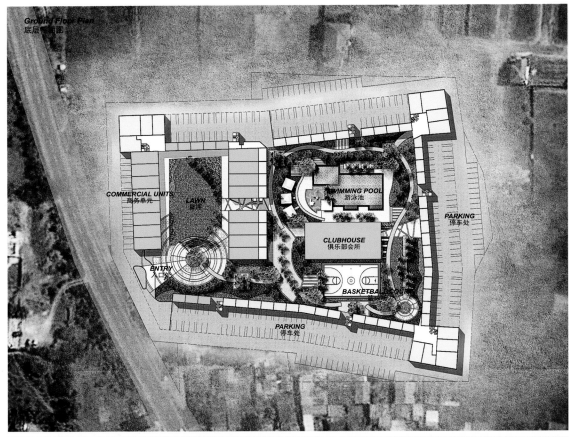

*Ground Floor Plan*
底层平面图

*2nd Floor Plan*
二楼平面图

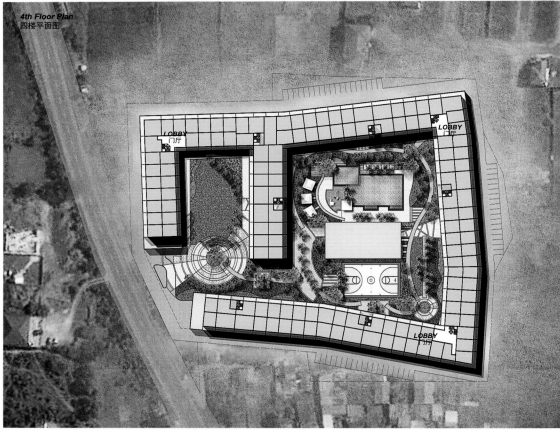

*4th Floor Plan*
四楼平面图

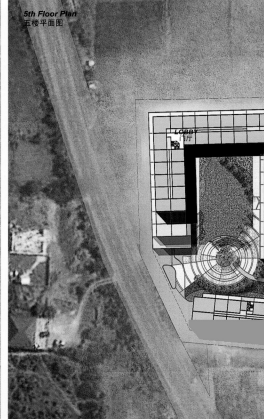

*5th Floor Plan*
五楼平面图

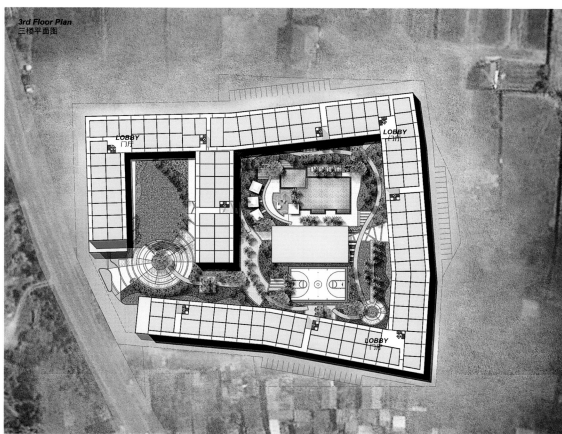

**3rd Floor Plan**
三楼平面图

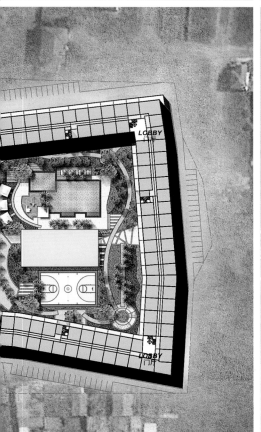

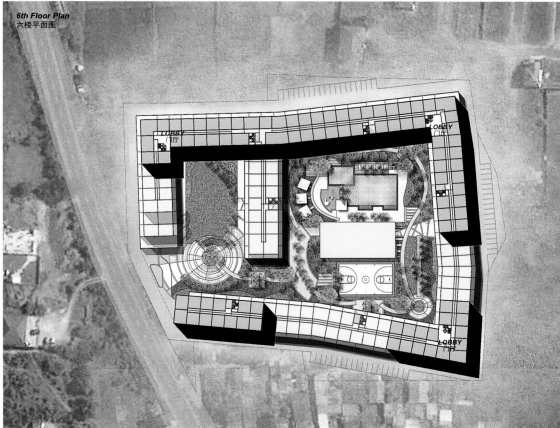

**6th Floor Plan**
六楼平面图

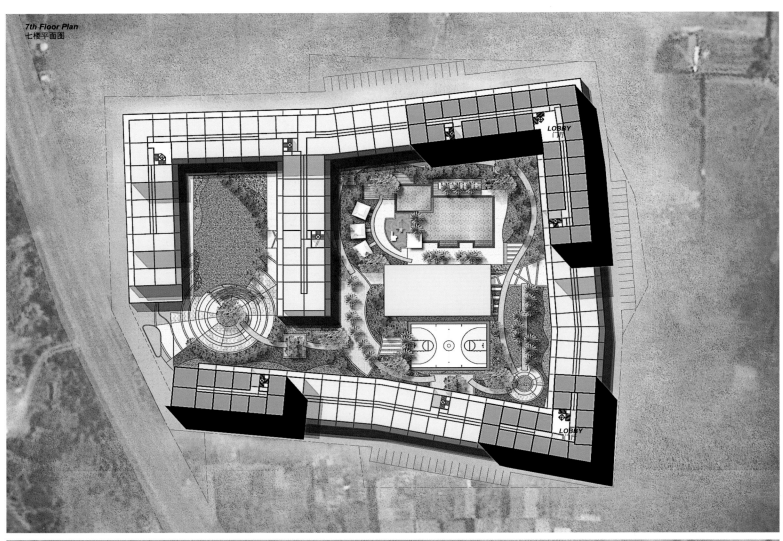

*7th Floor Plan*
七楼平面图

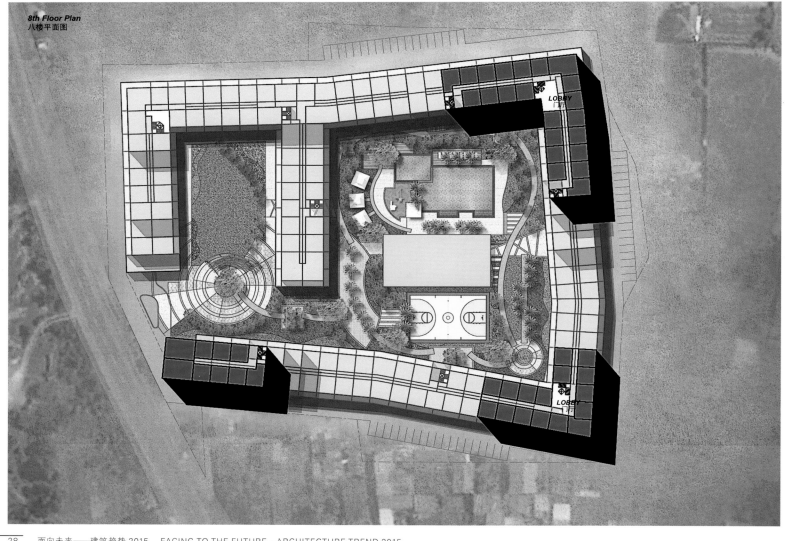

*8th Floor Plan*
八楼平面图

**Section**
剖面图

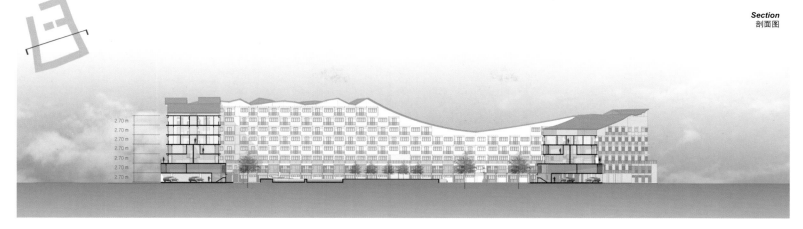

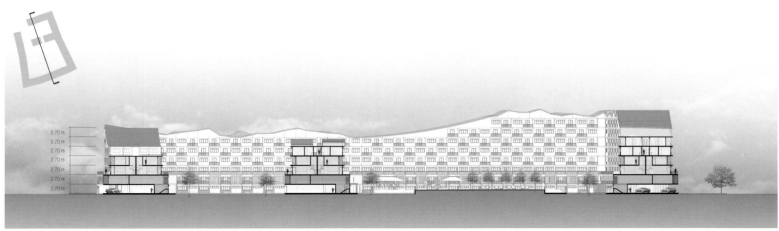

# 英国伯明翰图书馆
## Library of Birmingham, UK

设计单位：梅卡诺建筑事务所
开发商：伯明翰市议会
项目地址：英国伯明翰
建筑面积：35 000 m²
设计师：Mecanoo Architecten　Delft
摄影：Christian Richters

*Designed by: Mecanoo Architecten*
*Client: Birmingham City Council*
*Location: Birmingham, London*
*Building Area: 35,000m²*
*Designer: Mecanoo Architecten, Delft*
*Photography: Christian Richters*

叠加的
建筑体块
雕刻镂空的
建筑体面

Stacked
Building Block
Filigree
Building Decent

**项目概况**

英国伯明翰是一个拥有上百万人口的多元化城市，它具有多重身份，既包括文化的，也包括建筑方面；它是欧洲最年轻的城市，25岁以下的群体占到总人口基数的25%；它也是一个学生的城市，拥有50 000名学生，人数仅次于伦敦。伯明翰图书馆的筹建，将进一步丰富该地区的知识经济，它将被打造成为学习、信息和文化中心，成为连接该城市各年龄层、促进彼此交流的枢纽。在设计师们现代手法的演绎下，图书馆的功能不再仅局限于传授知识，它的影响力将超越其建筑的物理边界，得到进一步的扩张。

### 建筑设计

建筑由三个重叠堆放的体块构成，该设计不但为建筑本身增加了更多的层次感，而且通过体块的叠加，为图书馆创造了额外的户外花园空间，不仅丰富了建筑的软景观美感，而且花园的倾斜床与环绕伯明翰的平缓山坡相得益彰，彼此呼应。在伯明翰图书馆的入口处，图书馆的悬臂不仅是一个提供庇护的大棚，而且另外形成了一个宏伟的城市阳台，可欣赏广场上的活动和发生的事情。

伯明翰图书馆是一个透明的玻璃建筑，其精巧的雕刻镂空"皮肤"受到这个曾经的工业城市的工匠传统的启发。内部设计被形象地称为"街道的延伸"：人行道和自动扶梯动态地放置在图书馆的中心，在建筑物内的八个圆形空间之间形成连接。这些圆形大厅不仅在经由图书馆中起到了重要作用，而且还提供了自然采光和通风。作为屋顶鹰巢，其突出的地位使得可从广场看到这一微妙的房间。

伯明翰图书馆是环境评估法优秀评级的建筑，它采用了灰水系统和地源热泵技术。尽管图书馆是一个透明的建筑，但是，它通过建筑质量和前庭的缓冲能力维持能源效率。在下午的时候，外墙块内的遮阳物和反光材料在挡住恶劣的太阳光线的同时也允许自然光照射到内部。一楼受益于土壤的质量，它提供缓冲和隔离。圆形天井切出的方形创建一个受保护的室外空间，并使得日光深入建筑。该大楼将包括一个混合模式和自然通风的策略。外墙将响应外部条件，开口让新鲜空气进入和流出。软景观屋顶空间的加入将进一步提升建筑的功能条件。

## Project Overview

Birmingham is a multicultural British city of a little over one million people from very different backgrounds. It has many identities, both culturally and architecturally. It is not only Europe's youngest city with 25% of the population under 25 years old but it's also a student city with 50,000 students, second in student population only to London. The Library of Birmingham will become a centre of learning, information and culture that will help to foster Birmingham's knowledge economy. It is intended to become the social heart of the city; a building connecting people of all ages, cultures and backgrounds. The modern library is no longer solely the domain of the book—it is a place with all types of content and for all types of people. The library's influence will also extend beyond the physical boundaries of the building, its global digital presence allowing the public to access content from anywhere in the world.

## Building Design

The building consists of three stacked overlapping blocks, the design not only for the building itself adds more layering, and through the blocks' superposition, it additionally creates an opportunity for outdoor gardens to enrich the soft landscape of the building, while the sloping beds of the garden respond to the gentle slopes around Birmingham. The cantilever of the library is not only a large canopy that provides shelter at the common entrance of the Library of Birmingham and the REP, but additionally forms a grand city balcony with views of the events and happenings on the square.

The Library of Birmingham is a transparent glass building. Its delicate filigree skin is inspired by the artisan tradition of this once industrial city. The interior design is vividly called "street extension": travelers and escalators dynamically placed in the heart of the library forms connections between the eight circular spaces within the building. These rotundas play an important role not only in the routing through the library but also provide natural light and ventilation. Its prominent position as a rooftop aerie makes this delicate room visible from the square.

The Library of Birmingham is a BREEAM excellent rated building and incorporates grey water systems and ground source heat pumps. Although the Library is a transparent building, it maintains energy efficiency through the buffering capacity of the building mass and the atria. Sun shading and reflective materials within the façades block the harsh rays of the sun during the height of afternoon while allowing natural daylight into the interiors. The ground floor benefits from the mass of the soil which provides buffering and insulation. The circular patio cut out of the square creates a protected outdoor space and invites daylight deep into the building. The building will incorporate a mixed mode and natural ventilation strategy. The façade will respond to external conditions and openings will allow fresh air intake and outflow. The addition of soft landscaped roof spaces will further enhance the immediate surrounding conditions.

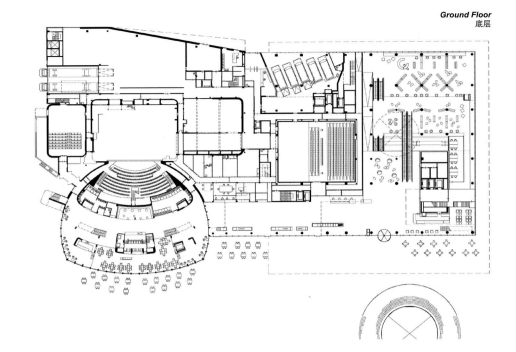

*Ground Floor* / 底层

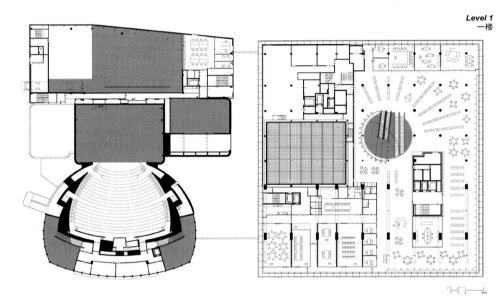

*Level 1* / 一楼

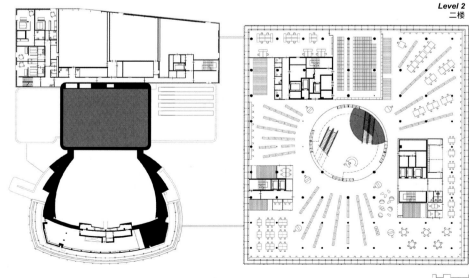

*Level 2* / 二楼

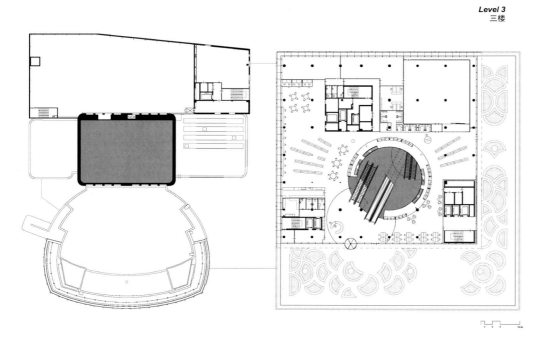

**Level 3**
三楼

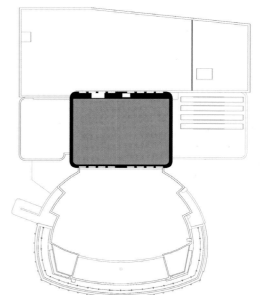

Level 4
四楼

Level 7
七楼

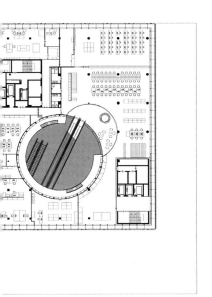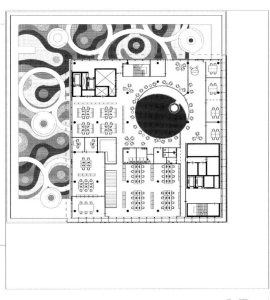

**Section**
剖面图

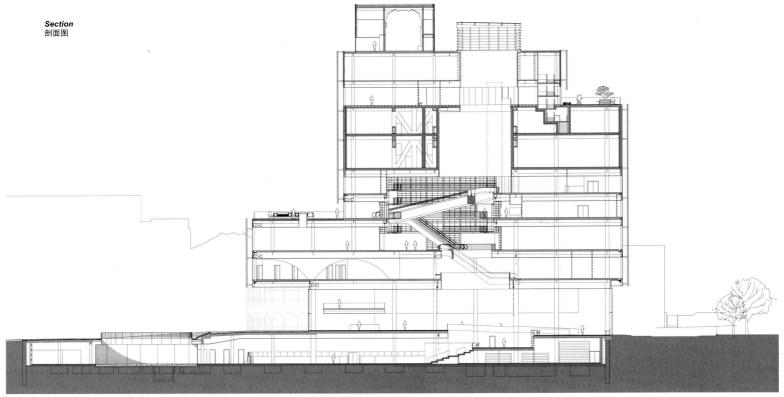

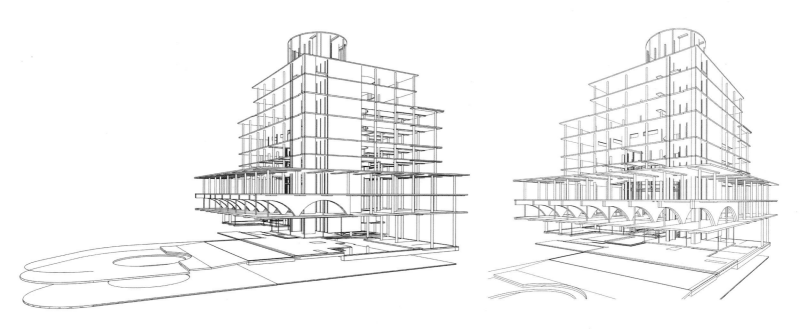

38　面向未来——建筑趋势 2015　FACING TO THE FUTURE—ARCHITECTURE TREND 2015

# 德国耶拿 Göpel 电子有限公司研发大楼
## Göpel GmbH, New Construction of a Research and Development Building

设计单位：wurm + wurm architekten ingenieure
开发商：耶拿 Göpel 电子有限公司
项目地址：德国耶拿
建筑面积：4 460m²
摄影：Ester Havlova

*Designed by: wurm + wurm architekten ingenieure*
*Client: Jena Göpel Electronics Co., Ltd.*
*Location: Jena, Germany*
*Building Area: 4,460m²*
*Photography: Ester Havlova*

石膏制冷天花板
通风片
错列式体量

Gypsum-Cooling-Ceiling
Ventilation-Flaps
Crosswise Layered Volumes

### 项目概况

这栋新的研发大楼将可提供 80 个工作岗位，它和同样由 wurm+wurm 设计的 Göpel 电子公司另两栋大楼处在同一小区，具有良好的交通优势。

### 建筑设计

由于建筑场地位于 Saale 的洪水多发区，设计将建筑结构进行抬升，并将停车场和庇护入口区设于建筑的下面。尽管建筑运用了复杂的几何体，但楼层平面还是以灵活的方式布置，第一层形成一块开阔的集合区，中间两层用作标准的功能空间，顶层则设有研讨室和经理办公室。

立面上，大块的金属板粘在可移动式轻金属组装框架上，形成建筑现代感十足的外观形象。另外，石膏制冷天花板能进行气候调节，供热装置可提供部分区域的供热，窗户边上的通风片还可提供自然通风，为建筑减少了大量的能源消耗。

◀ 错列式体量——建筑该体量的外观为 15m×40m 的矩形，是标准办公楼层的经典模式。楼层进行层层改造，每一层变成一个平行四边形，因此在顶层形成一个 X 形的结构。

层层堆叠的错列式体量由 16 根梯形柱子支撑着，形成一个垂悬部分，使消极空间形成了开阔的露台。

Crosswise layered volumes—the shape of the volume is developed from a rectangle with a size 15m x 40m, a classic pattern of a floor plan for standard offices. This floor plan is transformed from level to level. It becomes a parallelogram alternating on each level, so that by top view an X is created.

The staggered and crosswise layered volumes are supported 16 trapezoidal columns when there is an overhang. The negative-space which is generated by the delayed volumes is formed to open terraces.

## Profile

The new research and development building is designed for 80 further jobs. It is situated in the nearest neighbourhood of two other buildings which belong to Göpel electronic, also designed by wurm+wurm. There is excellent transport connection to motorway and railway.

## Architectural Design

Due to the situation of the construction site in the flood area of the Saale, it was necessary to elevate the structure. As it was impossible to reach the required number of parking places in the open space, the opportunity was provided to place the parking and a sheltered entrance area underneath the building. In spite of the buildings complex geometry the floor plan is usable in a flexible way. On Level 1 is seeded an open assembly area, the two levels in the middle are used as standard office-space and on the top level are sited seminar rooms and the executive.

On façade, big-sized metal sheets are fixed with adhesive-technology onto a movable assembled framework of light metal, so a modern image is created. Additionally, climate control is implemented by using gypsum-cooling-ceiling and heating-installation is applied by district-heating. Natural airing is affected by window-sided ventilation-flaps. Those measures save a lot of energy for the building.

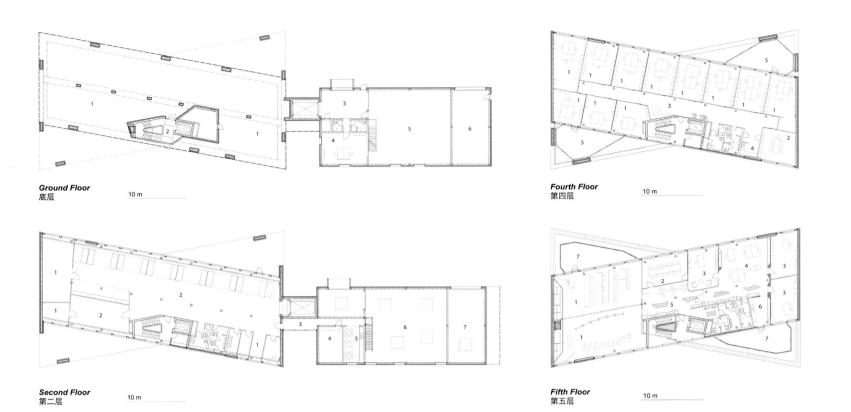

**Ground Floor**
底层  10 m

**Fourth Floor**
第四层  10 m

**Second Floor**
第二层  10 m

**Fifth Floor**
第五层  10 m

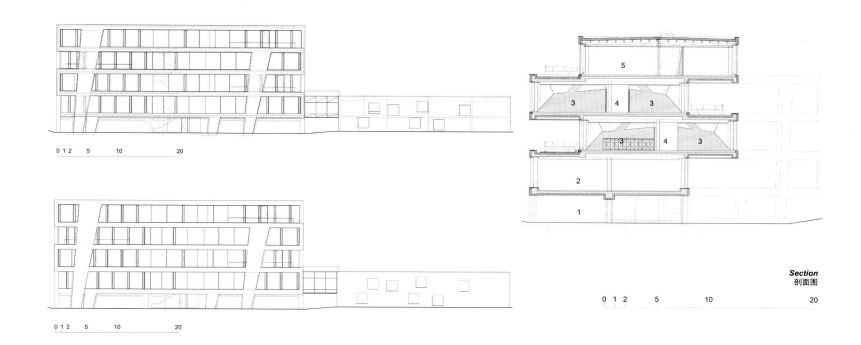

**Section**
剖面图

# 里约热内卢 MAR——艺术博物馆
## MAR—Art Museum of Rio de Janeiro

项目地址：巴西里约热内卢
建筑面积：11 240m²
摄影：Leonardo Finotti　Andres Otero

*Location: Rio de Janeiro, Brazil*
*Building Area: 11,240m²*
*Photography: Leonardo Finotti, Andres Otero*

**建筑概况**

该项目的挑战是要将三个拥有不同建筑特点的现有建筑进行合并,以容纳"里约热内卢博物馆"、"A Escola doolhar"学校以及周围的文化和休闲空间。

现有建筑、"Palacete Dom João"宫殿、警察大楼和里约旧的中心汽车站的连接,将成为里约热内卢具有历史意义的主要市区重建的一部分。

**建筑设计**

项目设计的第一步是建立一个流程制度,使博物馆和学校以集成和有效的方式来工作。因此,设计将在警察大楼的屋顶建立一个悬浮广场,这将结合所有的通道,并为文化活动和休闲度假集群一个酒吧区域。此外,警察大楼也将用于学校、礼堂、多媒体展区、行政管理区和综合楼员工区。

高晓,目前作为通至道路的通道,将变成整个综合楼的一个大门厅,并举办雕塑展览区。此通道将在两个建筑物之间进行控制,具有内置、开放和覆盖的空间特点。道路的选取框以及由该城市列出的传统元素将被用于厕所,仓库,装卸以及存储区域。

最后,作为该项目的主要标志,广场会有一个抽象的天线形式,流体和极轻的结构,类似水面波纹。一个建筑在充满了诗意的风格的同时,也是一个计算精确的现代化标志,这就是艺术博物馆所要传达的理念。

Building Merger
Cluster Reconstruction

建筑合并
集群重建

## Building Overview

Our challenge was to unite three existing buildings with different architectural characteristics to house the Museum de Arte do Rio, the school "A Escola do Olhar" as well as cultural and leisure spaces. The existing buildings, the palace "Palacete Dom João", the police building and the old central bus station of Rio, connected shall be part of the major urban redevelopment in the historic downtown of Rio de Janeiro.

## Building Design

The first step of the design was to establish a flow system allowing the Museum and school to work in an integrated and efficient manner. Therefore we proposed the creation of a suspended square on the police building rooftop, which will unite all accesses and host a bar and an area for cultural events and leisure. Consequently, the visitation will be from top to bottom. It was established that the palace, due to its large ceiling height and structure free plan should hold the exhibition areas of the museum.

The stilts, currently used as an access to the road, will turn into a large foyer for entire complex, and will hold the sculpture exhibition areas. Access will be controlled between the two buildings, characterizing this empty space as internal, open and covered. The marquee of the Road, heritage element listed by the City, will be used for lavatories, store and region of loading, unloading and deposits.

Finally as the main mark of the project, we suggested that the suspended square have an abstract and aerial form. A fluid and extremely light structure, simulating water surface waves. A poetic architectural character full of meaning, simple and at the same time modern in regards to the structural calculation, which is the concept the Art Museum conveys.

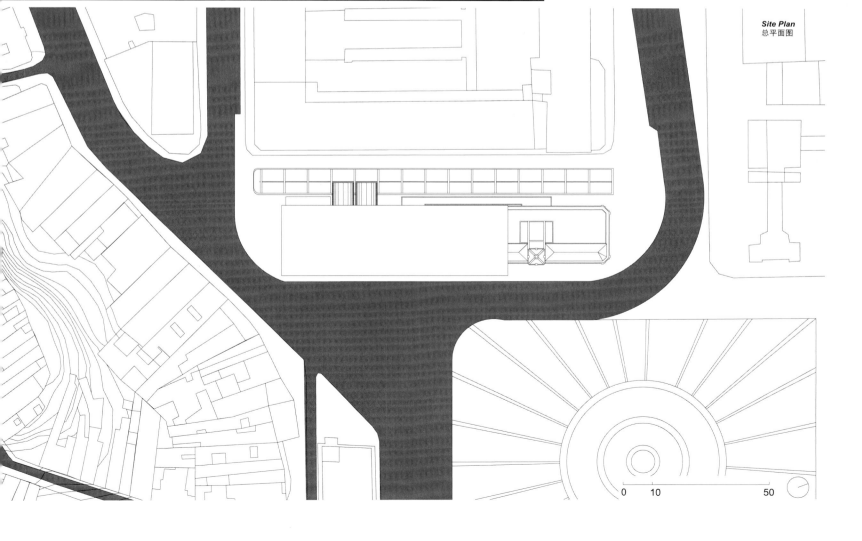

Site Plan
总平面图

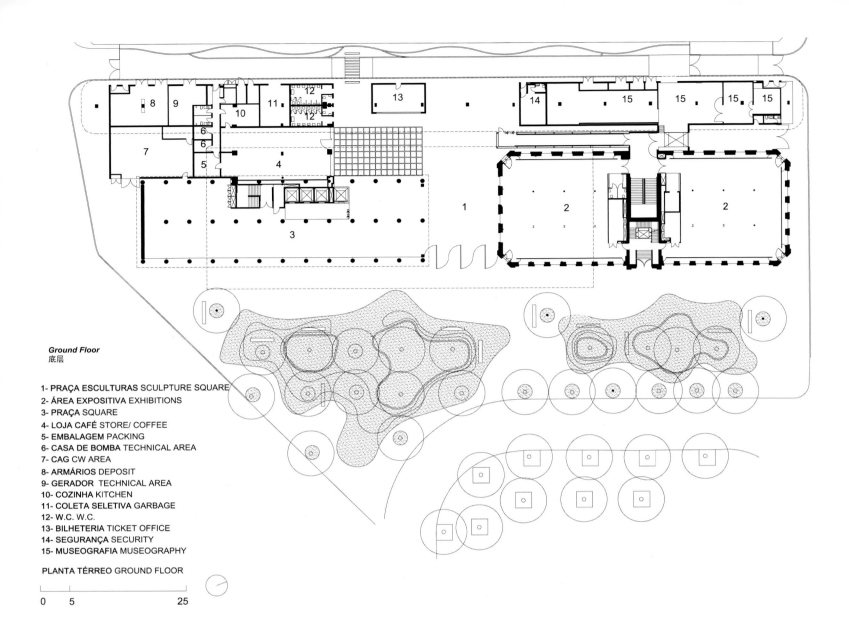

*Ground Floor*
底层

1- PRAÇA ESCULTURAS SCULPTURE SQUARE
2- ÁREA EXPOSITIVA EXHIBITIONS
3- PRAÇA SQUARE
4- LOJA CAFÉ STORE/ COFFEE
5- EMBALAGEM PACKING
6- CASA DE BOMBA TECHNICAL AREA
7- CAG CW AREA
8- ARMÁRIOS DEPOSIT
9- GERADOR TECHNICAL AREA
10- COZINHA KITCHEN
11- COLETA SELETIVA GARBAGE
12- W.C. W.C.
13- BILHETERIA TICKET OFFICE
14- SEGURANÇA SECURITY
15- MUSEOGRAFIA MUSEOGRAPHY

PLANTA TÉRREO GROUND FLOOR
0  5    25

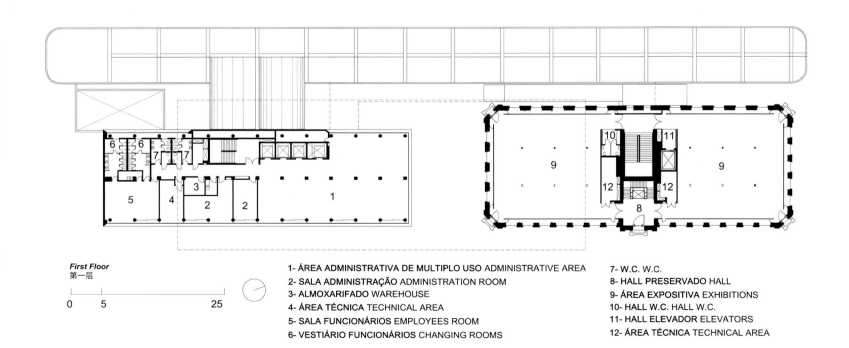

*First Floor*
第一层

0  5    25

1- ÁREA ADMINISTRATIVA DE MULTIPLO USO ADMINISTRATIVE AREA
2- SALA ADMINISTRAÇÃO ADMINISTRATION ROOM
3- ALMOXARIFADO WAREHOUSE
4- ÁREA TÉCNICA TECHNICAL AREA
5- SALA FUNCIONÁRIOS EMPLOYEES ROOM
6- VESTIÁRIO FUNCIONÁRIOS CHANGING ROOMS
7- W.C. W.C.
8- HALL PRESERVADO HALL
9- ÁREA EXPOSITIVA EXHIBITIONS
10- HALL W.C. HALL W.C.
11- HALL ELEVADOR ELEVATORS
12- ÁREA TÉCNICA TECHNICAL AREA

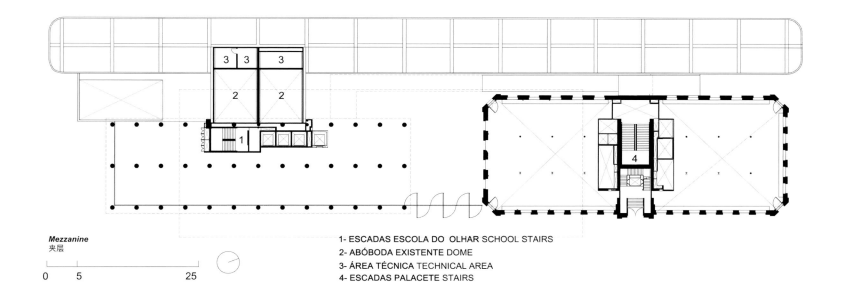
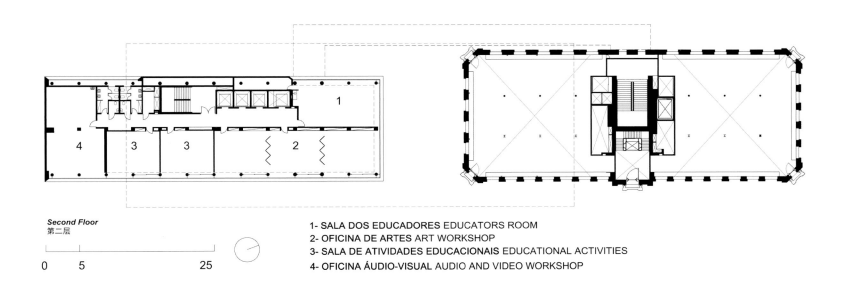
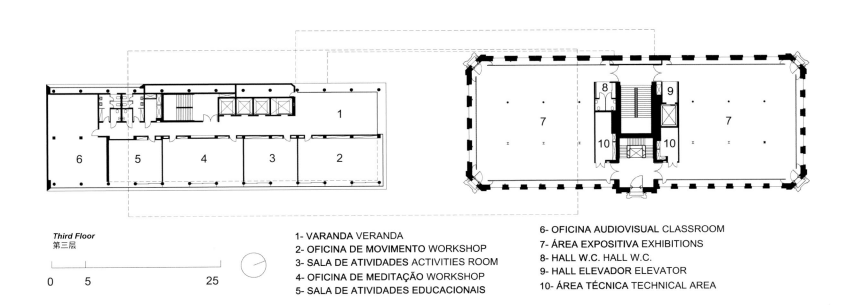

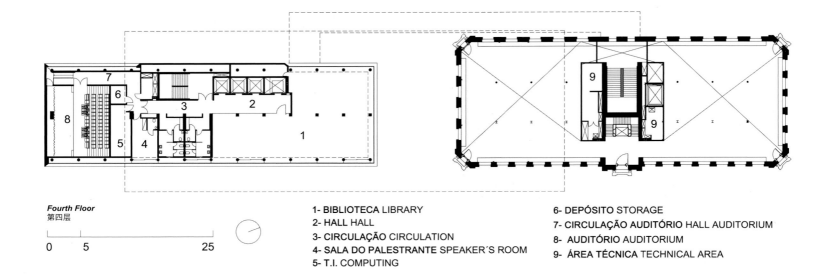

**Fourth Floor**
第四层

0    5    25

1- BIBLIOTECA LIBRARY
2- HALL HALL
3- CIRCULAÇÃO CIRCULATION
4- SALA DO PALESTRANTE SPEAKER´S ROOM
5- T.I. COMPUTING
6- DEPÓSITO STORAGE
7- CIRCULAÇÃO AUDITÓRIO HALL AUDITORIUM
8- AUDITÓRIO AUDITORIUM
9- ÁREA TÉCNICA TECHNICAL AREA

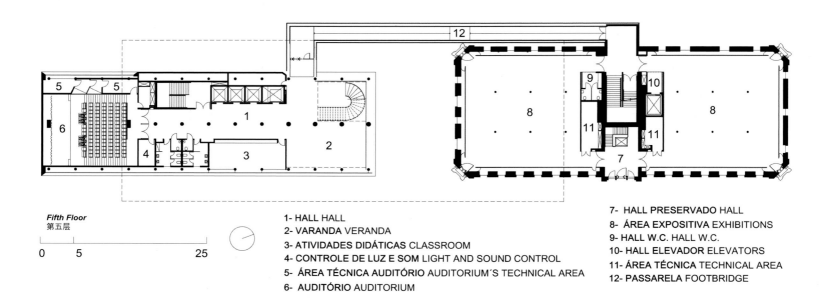

**Fifth Floor**
第五层

0    5    25

1- HALL HALL
2- VARANDA VERANDA
3- ATIVIDADES DIDÁTICAS CLASSROOM
4- CONTROLE DE LUZ E SOM LIGHT AND SOUND CONTROL
5- ÁREA TÉCNICA AUDITÓRIO AUDITORIUM´S TECHNICAL AREA
6- AUDITÓRIO AUDITORIUM
7- HALL PRESERVADO HALL
8- ÁREA EXPOSITIVA EXHIBITIONS
9- HALL W.C. HALL W.C.
10- HALL ELEVADOR ELEVATORS
11- ÁREA TÉCNICA TECHNICAL AREA
12- PASSARELA FOOTBRIDGE

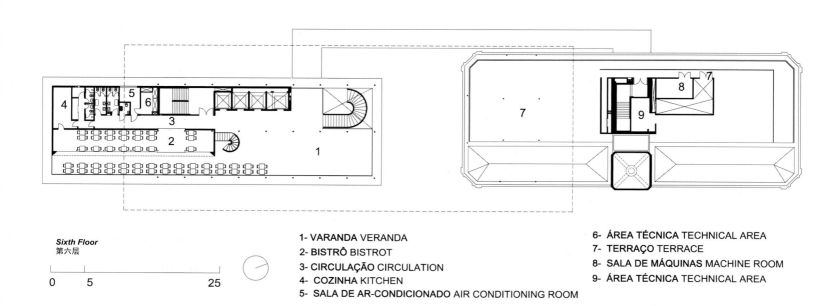

**Sixth Floor**
第六层

0    5    25

1- VARANDA VERANDA
2- BISTRÔ BISTROT
3- CIRCULAÇÃO CIRCULATION
4- COZINHA KITCHEN
5- SALA DE AR-CONDICIONADO AIR CONDITIONING ROOM
6- ÁREA TÉCNICA TECHNICAL AREA
7- TERRAÇO TERRACE
8- SALA DE MÁQUINAS MACHINE ROOM
9- ÁREA TÉCNICA TECHNICAL AREA

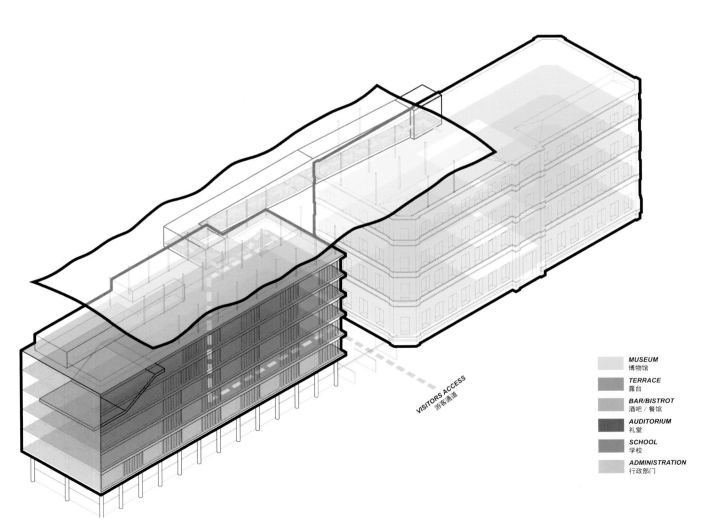

| | |
|---|---|
| MUSEUM | 博物馆 |
| TERRACE | 露台 |
| BAR/BISTROT | 酒吧/餐馆 |
| AUDITORIUM | 礼堂 |
| SCHOOL | 学校 |
| ADMINISTRATION | 行政部门 |

VISITORS ACCESS
游客通道

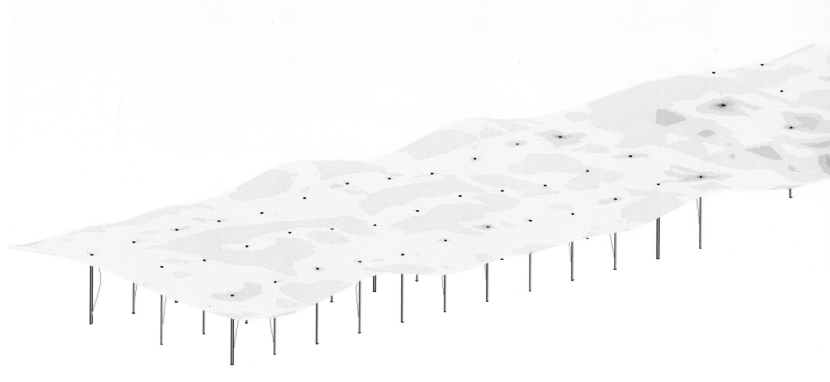

Dis 0.2cm
Max=0,8

220,19
200,00
175,00
150,00
125,00
100,00
75,00
50,00
25,00
0,0
-25,00
-50,00
-70,86

MXX, (kNm/m)
Automatic direction
Cases: 4 (COMB1)

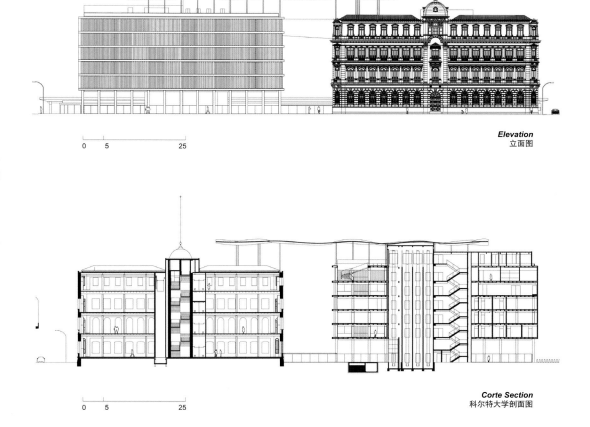

*Elevation*
立面图

*Corte Section*
科尔特大学剖面图

Lightweight Construction
Vertical Alignments
Steel Frame

轻型结构
竖向定线
钢架

# 奥地利格拉茨 NIK 办公大楼
## NIK Office Building

| | |
|---|---|
| 设计单位：Atelier Thomas Pucher | *Designed by: Atelier Thomas Pucher* |
| 合作单位：Bramberger 建筑师事务所 | *Collaboration: Bramberger Architects* |
| 项目地址：奥地利格拉茨 | *Location: Graz, Austria* |
| 建筑面积：1 400m² | *Building Area: 1,400m²* |
| 摄影：Andreas Buchberger | *Photography: Andreas Buchberger* |
| 设计团队：Thomas Pucher　Alfred Bramberger　Martin Mathy　Hans Waldhör　Ingmar Zwirn　David Klemmer　Boris Murnig　Christof Schermann　Sahar Arjomand Bigdely | *Project Team: Thomas Pucher, Alfred Bramberger, Martin Mathy, Hans Waldhör, Ingmar Zwirn, David Klemmer, Christof Schermann, Boris Murnig, Sahar Arjomand Bigdely.* |

**项目概况**

NIK 是为一个顶级品牌代理机构建立的一个独立式办公大楼,位于格拉茨莫尔河岸一个小型的绿色广场的中央。该项目获得了格拉茨一项邀请赛事的一等奖,预计将成为这座城市分散区域的一个标志性建筑。

**建筑设计**

建筑的首要问题是如何在原场址两层仓库的基础上增加一个 2.5 层楼的建筑结构,同时还要使建筑足够轻,能够建造在 20 世纪 80 年代的地下停车场的上面。设计的整个过程始终注意建筑的轻型结构,在建筑底下采用了一个复杂的分布网格,能够无形地扩散车库支撑结构上钢架构梁的集中负荷。

**Profile**

NIK is a freestanding office building created for a top branding agency, standing at the center of a small, green square on the banks of the river Mur. This project was the first prize of an invited competition in Graz, Austria. It is intended to be a landmark in a scattered area of the city.

**Architectural Design**

The chief problem to overcome was how to increase the number of floors by 2.5 (a double-story storage shed previously stood on the site), while keeping the new construction light enough to rest on the 1980s underground parking garage beneath. This was achieved by designing a consistently lightweight construction and also by using a complex distributor grid underneath, which invisibly spreads the concentrated loads of the steel frame girders over the garage's supports.

◀ 立面与造型——NIK 的形态塑造主要呈金色立方体的形态，体现出建筑朴素无华，简单大方的结构特点。泛着金色光泽的立面由清晰的竖向定线和玻璃的钢架结构组成，熠熠生辉的同时又显得平静而简单。厚实的立面在建筑内部也得到了很好的利用，形成一个结实、实用而且富有魅力的宛如雕刻般的建筑实体。

Façades and form—he form of the building is sculptural in simplicity: a stack of golden cubes. The steel-framed structure with clear vertical alignments and plenty of glass mediates between a high-density urban environment and an open green space. Calm and reduced, yet coated with a playful sheen of gold, the chunky façade blocks are also put into good use on the inside. The result is a construction that is solid, practical and inviting but at the same time manages to be aesthetically and literally light.

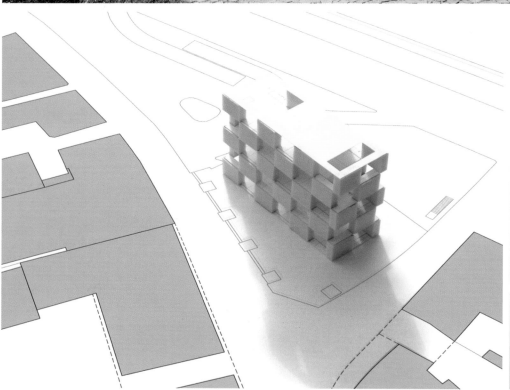

**Ground Floor Plan**
底层平面图

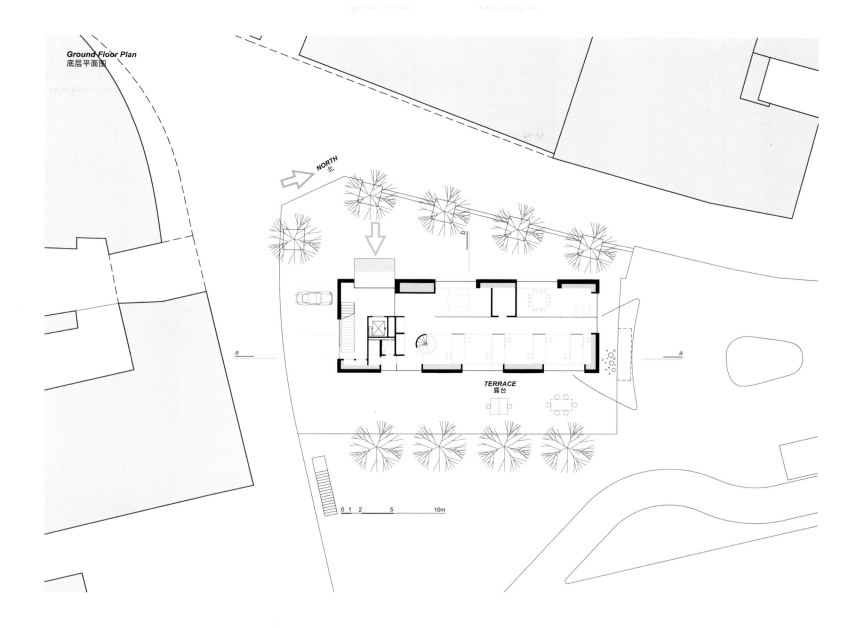

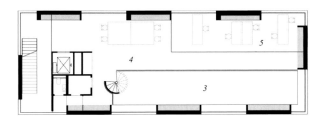

**Galerie**
画馆

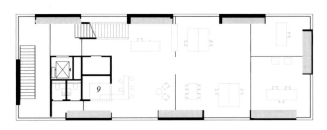

**2nd Floor**    **9. KITCHEN**
二楼    9. 厨房
          **10. OFFICE**
          10. 办公室

**1st Floor**    **6. RECEPTION**
一楼    6. 接待处
        **7. CONFERENCE ROOM**
        7. 会议室
        **8. OFFICE**
        8. 办公室

**3rd Floor**    **11. TERRACE**
三楼    11. 露台
        **12. CONFERENCE ROOM**
        12. 会议室
        **13. OFFICE**
        13. 办公室

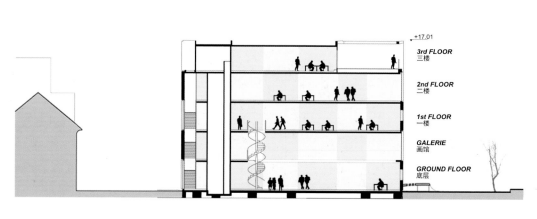

| | |
|---|---|
| 3rd FLOOR | 三楼 |
| 2nd FLOOR | 二楼 |
| 1st FLOOR | 一楼 |
| GALERIE | 画馆 |
| GROUND FLOOR | 底层 |

**Autonomous Building**
**Glass Sound Screen**
**Brick Façade**

自主建筑
玻璃屏风
砖砌立面

# 荷兰阿姆斯特丹 Furore 公寓大楼 A 栋
## Furore – Block A

设计单位：Kruunenberg Van der Erve Architecten
开发商：Stadgenoot
项目地址：荷兰阿姆斯特丹
建筑面积：6 128m²
摄影：Luuk Kramer

*Designed by: Kruunenberg Van der Erve Architecten*
*Client: Stadgenoot*
*Location: Amsterdam, Netherlands*
*Building Area: 6,128m²*
*Photography: Luuk Kramer*

**项目概况**

这栋引人注目的建筑优美地矗立在阿姆斯特丹心脏位置，虽然是一栋巨大的公寓大楼，却与周边的19世纪建筑非常协调。

**规划设计**

Furore大楼由3栋建筑体量构成，总体体现出该区域封闭式的建筑形式。这栋建筑还保留着自动结构，甚至表现出从20世纪20年代开始的线形城市规划特征。该双重性使得A栋的建筑师们将它与周边的小型建筑更好地联系起来。

另外，除了两层地下停车场，建筑的底层规划有4个商业空间和3个办公—住宅单元，上面7层是业主的住所和租房单元，这两类住宅均匀地散置在整栋综合体。

**建筑设计**

建筑底层5m高的基座散发出坚固性，是典型的自主建筑，基座上半部分的剖面成为一层住宅法式阳台的栏杆。所有的建筑单元都位于同一底座，看似一些独立的线条，实际上是由通道和一个玻璃声音屏风连接起来，这个屏风是用来转移交通噪声的。

由于立面上垂直的开窗和方形开口所形成的独特韵律，使建筑的砖砌表面显得格外优雅。大楼上面是高高的屋脊，大大的窗户通过砖砌结构中滑动式的玻璃板来阻隔交通噪声，为建筑正面带来了无限生机。

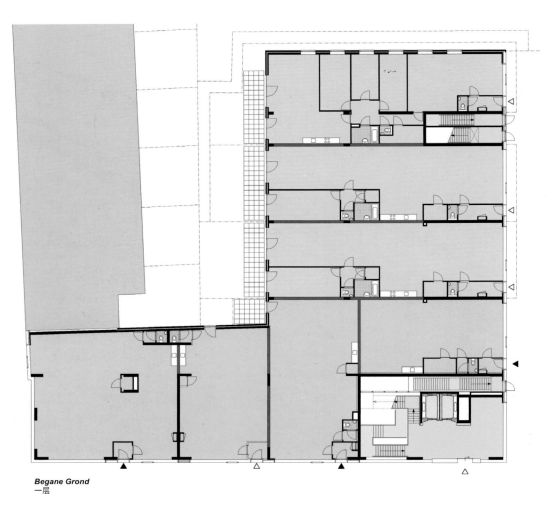

**Begane Grond**
一层

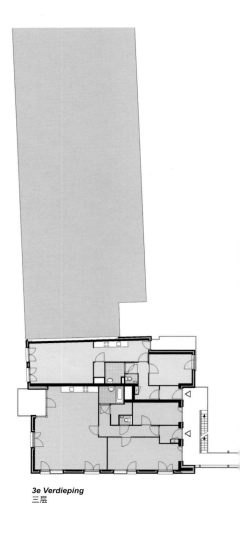

**3e Verdieping**
三层

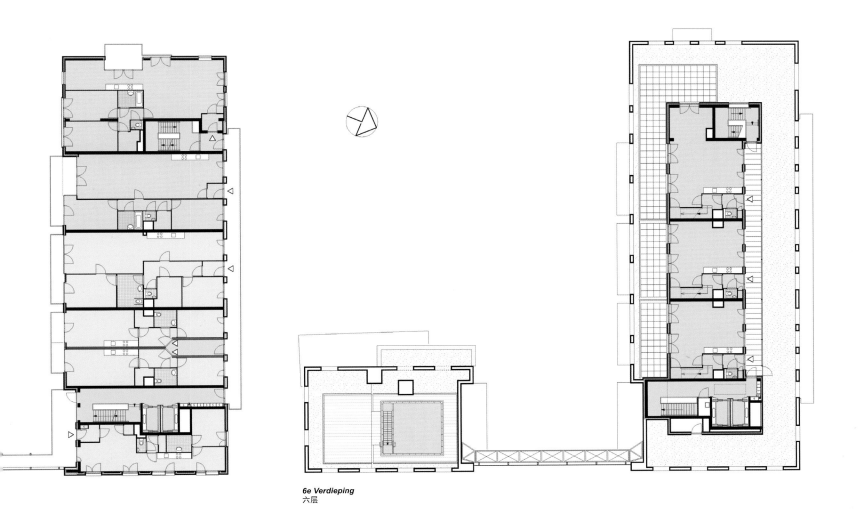

*6e Verdieping*
六层

## Profile

The notable building is located in the inner heart of Amsterdam, beautifully standing in harmony with the surrounding nineteenth century architecture although it is in fact a massive apartment complex.

## Planning Design

The combined three buildings form the Furore complex, which in its entirety is reminiscent of the closed building blocks of the area. Yet the buildings also retain characteristics of autonomous structures and even display the traits of linear urban planning from the 1920's. This double character has led the architects in Block A to make a better connection with the smaller buildings that surround it. Besides two levels of underground parking, four commercial spaces and three office-residence units are situated on the ground floor level of Block A. This is topped by seven floors of owner-occupied residences and rent-controlled units. The two types of residences are interspersed evenly through out the complex.

## Architectural Design

The five-meter-high base that decorates Block A at the ground level emanates a solidity that is typical of an autonomous building. The upper section of this base serves as a balustrade for the French balconies of the first floor residences. Both building units, which stand on the shared base, resemble independent lines but they are connected by walkways and a glass sound screen erected to deflect traffic noise.

The brick surface of the two buildings is graced by the rhythm of vertical fenestration and square openings cut into the façade where it continues up to form a high roof ridge. Large windows which are shielded from the traffic noise by sliding glass panes set flush to the masonry. The fact that the façade openings form an oblique rather frontal composition enlivens the face of this complex.

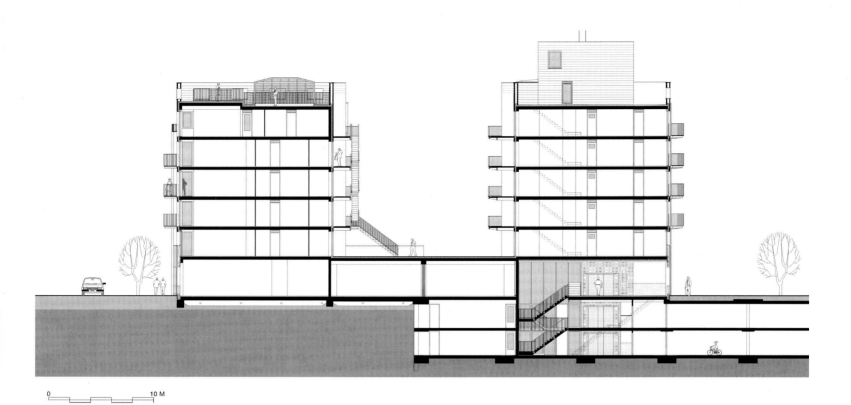

# 瑞士苏黎世 Cocoon
## Cocoon

设计单位：Camenzind Evolution
开发商：Swiss Life
项目地址：瑞士苏黎世
建筑面积：1 900m²
设计团队：Stefan Camenzind  Marco Noch  Susanne Zenker
摄影：Camenzind Evolution   Romeo Gross

Designed by: Camenzind Evolution
Client: Swiss Life
Location: Zurich, Switzerland
Building Area: 1,900m²
Project Team: Stefan Camenzind, Marco Noch, Susanne Zenker
Photography: Camenzind Evolution, Romeo Gross

**项目概况**

Cocoon 位于苏黎世 Seefeld 地区一个美丽的山坡上，得天独厚的位置优势源于 Cocoon 与一片绿洲紧密地贴在一起，可以眺望不远处的湖景和山景，三面被巨大的古树环抱，像一个美丽的大公园。

**建筑设计**

这个椭圆形的建筑宛如一个从地上升起的优雅、盘旋而上的雕塑。建筑正立面的装饰元素有意识地对整体结构增添了一些微妙和复杂感。

该建筑被包裹在很精细的类似鳞片状的不锈钢丝网内，这种金属外表皮以柔和的线条沿着宽阔的螺旋体优雅上升，与屋顶露台被开放的外观框架所连接，既可以保证隐私，又透露出一种约束的优雅，同时还形成了一个硬朗简洁的外观。

整体设计使得建筑在白天看起来是封闭式的，朝向中庭，在夜间则像是一个透明的光辉灯塔。

螺旋状
不锈钢丝网表皮
流线体系

Spiral
Stainless Steel Mesh
Circulation System

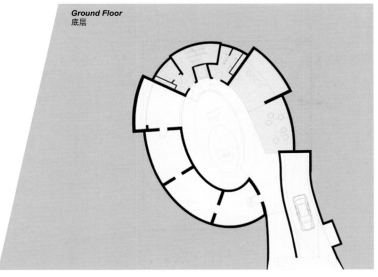

*Ground Floor*
底层

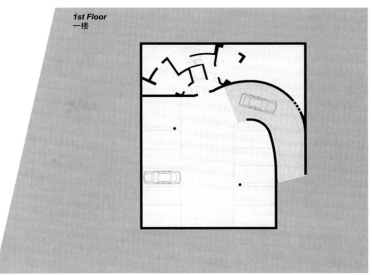

*1st Floor*
一楼

*Ground Floor*
底层

◀ 内部流线体系——根据螺旋状的概念，Cocoon 可被看作是一种"通信景观"。所有的空间都围绕着一个灯光辉煌的中庭，沿一个斜坡冉冉上升，这种连续的螺旋空间避免了传统水平空间分隔带来的沟通障碍，充分发挥了互动和合作的可能性。在内部，椭圆的扩展和螺旋彼此之间的交错，各种不同的要素如电梯、螺旋坡道、弧形和楼梯井等构成了清晰的内部空间结构，多功能的流线体系，使得内部的空间结构与周围环境进行了良好的互动。

Internal flow system—according to the concept of the spiral, Cocoon can be regarded as a form of "communications landscape". All the space is around a light-flooded atrium, gently rising along a slope; this kind of continuous spiral space avoids the communication barriers brought by the separated traditional horizontal space, fully playing the possibility of interaction and cooperation. Internally, the expansion of the ellipse, interleaved between each coil and various different factors such as elevators, spiral ramps, arcs and stairwells constitute the building interior structure with clear and multi-functional streamline system, which makes the internal space structure get a good interaction with the surrounding environment.

## Profile

Cocoon is located on a beautiful hillside of Seefeld District in Zurich, which enjoys excellent lake and mountain views. The location's distinctive flair stems from the exceptional park-like setting – a green oasis into which Cocoon snugly nestles. The building is flanked on three sides by mighty, age-old trees.

## Architectural Design

The elliptical structure reads as a freestanding sculptural volume that gracefully spirals from the park. The façade assembly consciously adds a note of subtlety and sophistication to the overall composition.

The building is wrapped in a fine, almost scaly veil of stainless steel wire mesh. This curtain curls elegantly upwards in soft lines along the expanding spiral, its junction with the roof terrace accentuated by an open façade frame. The stainless steel mesh enveloping the building combines visual privacy with restrained elegance, while establishing a strong and unmistakable presence.

The shrouded, sculptural stand-alone building, introverted during the daytime as it looks inwards towards the atrium, is recast in the evening hours as a transparent shining beacon.

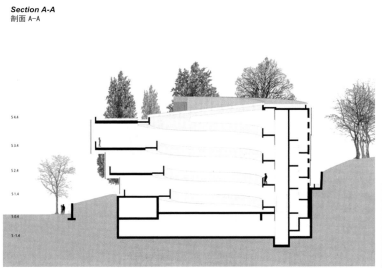

**Section A-A**
剖面 A-A

**Section B-B**
剖面 B-B

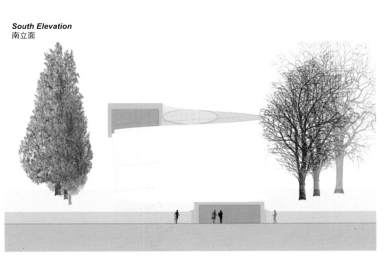

**South Elevation**
南立面

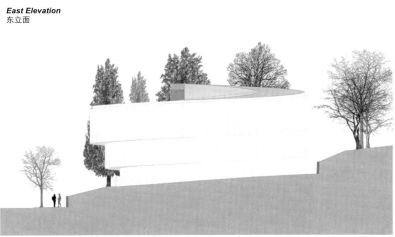

**East Elevation**
东立面

Geometric Model
"Trefoil"
Cube Core

几何模型
"三叶草"
核心筒

# 德国斯图加特奔驰博物馆
## Mercedes-Benz Museum

| | |
|---|---:|
| 设计单位：UN Studio | *Designed by: UN Studio* |
| 开发商：DaimlerChrysler Immobilien, Berlin | *Client: DaimlerChrysler Immobilien, Berlin* |
| 项目地址：德国斯图加特 | *Location: Stuttgart, Germany* |
| 建筑面积：35 000m² | *Building Area: 35,000m²* |
| 摄影：Christian Richters  Brigida Gonzalez | *Photography: Christian Richters, Brigida Gonzalez* |

**项目概况**

这座博物馆建筑位于梅赛德斯总部工厂旁边，不仅满足了越来越多的参观者的需求，也将是奔驰品牌的推广工具。馆址四周环绕着多车道公路，在 Neckartal 城区内从很远的位置就可以看到这座建筑。

**建筑设计**

建筑的几何模型是基于在三叶草的基础之上，设计以一个双螺旋体为出发点，借助弯曲旋转的形态直到形成一个双螺旋结构，而平面则形成一个边长大约 80m 的三角形。

设计在螺旋体内插入各个层面，使其环绕如同三叶草形的中庭，层面间保持一定的高差成为建筑的各个楼层，以这种方式在螺旋结构中设计了两种展区层："传奇之旅"展区层和日光照明的"典藏之旅"展区层。参观者乘坐中庭的电梯到顶层，然后通过两条互相交错的螺旋形斜道从上而下参观。

"传奇之旅"展区被一个环形梁所包围，环形梁以环绕型斜面的形式由箱梁和立式外墙组成，并由承重能力特别强的结构混凝土制成，墙厚达 50cm。这个斜面连接核心筒，支撑在外立面柱上，作为曲线形的连续梁承受垂直载荷。

整个设计从复杂的几何体出发，通过一系列的技术设计，为传奇的汽车品牌形成一栋全新的标志性大楼。

## Project Overview

The museum building is located next to the headquarters of the Mercedes factory, which is not only meet the needs of a growing number of visitors, also will be the tool for Mercedes-Benz brand promotion. Location is surrounded by the multi lane highways, and it can be seen in a position very far from the city within Neckartal City.

## Architectural Design

The geometric model employed is based on the trefoil organization. The design takes the two spiraling structures as the starting point. The curved rotation creates a double spiraling structures and the plane forms a triangle with about 80m side length.

Several levels are inserted into the spiral structures, surrounding the trefoil-shaped atrium. Certain dispersion is kept between different levels. Thus two exhibition areas are created within spiraling structure: "Legend rooms" and the bright "Collection rooms". Visitors travel up through the atrium to the top floor from where they follow the two main paths that unfold chronologically as they descend through the building.

The "Legend rooms" are surrounded by an annular girder composed of box beam and upright wall in the form of encircling slope. The box beam is made of structural concrete of strong bearing capacity; the wall is 50cm thick. The slope connects to the core tube, supporting on the front column and bearing vertical load as curved continuous beam.

The whole design starts from the complex geometry. By a series of technique designs, a brand new iconic building is created for the legendary automobile brand.

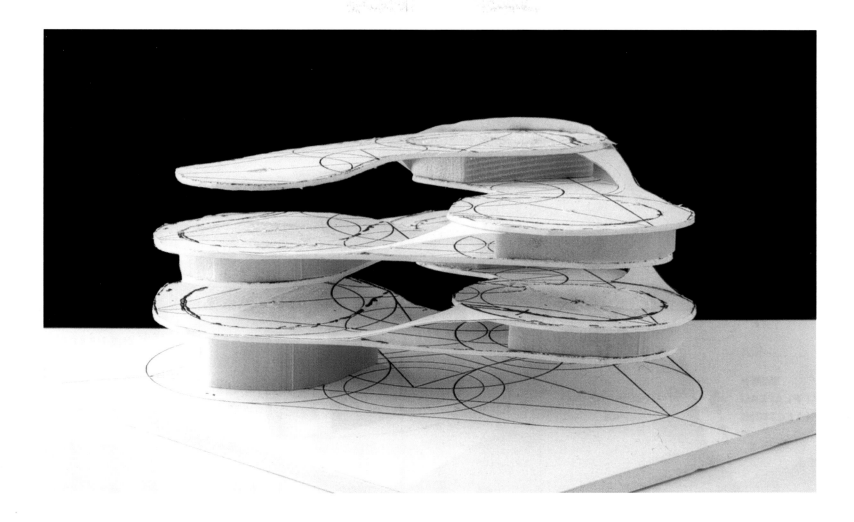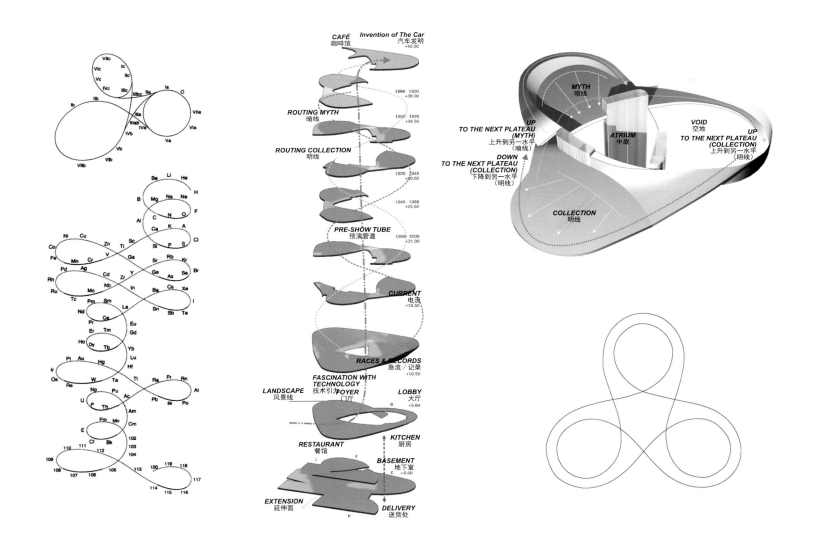

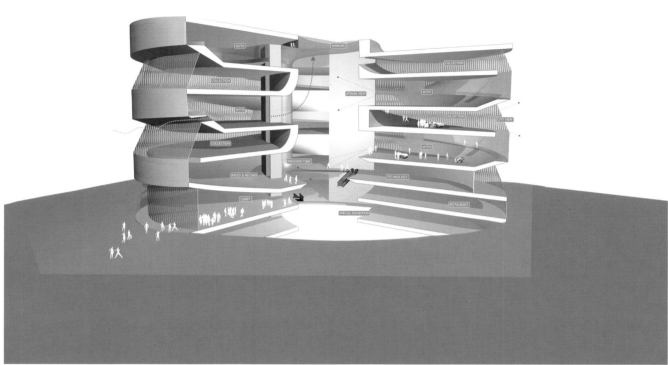

# 澳大利亚悉尼 Blakehurst 俱乐部
## The Blakehurst Club

设计单位：Tony Owen Partners
项目地址：澳大利亚悉尼

*Designed by: Tony Owen Partners*
*Location: Sydney, Australia*

Geometric Roof
3-Dimensional
Digital Modelling
Open Plan Space

几何屋顶
3D 建模
开放式空间

**项目概况**

凯尔湾保龄球俱乐部坐落在悉尼南部郊区 Blakehurst 风景壮观的海滨区域，拥有优美的景观和良好的位置优势，成为区域内社交生活的重要场所。原 Blakehurst 俱乐部建于 1962 年，在发展的历程中，俱乐部成为了一个有着强大的会员数量和赞助收入的社会区域。

**几何设计**

建筑的屋顶成为整个设计中的一大亮点，设计通过大量利用 3D 数字建模软件来探索几何形态的不同选择，最终选用分支的"L 系统"几何体来建造平坦屋顶。该几何体在白天不同时刻都能反射光线，呼应着户外水景中的迷人景色。

设计选择将屋顶抬升来形成一个空间，使其能最大程度地发挥屋顶形状及建筑位置的潜力，并且设计将所有的空调都安装在屋顶周边，最大限度地将天花板体量创造成一个与户外相连接的明亮又通风的空间。

**功能布局**

建筑底层为俱乐部更衣室和行政区域，还有功能中心及餐馆。重新设计的顶层则为白色的圆形酒吧，成为空间的焦点。休息区、小餐馆等其他区域都围绕着圆形酒吧呈辐射状分布，以流动式的形态延伸到户外水景上的巨大阳台，这一开放式的空间构成，使全景效果得到最大化利用。

▶ 设计特色——俱乐部的墙面均由透明玻璃材质组成，在夜色朦胧之时，建筑内清亮的灯光透过玻璃散发出来，好似天未破晓时隐约的月，在空旷地面的衬托下显得格外柔美，将设计者在设计建筑时极力打造的空间感最大化地体现了出来。

Design characteristic—the metope of the club are composed of transparent glass material, when the night is dim, the clear light of the buildings will come out through the glass, and is like a day when the moon is not a hint of dawn, against the background of open ground is very gentle, which will maximize to reflect out the dimensional feeling when the designer designed buildings.

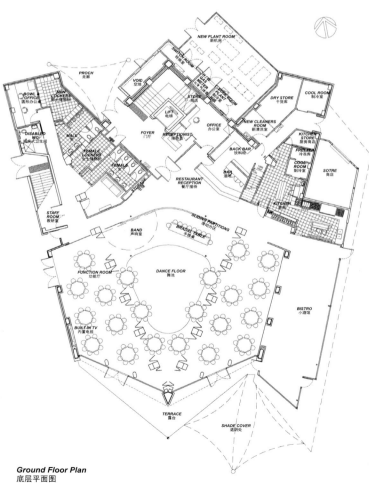

**Ground Floor Plan**
底层平面图

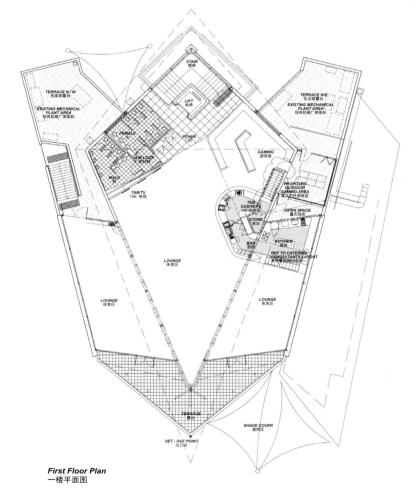

**First Floor Plan**
一楼平面图

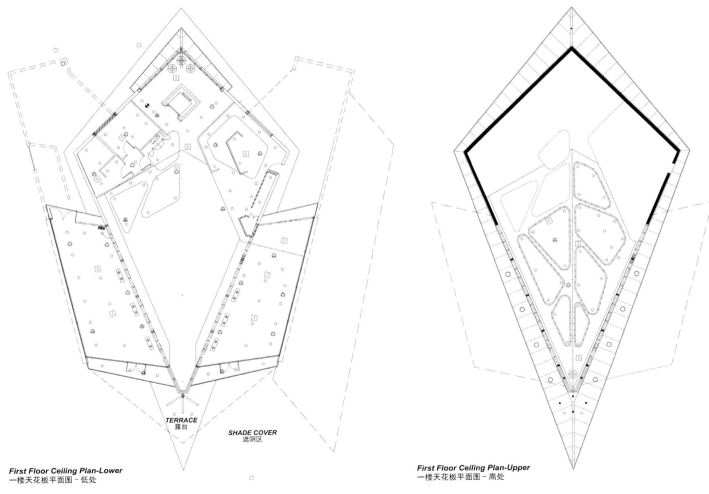

TERRACE
露台

SHADE COVER
遮阴区

**First Floor Ceiling Plan-Lower**
一楼天花板平面图 - 低处

**First Floor Ceiling Plan-Upper**
一楼天花板平面图 - 高处

## Profile

The Kyle Bay Bowling Club is located on a spectacular waterfront in Blakehurst in Sydney's southern suburbs. With great views and location, the club is an important place of social life in the area. The original Blakehurst Club was built in 1962. This design featured a unique diamond shaped parabolic roof. The club became a fixture in the social life of the area and had a strong membership and patronage.

## Geometric Design

Roof of the building become a highlight of the whole design. The architects made extensive use of 3-dimensional digital modelling software to explore different options for the roof geometry. Eventually a branching "L-system" geometry was chosen for the coffered roof. This geometry reflects light throughout different times of the day to create an effect that echoes the play of light on the water outside.

## Function Layout

The ground floor of the building contains the club change rooms and administrative facilities as well as a function centre and a restaurant. The redesigned upper level has a central white circular bar which is the focus of the space around which the other areas radiate. This area consists of a large open plan space to take maximum advantages of the panoramic views. The upper level consists of the lounge and bistro areas which flow onto the large outdoor balconies on the water.

**West Elevation**
西立面
SCALE 1:100
比例 1:100

**South Elevation**
南立面
SCALE 1:100
比例 1:100

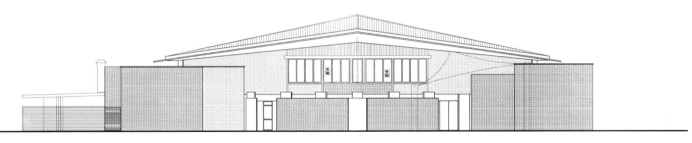

**East Elevation**
东立面
SCALE 1:100
比例 1:100

**North Elevation**
北立面
SCALE 1:100
比例 1:100

# 昂赞图书馆
## Library in Anzin

建筑师：Dominique Coulon et associés
项目指导：Olivier Nicollas
开发商：昂赞之城
建筑面积：1 750 m²

*Architectes: Dominique Coulon et associés*
*Project Director: Olivier Nicollas*
*Client: City of ANZIN*
*Building Area: 1,750m²*

### 建筑设计

该建筑体块是由独立的、自由的几何体块构成，整个建筑给人营造了一种诗情画意的感觉。透明度的精细领域显示其内容：在建筑内部，拥有丰富的、均匀的光线，空间是开放的、流动的，为阅览者提供了最佳的灵活性，高大的差距所产生的照明效果在空中呈现出流动性。多媒体图书馆覆盖着大片白色的面纱反射光线，该建筑宣称它的明亮程度就像是一张折纸一样，连续的褶皱和襟翼重复这一形象。它是白色的，几乎是微不足道的，就像一个概念的单纯的投影，但它充满了构成它超出了其物理极限的生活。

### Building Design

The volumes are independent and geometrically free, giving the whole a wonderfully poetic feel. The deliberate areas of transparency reveal its content: on the inside, there is abundant, uniform light. The space is open and fluid, offering optimal flexibility. The lighting effect produced by the tall gaps that appear to float in space. The multimedia library is covered with large white veils that reflect the light. The building asserts its lightness, like an origami. The successive folds and flaps repeat this image. It is white, almost immaterial, like the mere projection of a concept, yet it is brimming with the life that constitutes it beyond its physical limits.

Geometry
Architecture
Light Volume

几何体建筑
体量轻盈

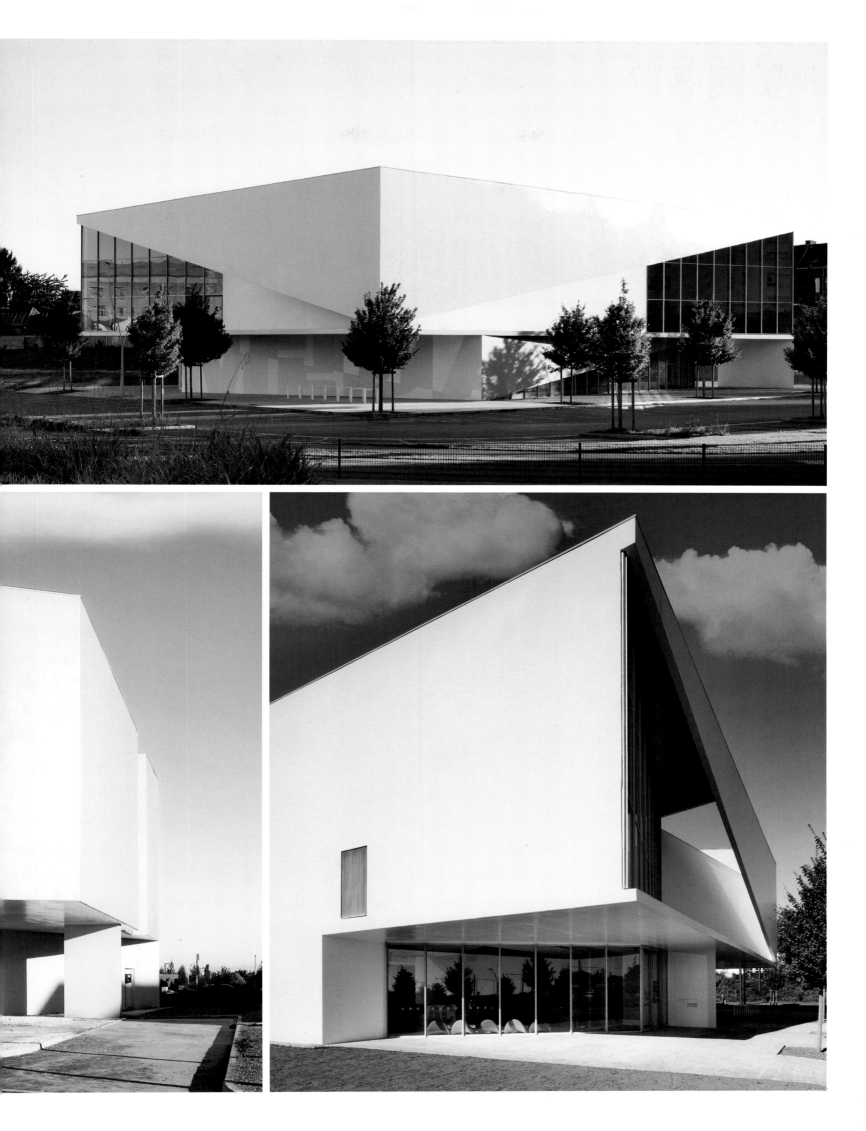

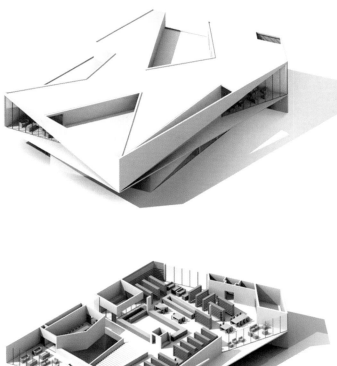

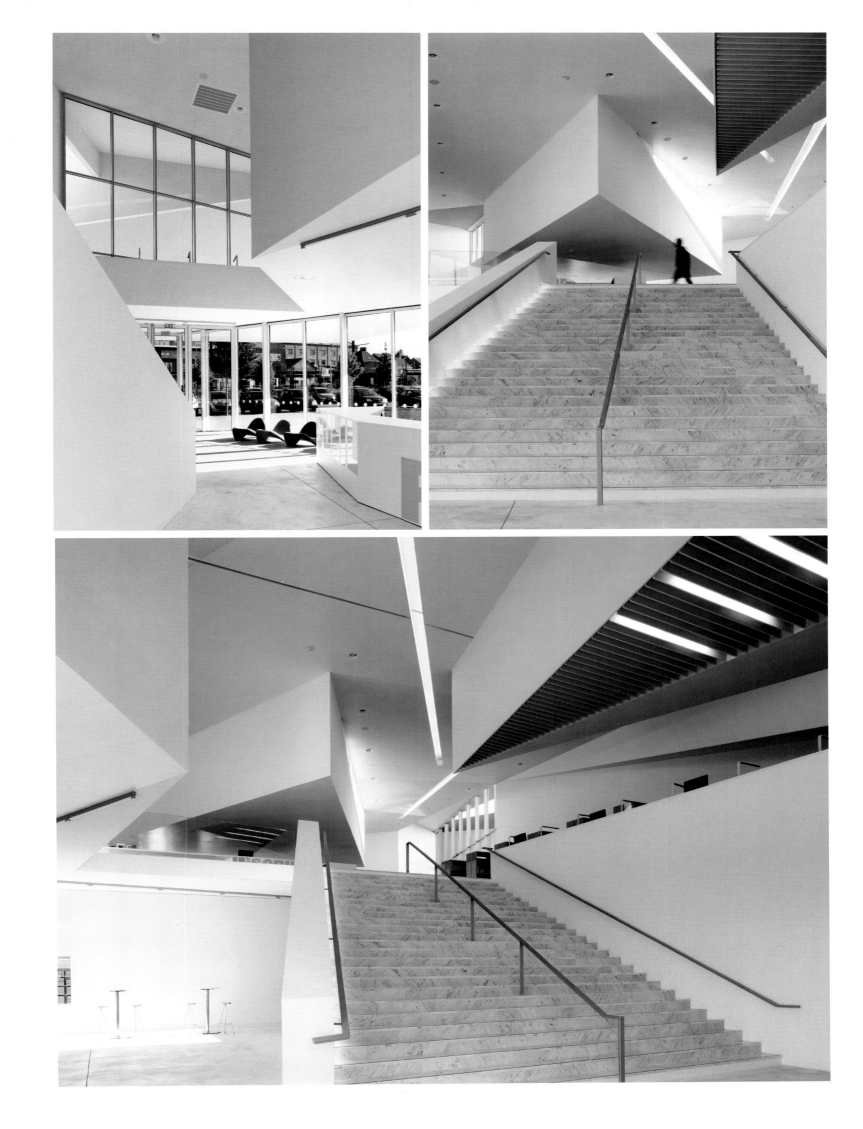

Larch Wood
Self-Supporting
Structure

落叶松木
自支撑结构

# 俄罗斯莫斯科电视秀亭台
Gazebo for TV Show

设计单位：Za Bor Architects
项目地址：俄罗斯莫斯科
建筑面积：42m²
摄影：Peter Zaytsev

*Designed by: Za Bor Architects*
*Location: Moscow, Russia*
*Building Area: 42m²*
*Photography: Peter Zaytsev*

**项目概况**

该项目位于莫斯科典型的郊区中，实际上是为了参加一档名为《乡村脱口秀》的电视节目而产生。独特的建筑形态，使其不论夏天还是冬天，都是一个遮风挡雨的好地方。

**建筑设计**

由于业主喜欢烧烤以及园林植物，于是建筑师采用白色落叶松木制成 14 个面，并将它们拼接起来，形成一个海浪似的、螺旋状的自支撑结构，就像是一只遗失在美丽花园的海螺。建筑共包含有用餐区和烧烤区，整体透明开放，激发了人与自然的关联。

虽然建筑结构形体复杂，但是设计以具有中性色彩且天然的落叶松为主要材料，部分用钢材和砖加以辅助，使得建筑既强有力地展现了项目本身，又与周边美丽的场地环境相融。不论是夏季葱郁的绿，还是秋天绚丽的黄和红，这座造型复杂的精巧建筑都将莫斯科郊区田园般的景色衬托得更为优雅迷人。

**Profile**

Located in a typical suburban area of Moscow, the project has been developed specially for popular TV show "The Village Talks". Its unique form makes it an excellent place to protect from wind and rainfall not only in summer but in winter as well.

**Architectural Design**

Because the owner enjoys cooking on the grill and garden trees, the architects make a small-size self-supporting structure consisting of 14 planes made of larch white-tinted wood. The gazebo has the helical structure resembling a sea wave, like a sea snail lost in a lovely garden. The gazebo includes a dining area and a barbeque area. Transparency and openness of the construction inspires a contact between man and nature.

Although the project is a complex structure consisting of 14 flat segments, the architects use neutral colors and natural larch wood, which helps, to present an object effectively and emphasize its structural features. From the lush green in summer to yellow and red in autumn, the complex and aggressive form sets off the elegance and charm of pastoral Moscow suburbs.

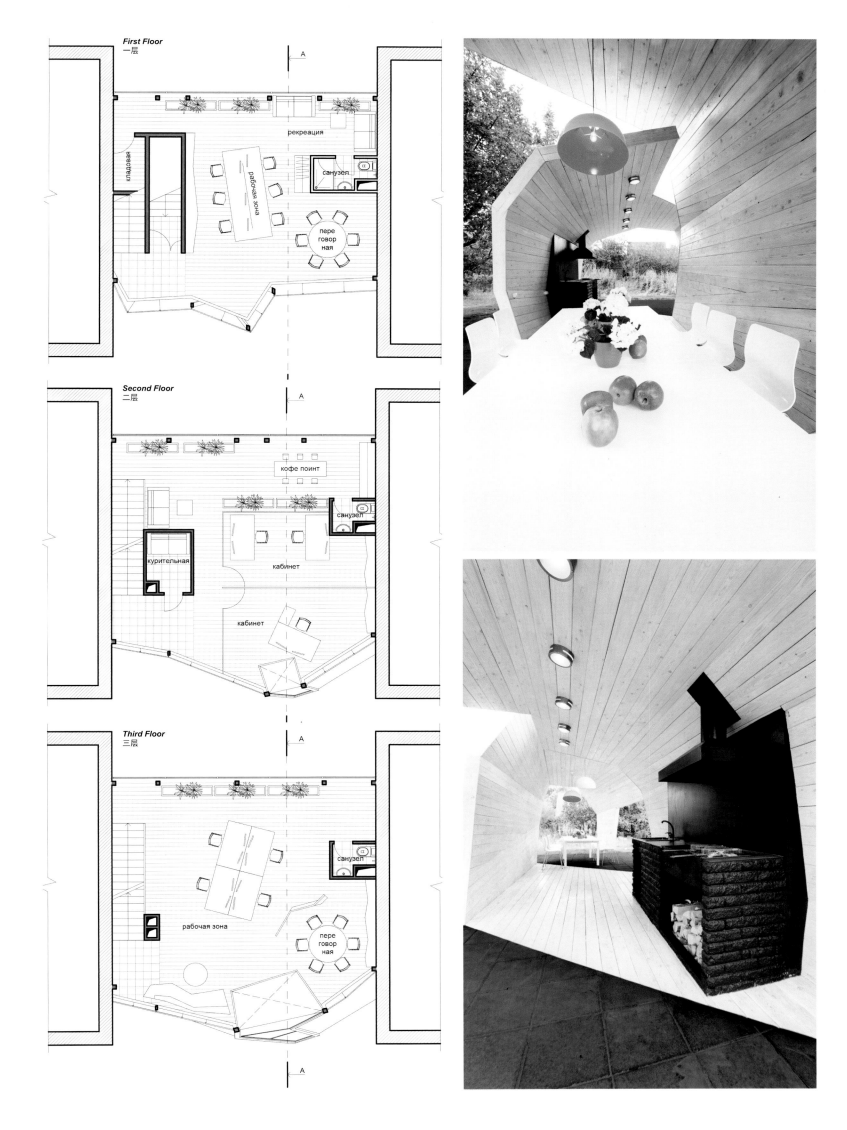

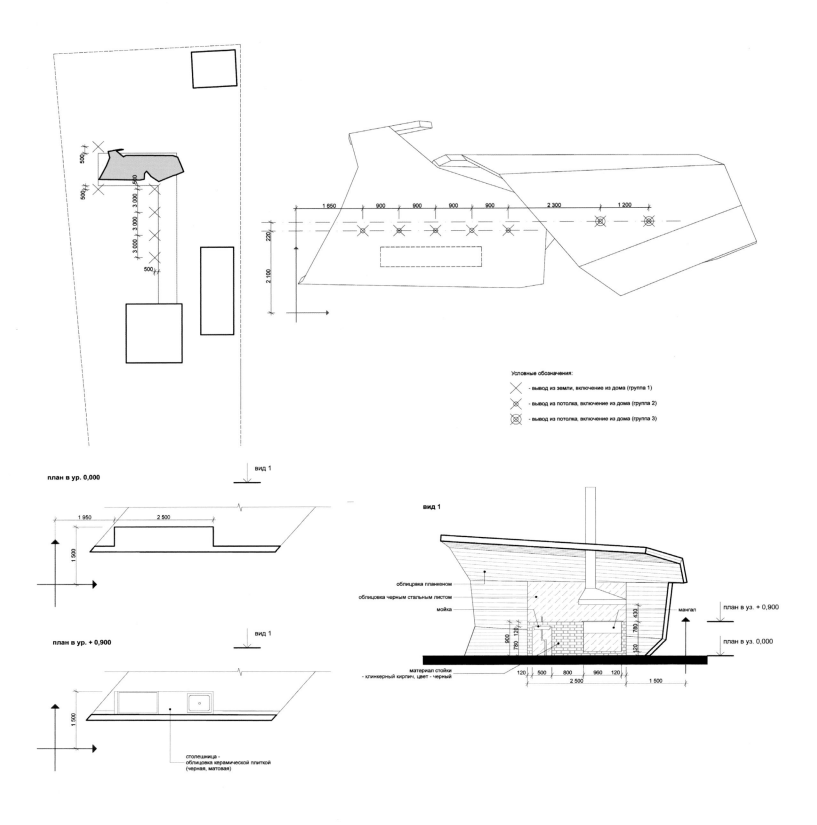

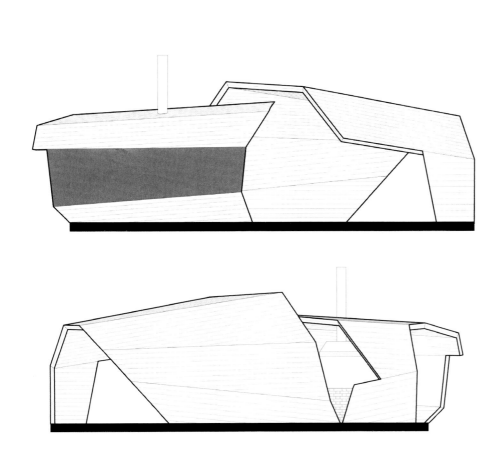

Section
剖面图

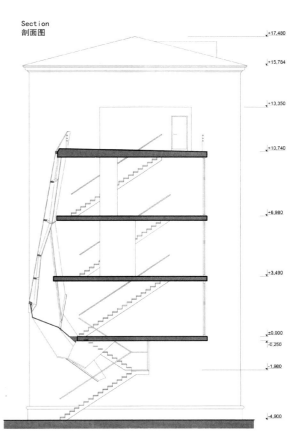

# 德国纽伦堡 Sipos Aktorik GmbH 生产车间
## A Production Hall of a Sipos Aktorik GmbH

设计单位：wurm + wurm architekten ingenieure
开发商：Sipos Aktorik GmbH
项目地址：德国纽伦堡
建筑面积：7 700m²
摄影：Ester Havlova

Designed by: wurm + wurm architekten ingenieure
Client: Sipos Aktorik GmbH
Location: Nuremberg, Germany
Building Area: 7,700m²
Photography: Ester Havlova

# 紧凑体量 屋顶天窗 不锈钢板瓦

# Compact Volume Skylight Shingles of Stainless Steel

**项目概况**

该项目位于纽伦堡附近的阿尔道夫镇，Sipos Aktorik GmbH 这一生产车间带有服务区、住宅小区和办公空间，直接依附于生产区，并以良好的交通网络与各个分区联系起来。

**建筑设计**

新大楼设计成紧凑型的体量，融入到阿尔道夫的丘陵景观中，其空间更像是一个巨大的宅邸，而不像是工厂。

生产车间屋顶设有天窗，整个屋顶造型就像一个酒吧，使车间内可得到丰富的自然采光，建筑夺目的颜色也增强了室内的光线效果。生产车间的钢结构为模块化制成，制造区域可随意膨胀成双倍大，装配区和两层办公区由高高的玻璃墙进行功能分隔，使得不同工作环境间可进行内部联系。管理区域嵌入两层晕光，通过巨大的玻璃滑动门可从办公空间到达这些区域，打破了建筑室内外的界线。

建筑前立面采用不锈钢板瓦制成，从而形成了该建筑清晰的外观形状。创新的制冷金属天花板系统则使办公空间保持适宜的温度，非常舒适宜人。

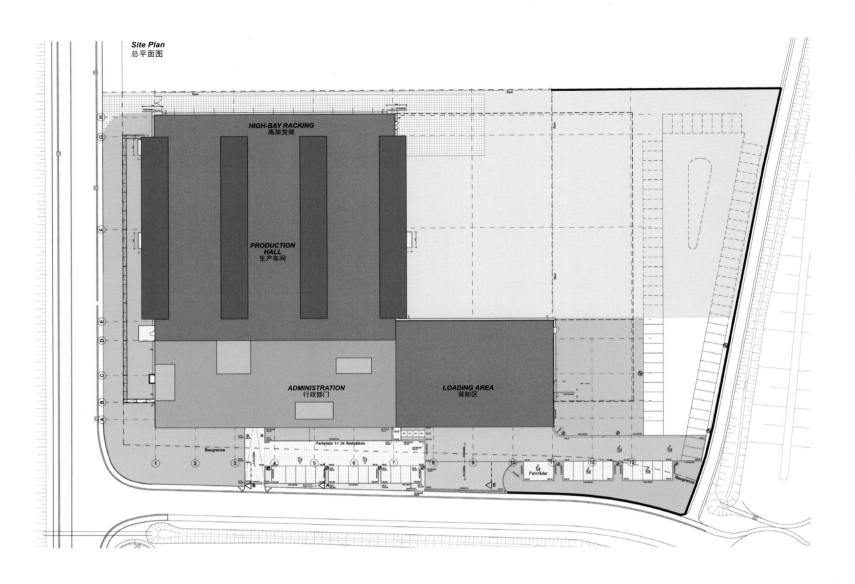

*Site Plan*
总平面图

## Profile

This project is located in Altdorf, near Nuremberg. This construction of a production hall includes service, development and office space which are attached to the production area and linked to the various divisions by a good network.

## Architectural Design

The new building of the Sipos Aktorik GmbH is designed as a compact volume, embedded into the easily hilly landscape at Altdorf, which in its spatial effect is more similar to a large mansion than a factory.

The roof of the production hall is occupied with skylights shaped as a bar; daylight reaches the working stations. The light impression of inside is still strengthened by the bright color of the construction. The steel structure of the production hall is modular developed, whereby the manufacturing area is optionally expandable up to the double size. The functional separation between assembly area and the two-story office area is abrogated by high glass walls. This conveys internal communication between these different working environments. Partly two-storey halation is incised into the management area, which is accessible by large glass sliding doors from the office space; outside and inside become interwoven with one another.

Shingles of stainless steel are selected for front material, which support the clearness of the shape of the building. The office space is kept pleasant at a moderate temperature by an innovative cooling system with a cooling ceiling of metal.

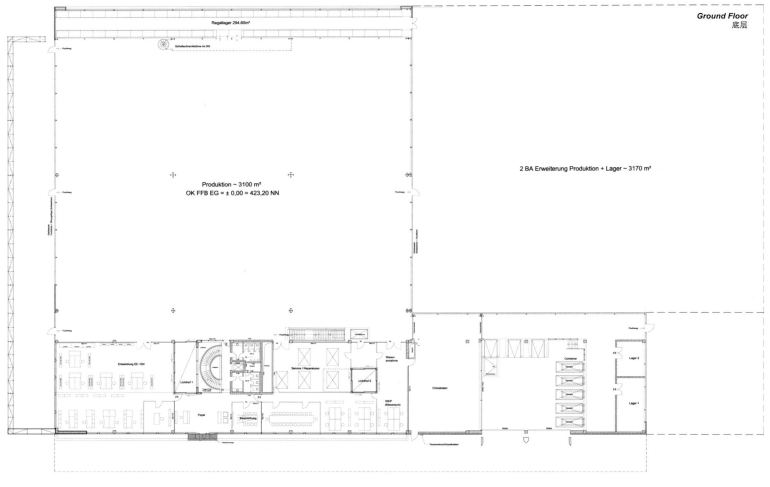

*Ground Floor*
底层

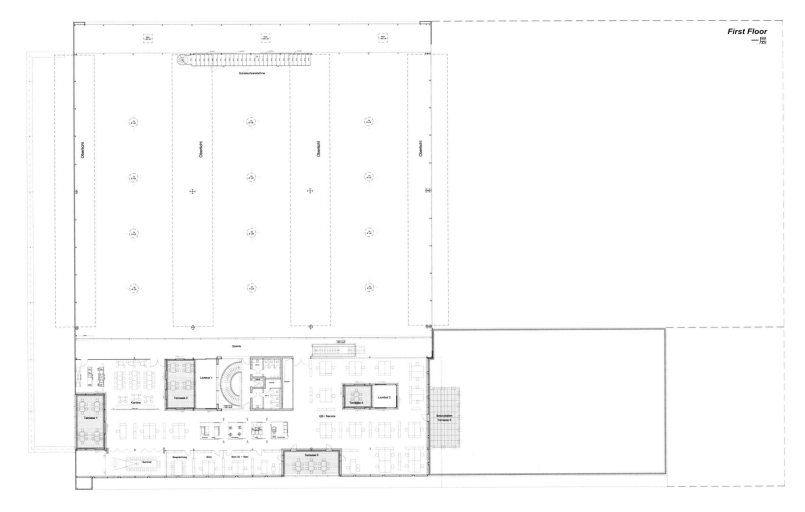

**First Floor**
一层

**Section A-A**
剖面 A-A

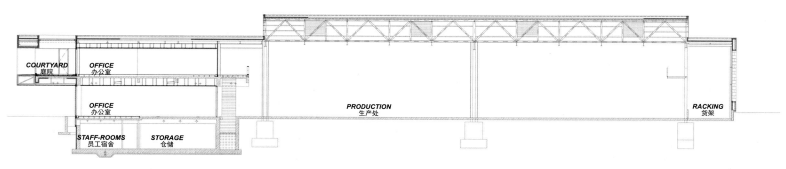

| COURTYARD 庭院 | OFFICE 办公室 | | | | |
| --- | --- | --- | --- | --- | --- |
| | OFFICE 办公室 | | PRODUCTION 生产处 | | RACKING 货架 |
| | STAFF-ROOMS 员工宿舍 | STORAGE 仓储 | | | |

**10 METER**
10 米

**East Elevation**
东立面

**West Elevation**
西立面

**North Elevation**
北立面

**South Elevation**
南立面

A Little City
Curved Concrete Walls
A Great Veranda

微缩城市
弧形混凝土墙
大型游廊

# 巴西里约热内卢艺术城
## Cidade das Artes

设计师：Christian de Portzamparc
开发商：里约热内卢市政厅文化秘书处
项目地址：巴西里约热内卢
建筑面积：90 000m²

Designer: Christian de Portzamparc
Client: City Hall of Rio de Janeiro, Secretaria Municipal das Culturas
Location: Rio de Janeiro, Brazil
Building Area: 90,000m²

**项目概况**

艺术城位于里约热内卢巴哈德提虎卡区域的一个14km平原的中心，绵延于海洋与山脉之间。该场地由穿过该地区的两条高速公路构成，位于新城心脏地带的艺术城将成为该区有力的城市标志和公共空间。

**建筑设计**

该建筑是一个小型的城市，容纳在一个大型结构中，经抬高建立在离地面10m以上的一个大型露台之上。这一露台是一个公共空间，从这里人们不仅可以看到山脉和海洋等周边迷人的景色，还能作为一个聚集地，为所有设施提供通道。

建筑屋顶和露台这两个水平板之间，设计有大型弧形混凝土墙，使其在体量与空隙部分的相互作用下形成大厅空间的同时，还与Siera Atlantica山的美丽弧线和海岸线遥相呼应。

艺术城设计将聚集各种各样的场所，不仅包括一个可以转换为歌剧厅和剧院的音乐厅，还包括众多电影院、舞蹈室、排练室、展览空间、餐厅以及多媒体图书馆等，使建筑不仅是一座大型房屋，更是一个飘浮于城市上方的大型游廊，展示着城市的无尽魅力。

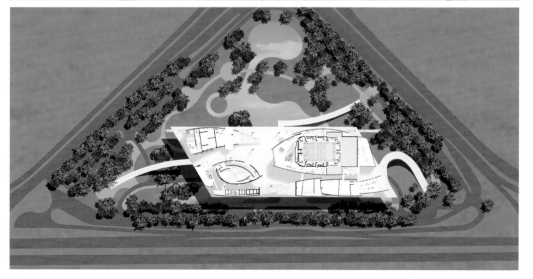

## Profile

The Cidade das Artes is situated between sea and mountain, in the center of 14km of plain. The site is structured by two highways that cross the district. In the center of this cross, the Cidade das Artes will be in the very heart of the new city, becoming a strong urban mark and public space.

## Architectural Design

The building is a little city contained in one big structure raised and established on a vast terrace ten meters above ground, from which one will see the mountain and the sea. This terrace is the public space; it is the gathering place that gives access to all facilities.

Between the two horizontal plates of the roof and the terrace are set the large curved concrete walls that contain the halls in interplay of volumes and voids. The architecture echoes the beautiful curves of Siera Atlantica Mountains and the line of the sea.

Cidade das Artes will gather a large variety of places: a concert room which is convertible in room of opera and in theatre, movie theatres, dance studios, numerous rehearsal rooms, exhibition spaces, restaurants, and a media library. The Cidade das Artes is seen as a large house, a great veranda above the city, homage to an archetype of Brazilian architecture, representing infinite charm of this city.

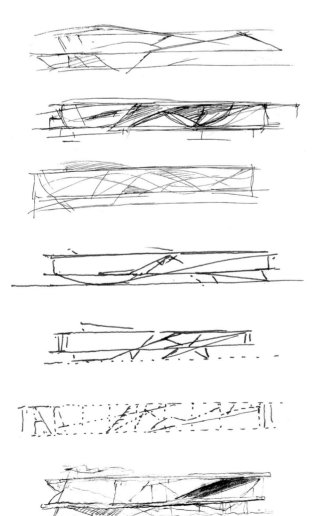
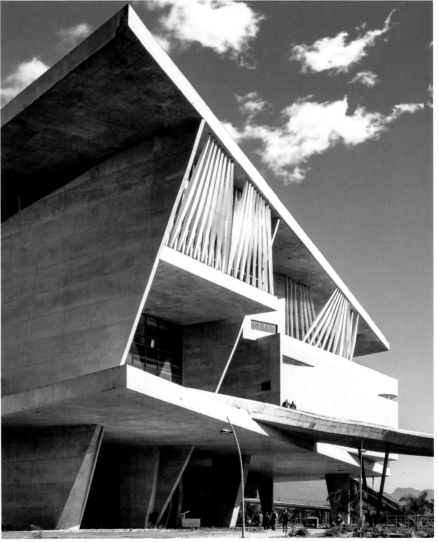

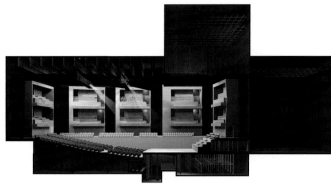
*Philharmonie*
交响乐团

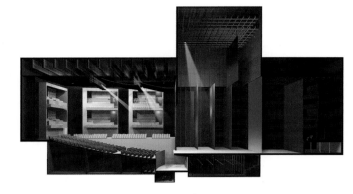
*Opera*
歌剧院

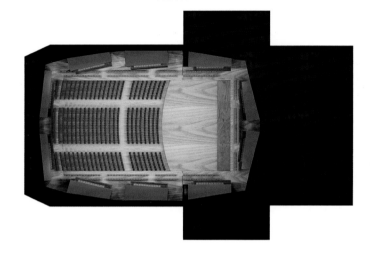

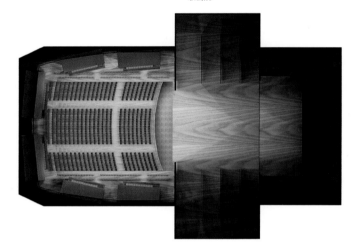

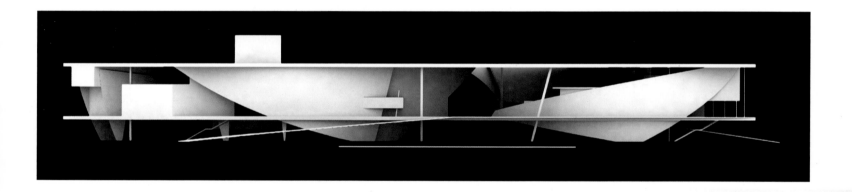

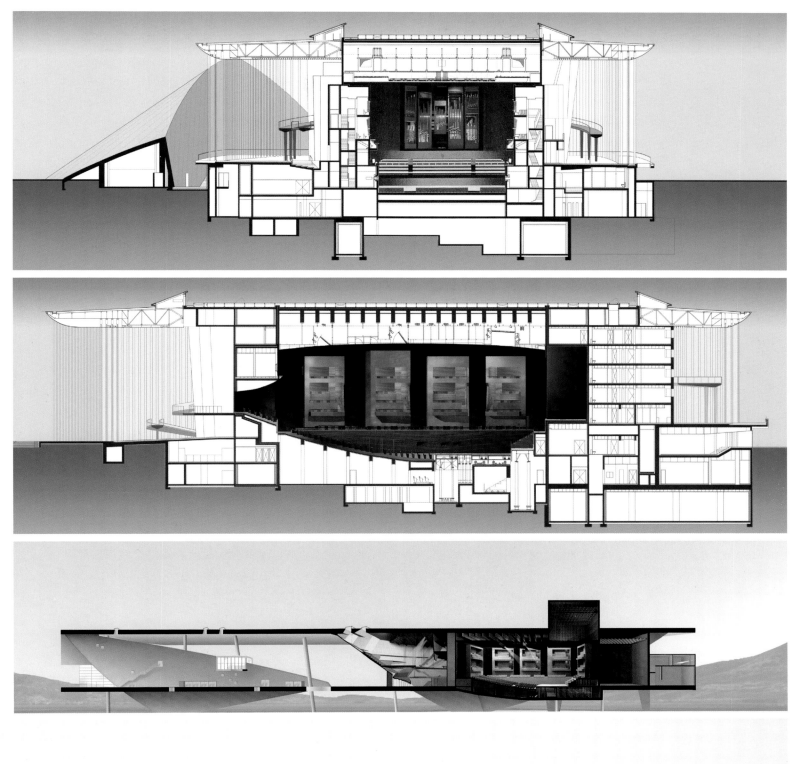

128　面向未来——建筑趋势 2015　FACING TO THE FUTURE—ARCHITECTURE TREND 2015

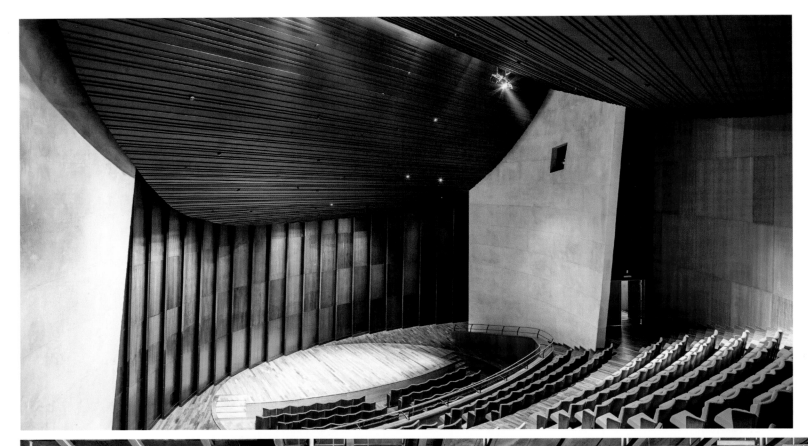

# 巴西圣保罗 Alphaville 俱乐部
## Alphaville Club

设计单位：Forte, Gimenes & Marcondes Ferraz Arquitectos
开发商：Alphaville Piracicaba club
项目地址：巴西圣保罗
建筑面积：930m²
合作者：Marília Caetano    Ana Paula Barbosa    Dante Furlan
　　　　Mônica Harumi    Bruno Milan    Flávio Faggion
摄影：Fran Parente

Designed by: Forte, Gimenes & Marcondes Ferraz Arquitectos
Client: Alphaville Piracicaba Club
Location: Sao Paulo, Brazil
Building Area: 930m²
Collaborators: Marília Caetano, Ana Paula Barbosa, Dante Furlan, Mônica Harumi, Bruno Milan, Flávio Faggion
Photography: Fran Parente

## 项目概况

该项目位于圣保罗皮拉西卡巴，是集舞厅、酒吧、健身房以及网球场、足球场、水上公园等为一体的大型综合俱乐部，旨在为封闭式的公寓居民提供一处娱乐、休闲的场所。

## 建筑设计

主大楼是一栋 7m 宽、56m 长、巨大的、呈连绵带状的线形建筑，它主要由体量极为轻盈的白色金属材质构成。其体量一侧固定在后勤区不透明的体量中，另一侧向游泳池完全敞开，楼梯和游泳池之间为木质甲板和石质的日光浴场。另外，与这一建筑相连的是较低场地上的网球场，因此，建筑屋檐的末端则成为观看网球赛的观众席。

屋檐的天花板完全由整齐的木条制成，与白色金属结构形成鲜明对比，在木质天花板的"裂缝"之间设计有LED软管，在夜色中将该处衬托得极为醒目。

和社交区域一样，俱乐部入口屋檐也为白色金属结构与木质天花板的设计，简单的造型给居民和游客一种惬意的感受，在娱乐运动的同时，体会休闲时间的浓浓幸福感。

## Profile

Located in Piracicaba, São Paulo, the Alphaville Club is a large comprehensive club including a ballroom, bar, fitness room, tennis court, soccer field and a generous water park. The project aims to create pleasant leisure spaces for residents of a future closed condominium.

## Architectural Design

The main building is a large linear construction, 7m wide and 56m long, like a continuous ribbon, composed by a very light white metallic structure. This volume is anchored in the opaque volumes of the support spaces on one side and, on the other side, fully opened to the swimming pools area, which are intercalated by wood decks and stone solariums. Continuing this slim construction, there is a tennis court, in a lower landing. Thus, the end of the marquee works almost as an audience, a grandstand for tennis matches.

The ceiling of this marquee contrasts with its fully white structure; it is entirely made of regular slatted wood. In between these "gaps" of the wood ceiling, LED hoses are installed randomly, highlighting the wood ceiling.

Similar to the social area, the entrance marquee of the condo has white metallic structure, wood ceiling and features itself as a simple construction which receives the resident and visitor in a cozy way. The project sought to create a coherent set of constructions aiming at the users' well-being in their leisure moments.

Linear Construction
Light Metallic Structure
Special Marquee

线形建筑
轻金属结构
特色屋檐

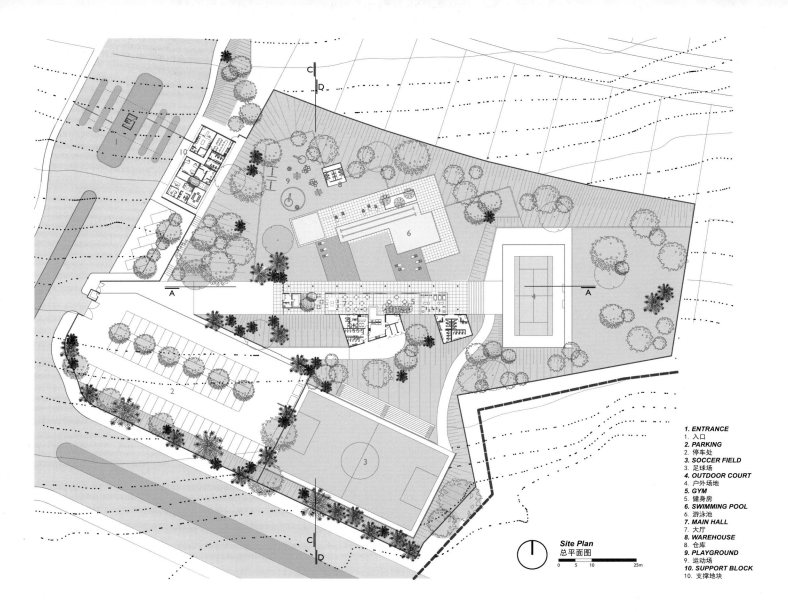

**Site Plan**
总平面图

1. **ENTRANCE**
1. 入口
2. **PARKING**
2. 停车处
3. **SOCCER FIELD**
3. 足球场
4. **OUTDOOR COURT**
4. 户外场地
5. **GYM**
5. 健身房
6. **SWIMMING POOL**
6. 游泳池
7. **MAIN HALL**
7. 大厅
8. **WAREHOUSE**
8. 仓库
9. **PLAYGROUND**
9. 运动场
10. **SUPPORT BLOCK**
10. 支撑地块

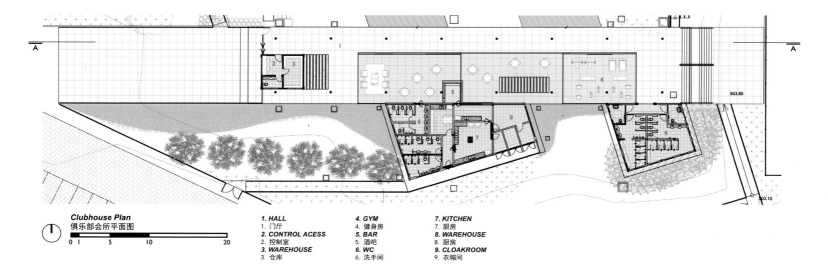

**Clubhouse Plan**
俱乐部会所平面图

0  1   5   10   20

1. HALL — 1. 门厅
2. CONTROL ACCESS — 2. 控制室
3. WAREHOUSE — 3. 仓库
4. GYM — 4. 健身房
5. BAR — 5. 酒吧
6. WC — 6. 洗手间
7. KITCHEN — 7. 厨房
8. WAREHOUSE — 8. 厨房
9. CLOAKROOM — 9. 衣帽间

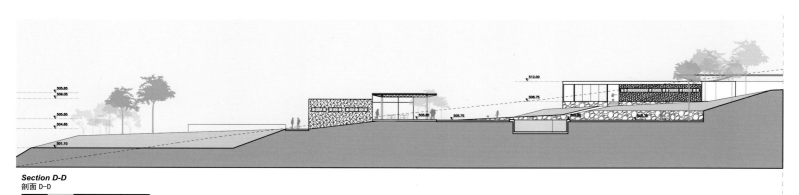

**Section D-D**
剖面 D–D

0  5  10  25m

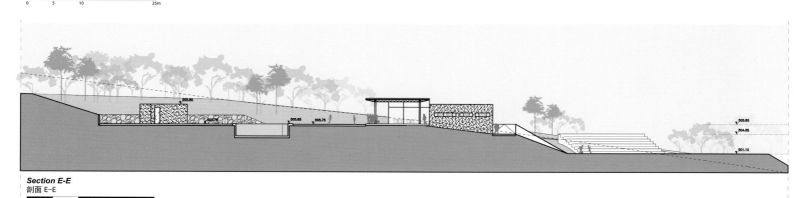

**Section E-E**
剖面 E–E

0  5  10  25m

# 荷兰阿尔梅勒羊厩
## Sheep Stable Almere

设计单位：70F architecture
项目地址：荷兰阿尔梅勒
总建筑面积：400m²
摄影：Luuk Kramer

Designed by: 70F architecture
Location: Almere, Netherlands
Gross Floor Area: 400m²
Photography: Luuk Kramer

**项目概况**

位于荷兰的阿尔梅勒市拥有约 80 只绵羊，而该建筑则是为这些绵羊设计的一处居所，方便当地管理局对这群绵羊的集中管理和安置。

**建筑设计**

整座建筑大部分以木头作为材料，结构使用了松木，外部覆盖层使用了雪松面板材料，而弯曲的大梁则采用钢构来强调室内的管状造型。管状的羊厩共有三个开口供日光照入，最低的一个开口设置在接近地面的高度，让人们在羊厩关闭时也能看到羊厩内的情况。

另外，在长立面末端的细节设计对整个羊厩的建筑设计显得尤为重要，突出了建筑横剖面的三角造型，立面也由此慢慢地变为屋顶。这座建筑不仅是绵羊的居所，还成为一般民众或学校课程可参观的场所，使人们对羊群有了更深的认知，在人与动物的情感交流中，学会爱和保护。

Tube-like Shape
Homogenous Cross Section
A-symmetrical Cross Section

管状
三角横剖面
不对称剖面

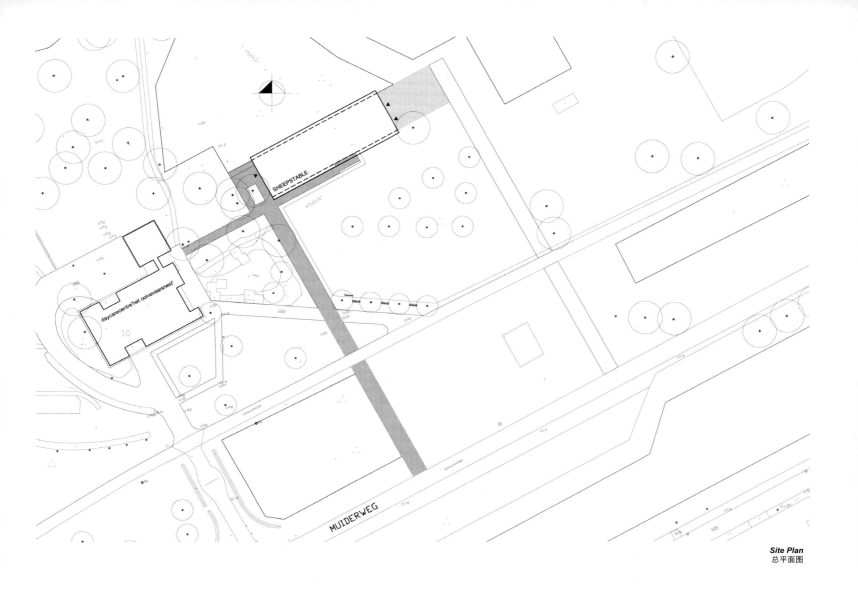

*Site Plan*
总平面图

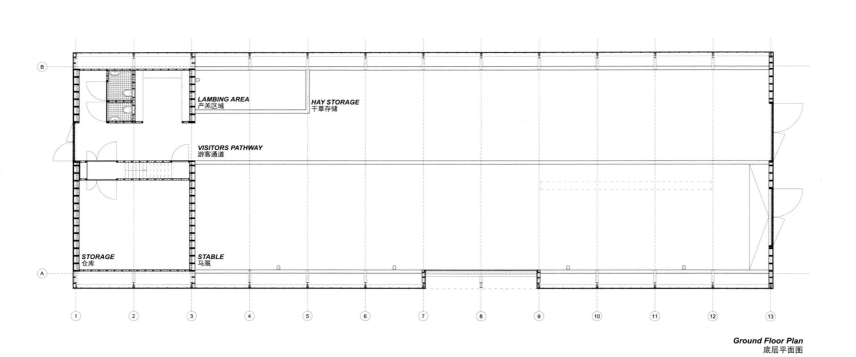

*Ground Floor Plan*
底层平面图

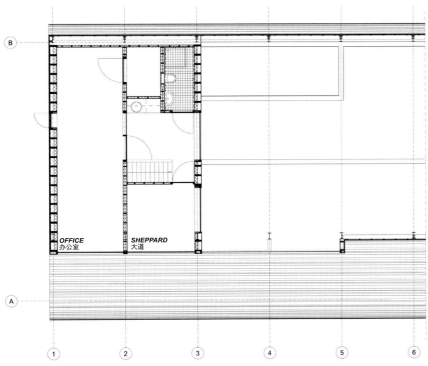

*First Floor Plan*
一层平面图

### Profile
The city of Almere has a sheep population of about 80 sheep. This stable is designed for the sheep, in order to keep them under control.

### Architectural Design
The construction (pine) and cladding (Western Red Cedar) are made of wood. Only the curved girders are made of steel to emphasize the tube-like shape of the interior. The tube has three strategically placed daylight openings, of which one is close to the floor level, so people can look inside even when the building is closed. All vertical walls in the stable and office are clad with beech plywood.

The detailing of the corner of the building is extremely important for the overall experience of the architecture of this building. It emphasizes the cross sectional shape of the building, and finishes the long façade of the building, which starts as a façade and slowly becomes roof. The stable is designed to make it possible for the public or school classes to visit the building and experience the keeping of sheep up close. In the communication with sheep, man can better understand the sense of love and protection.

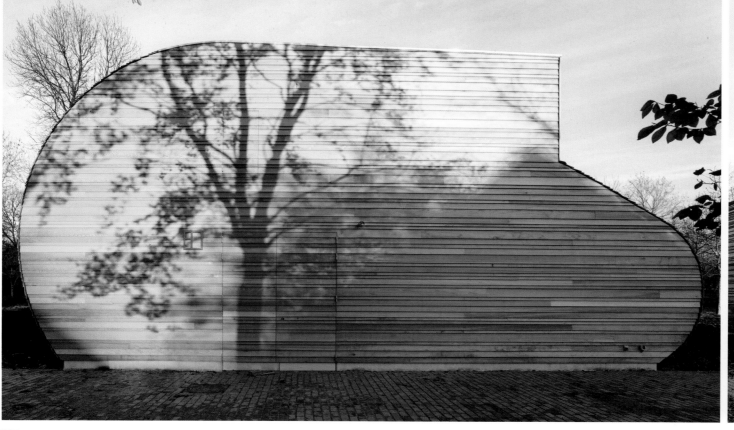

◀ 不对称剖面——建筑师将羊厩设计成简单的管状，但其剖面则处理成一个不对称的造型。绵羊居住的部分较低，较高的部分设在公共道路和干草储存室之上，这样既能储存更多的干草，又使得室内的空气能自然流动。羊厩下方的两道开口，更增强了这一效果。A-symmetrical cross-section—the stable is designed with an a-symmetrical homogeneous cross-section. The part of the building where the sheep reside is relatively low; the high part is situated above the (public) pathway and the hay storage section, making it possible to store a maximum amount of hay. This shape also creates a natural flow for the air inside the building, which is refreshed by two slits at the foot of each long side of the building.

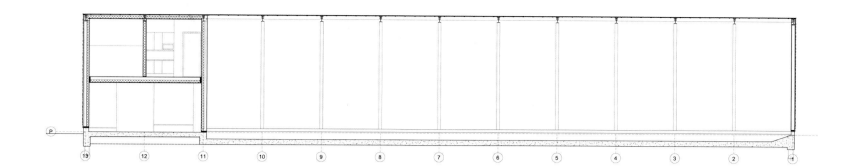

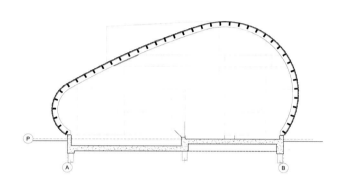 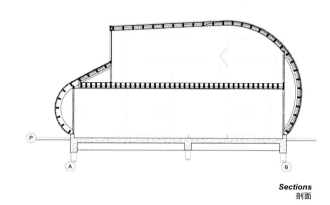

*Sections*
剖面

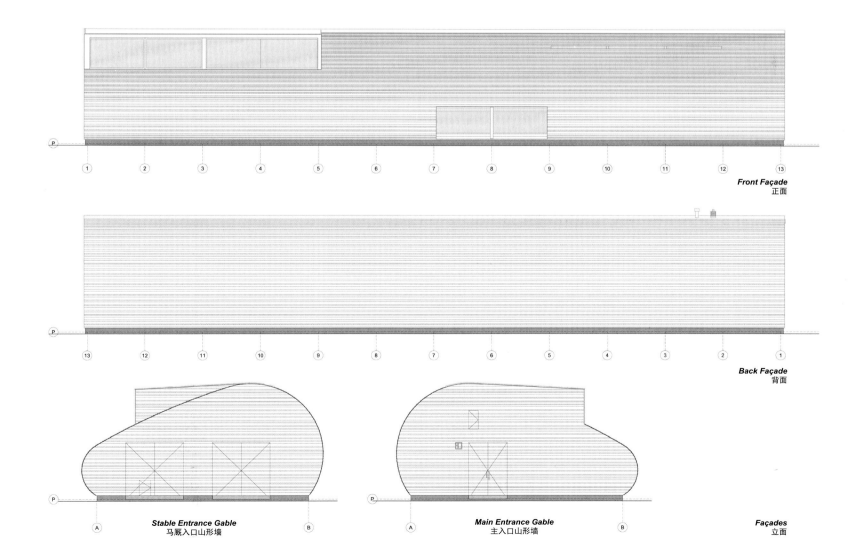

*Front Façade*
正面

*Back Façade*
背面

**Stable Entrance Gable**
马厩入口山形墙

**Main Entrance Gable**
主入口山形墙

*Façades*
立面

Mullions Skin
Metallic Punched Panels
Wooden Framework

竖向框立面
金属穿孔面板
木框架结构

# 法国巴黎 Asnieres-Sur-Seine 学校体育馆
## Gymnase Scolaire, Asnieres-Sur-Seine

设计单位：Ateliers O–S architectes
开发商：Asnieres–Sur–Seine 城
项目地址：法国巴黎
建筑面积：1 001m²
摄影：Cecile Septet

*Designed by: Ateliers O-S architectes*
*Client: Asnieres-Sur-Seine City*
*Location: Paris, France*
*Building Area: 1,001m²*
*Photography: Cecile Septet*

## 项目概况

该体育馆位于巴黎的阿涅尔河畔,高架铁路的两侧,一个 30m 宽、250m 长的狭窄地块上。这一沿着铁路路堤延伸且南面向塞纳河敞开的场地条件,指引着建筑的设计方向。

## 建筑设计

建筑外立面采用了有节奏的竖向框,整体统一的基础上局部又有所不同,笔直的立面线条从铁路路堤中拔地而起,使得整座建筑宛如一个巨大岩石。

沿着街道的体育大厅在城市和体育馆之间形成强烈的视觉连接,大厅周边围绕着更衣室、设施房和其他房间。屋顶上的一个开口为室内带来弥漫的光线,补充着立面巨大窗户带来的自然光线。

建筑沿街立面由两层构成,底层的落地玻璃是体育馆向该城市开放的唯一区域,为街道带来勃勃生机。巨大的高层由金属穿孔面板构成,垂直的灯光使建筑显得格外醒目。另外,建筑内部结构由木框架构成,并且内部挑高很高,既保持了室内空气通畅,又营造出一个明亮通透的内部空间。

## Profile

"Gymnase Scolaire" is located in a narrow site, 30m wide and 250m long. The site opens to the south on the river La Seine, along a railway embankment, which directs the design.

## Architectural Design

The building's skin maintains a rhythmic pattern of structure and mullions, unified in its entirety and different partly. The gymnasium emerges from the railway embankment as a rock with straight edges.

The sports hall is located along the street, creating a strong visual link between the city and the gymnasium. Around are organized the changing rooms, the facilities and other additional rooms. An opening on the roof brings diffuse light, completing the light coming from the wide windows on the façades.

The street façades are composed by two levels: a glassed low level that brings life to the street at pedestrian scale, and a high massive level composed of metallic punched panels that highlight the building with vertical lights. The inner structure is made of a wooden framework leaning over a concrete basement along the railway. The high laminated timber beams appears on the roof, giving rhythm and depth.

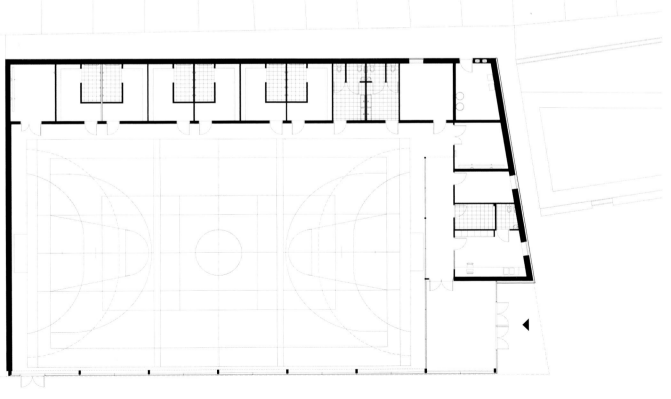

RUE MARIE CURIE

**Section**
剖面图

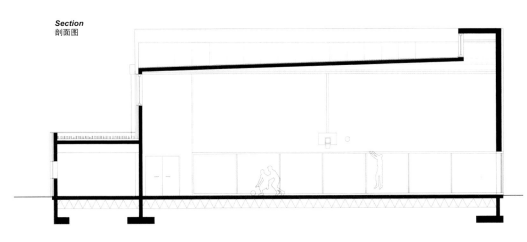

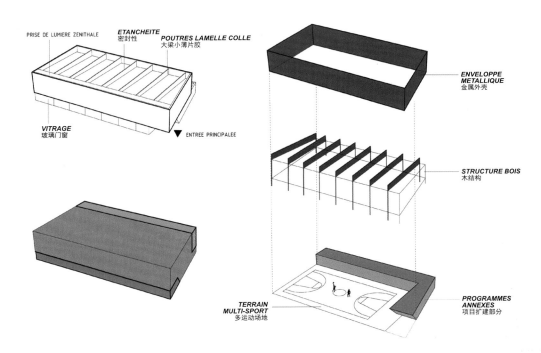

PRISE DE LUMIERE ZENITHALE / **ETANCHEITE** 密封性 / **POUTRES LAMELLE COLLE** 大梁小薄片胶 / **VITRAGE** 玻璃门窗 / ENTREE PRINCIPALEE

**ENVELOPPE METALLIQUE** 金属外壳

**STRUCTURE BOIS** 木结构

**TERRAIN MULTI-SPORT** 多运动场地

**PROGRAMMES ANNEXES** 项目扩建部分

# 荷兰阿姆斯特丹 A8 高速路公园
## A8ernA

设计单位：NL Architects
开发商：Gemeente Zaanstad
项目地址：荷兰阿姆斯特丹
建筑面积：37 500m²
设计团队：Pieter Bannenberg　Walter van Dijk　Kamiel Klaasse
　　　　　Erik Moederscheim　Sarah Möller　Annarita Papeschi
　　　　　Michael Schoner　Wim Sjerps　Crystal Tang

Designed by: NL Architects
Client: Gemeente Zaanstad
Location: Amsterdam, Netherlands
Building Area: 37,500m²
Project Team: Pieter Bannenberg, Walter van Dijk, Kamiel Klaasse, Erik Moederscheim, Sarah Möller, Annarita Papeschi, Michael Schoner, Wim Sjerps, Crystal Tang

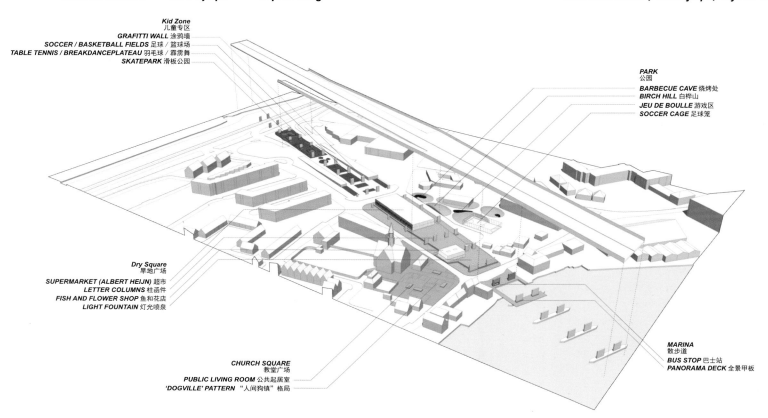

## 项目概况

Koog aan de Zaan 是阿姆斯特丹附近一个美丽的小村庄,坐落在 Zaan 河边。20 世纪 70 年代早期,这里新建了一条高速公路,而该项目就是试图在保留小镇两侧之间联系通道的基础上,使公路下面的空间变得活跃起来。

## 建筑设计

溜冰公园类似于一个"米老鼠"形态的碗状物,这个泳池就像是被掏空的水滴一样,坐落在高速路下。高速路旁边是一个小公园,里面有一些小丘陵,强化了绿的气息,还设有一个"烧烤穴"和一个足球场。

教堂前广场中多余的绿化都被清理,使其变得更具吸引力和实用性。在广场上,一种浅色的砖强化了原来的城市平面,并且用一种 Dogville 式的方式清晰地表达了这里原来房屋的配置。在高速路下挖出迷你港口,使水面能与主街道相连,一道堤坝可方便人们走到河道中的第一对柱子处,这一全景平台可观看到河面美妙的景色。

## Profile

Koog aan de Zaan is a sweet little village near Amsterdam. It is located at the river Zaan. In the early seventies a new Freeway was constructed. The project is an attempt to restore the connection between both sides of town and to activate the space under the road.

## Architectural Design

The sophisticated skate park features a "Mickey Mouse" shaped bowl. The pool is a kind of excavated blob that sits under the highway. Next to the highway is a small park with some hills that intensify the experience of the greenery. Carved out from these are a "barbecue cave" and a soccer cage.

Redundant greenery was removed from the square in front of the church that as such became much more attractive and usable. On the square, the original city plan is highlighted in a lighter brick and articulates in a Dogville kind of way the configuration of houses that used to be here. By introducing the mini harbor that is excavated from the land under the highway the water connects to Main Street. A jetty allows access to the first two columns in the stream. The Panorama Deck features wonderful views over the river.

**Bowel**
**"Excavated Blob"**
**Innovative Park**

碗状物
"水滴"
创意公园

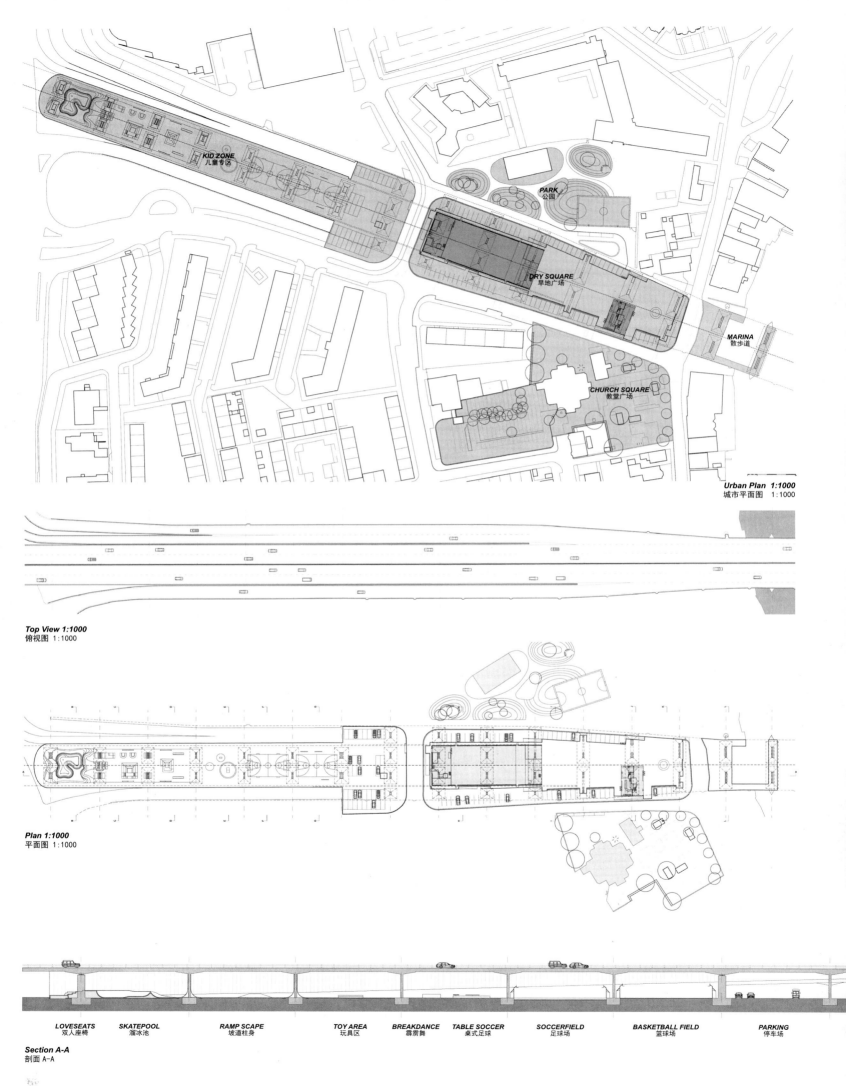

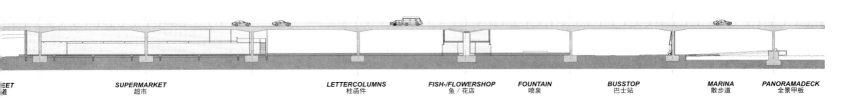

| EET | SUPERMARKET | LETTERCOLUMNS | FISH-/FLOWERSHOP | FOUNTAIN | BUSSTOP | MARINA | PANORAMADECK |
| 道 | 超市 | 柱函件 | 鱼／花店 | 喷泉 | 巴士站 | 散步道 | 全景甲板 |

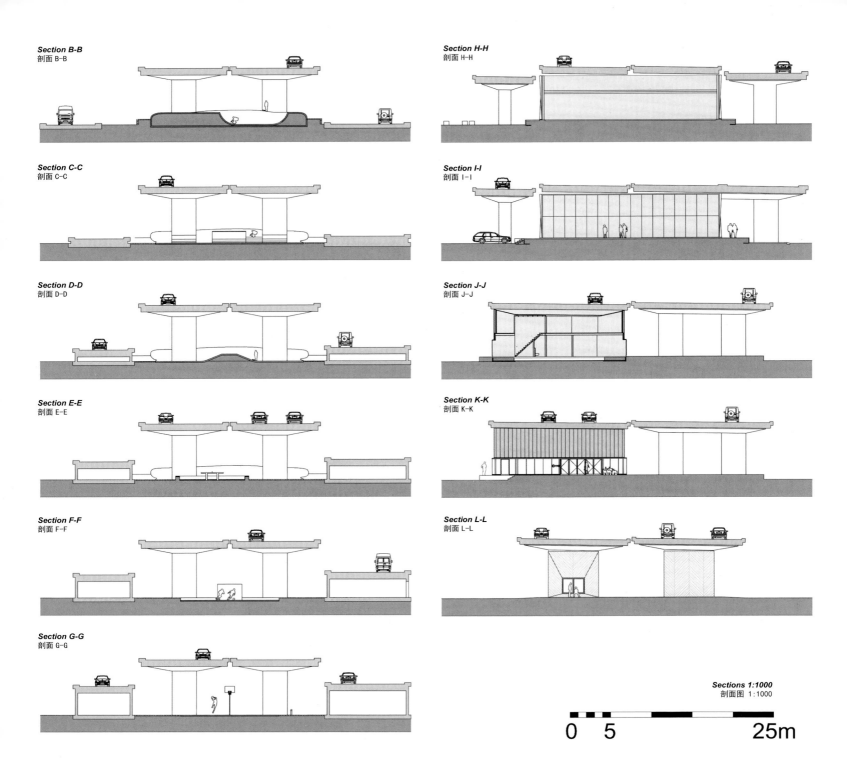

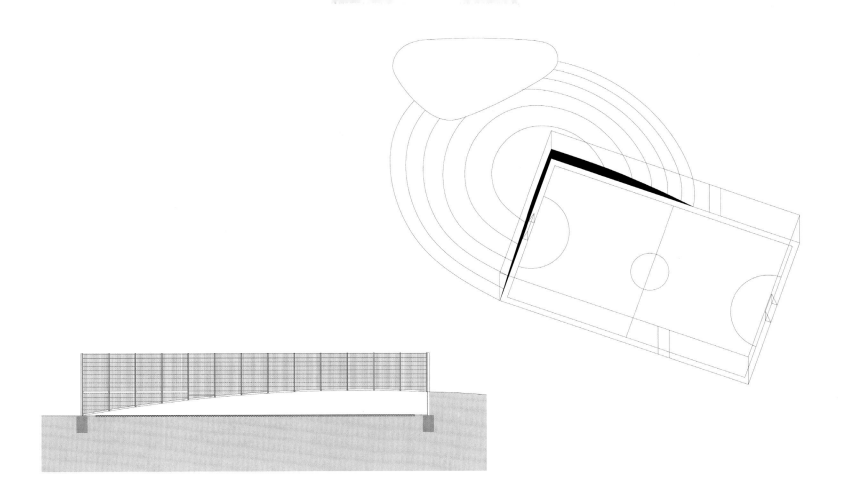

**Detail Soccer Cage 1:100**
足球笼详图 1:100

0　1　　　5m

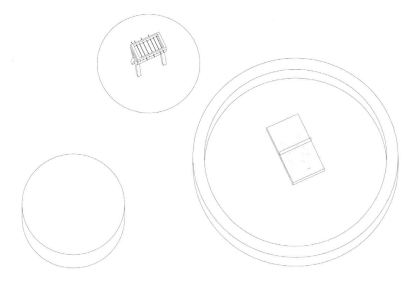

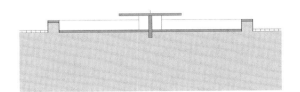 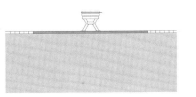 

**Detail Play Zone 1:100**
游乐区详图 1:100

0　1　　　5m

# 2015 年世博会奥地利馆
## Austrian Pavilion in Expo, 2015

| | |
|---|---|
| 设计单位：Paolo Venturella Architecture | Designed by: Paolo Venturella Architecture |
| 设计团队：Fabrizio Furiassi　Manuel Tonati　Angelo Balducci | Design Team: Fabrizio Furiassi, Manuel Tonati, Angelo Balducci |
| 委托人：2015 年展览会 | Client: 2015 Exhibition |
| 建筑面积：960m² | Building Area: 960m² |

# 可持续发展 蔓延的房子 奥地利木屋式结构
# Sustainable Development Spread House Austrian Chalet-style Architecture

**项目概况**

该项目的设计理念主要是"可持续发展"——通过"蔓延的房子"这一建筑实体,清楚地表明如何架构真正的"可持续"的内涵:自然光是可控制的。

**建筑设计**

该展馆的主要概念是创建两个不同的空间。一个用于展览,另一个被设计成为一个大温室,在温室的花园里,人们可以直接采摘食物来吃。虽然两个空间的形状是相似的,但在功能上则完全不同。一个是用玻璃制造,减轻太阳的自然光照;而另一个是覆盖冷却液材料以获得更暗的空间,在此空间可以展示展览板和地方办事处。从景观的审美观点来说,该建筑借用了奥地利木屋和温室的典型形状,花园面向太阳被设置在南侧,它与酒吧和餐厅相连,以创造一个人们可以很容易地吃他们自己摘取的食物的地方。展览空间为避免阳光直射,被设置在北侧,而且展馆部分被抬升以容纳该地区的人流。

**Plan 00**
平面图 00

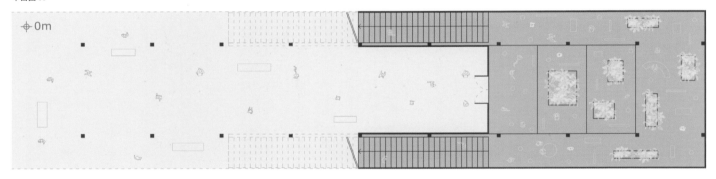

**Plan 01**
平面图 01

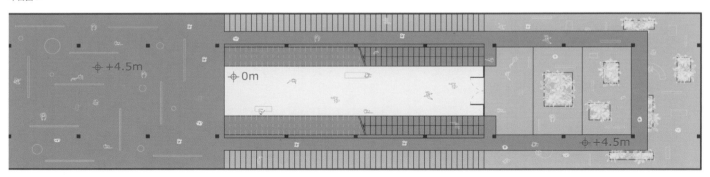

**Plan 02**
平面图 02

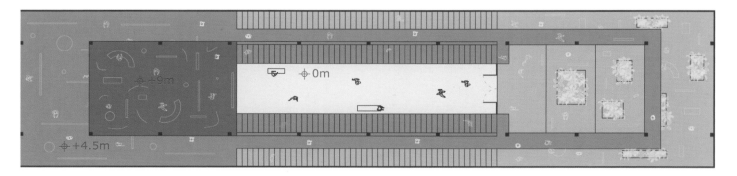

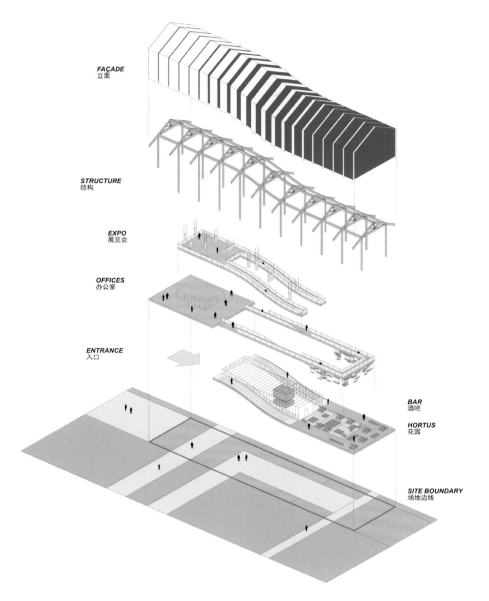

## Project Overview

The main concept of the project is "sustainable development"—the "Rampant house" clearly show the message of how an architecture can really be "sustainable" the natural light can be controlled.

## Building Design

The main concept of the pavilion is to create two different spaces. One is for the exhibition and another one for a big greenhouse. They are connected to allow people to walk in a unique single space. The greenhouse is the place where people can pick up and eat food directly from the hortus. The two spaces are similar in shape but totally different in their function. One is made in glass, lightened with natural light of the sun, and the other is covered with stoff material for a darker space where is possible to show panels of the exhibition and places for offices. From an aesthetic point of view is used the typical shape of the Austrian wood house and of the greenhouse. The Hortus is placed on the south side facing the sun. It is connected with the bar and restaurant in order to creating a place where people can easily eat the vegetables picked up by themselves. The exhibition space is placed on the north side to avoid direct sunlight. Moreover this part of the pavilion is lifted up to allow the pedestrian flow on the area.

*Elevation*
立面图

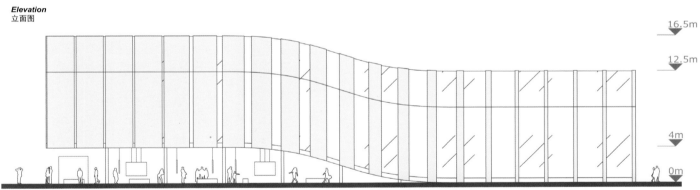

*Section*
剖面图

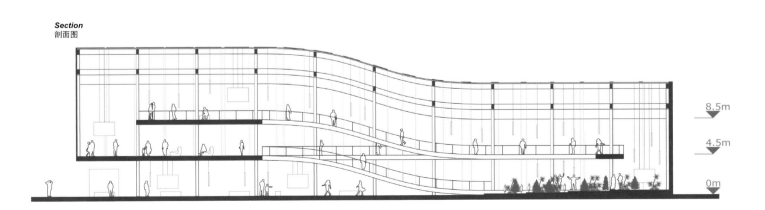

# 广州杨家声设计顾问有限公司办公室
## Ben Yeung Office

设计公司：广州杨家声设计顾问有限公司
开发商：广州杨家声设计顾问有限公司
项目地址：广东省广州市荔湾区西增路 63 号原创元素创意园 D5,D6 写字楼
总建筑面积：1 300m²

Designed by: ben yeung & associates ltd.
Client: ben yeung & associates ltd.Guangdong
Location: Guangzhou, Guangdong
Gross Floor Area: 1,300 m²

Loft Style
Open-plan Office Space
Natural Lighting Design

loft 风格
开敞式办公空间
自然采光设计

**项目概况**

Ben Yeung Office 坐落在广东省广州市荔湾区西增路63号原创元素创意园，建筑面积 1 300m²，建筑主要采用 loft 风格设计，保留浓厚的工业味道，同时以简约前卫、绿色环保的理念为团队打造现代舒适的开敞式办公空间。

**建筑设计**

办公室在旧有厂房空间的基础上利用钢结构加固支撑，以原始的砖墙作为空间间隔。大门与大堂接待处等主要空间用空旧酒瓶作为装饰，呼应原有的啤酒厂文化。硬朗的质感，融合柔和的原木设计元素，简约而协调。

屋顶自然采光设计运用于大堂接待前厅、开放式图书馆阅读区及办公区域，充沛的阳光让整个空间更加明亮通透；顶部的自然通风调节室内温度，保持空气流通；垂直绿化墙更成为办公室内的天然氧吧。这种亲近自然的设计，最大限度地实现空间的自身节能，打造健康环保的办公环境。

办公空间间隔采用开敞式设计，没有整幅墙壁和门扇的阻隔局限，营造亲近和睦的办公气氛，让团队活跃于沟通交流，以高质量的工作环境让团队专注于优化项目成果。

***First Floor Plan***
一层平面图

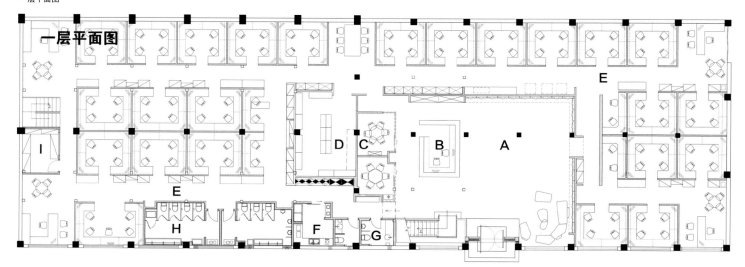

## Project Overview

Ben Yeung Office is located in the Liwan District, Xizeng Road, and original elements of Creative Park No. 63, Guangzhou, Guangdong, with building area of 1,300 square meters. The building mainly uses loft-style design to retain a strong industrial flavor, while minimalist avant-garde and the concept of green create a modern and comfortable open-plan office space for the team.

## Building Design

The office uses steel which based on the old factory space for the reinforcement of the support and makes the original brick as a spatial interval. The main door and the reception hall spaces use old empty bottles as decoration, echoing the original brewery culture. Tough texture merged with soft wood design elements is simple and harmonious.

The natural lighting design of the roof is used for lobby reception, open library reading area and office area, with plenty of bright sunshine to make the whole space more transparent; natural ventilation at the top to adjust the room temperature, keeping the air circulating; vertical green wall is more become a natural oxygen bar in the office. This close nature of design is maximize to realize space itself energy saving, creating a healthy and environmental protection working environment.

Office space intervals use open type design without the barrier and limitation of the whole walls and doors, creating an close and harmonious office atmosphere, so the team is active in communication, with high-quality work environment for the team to focus on optimizing the results of the project.

*Second Floor Plan*
二层平面图

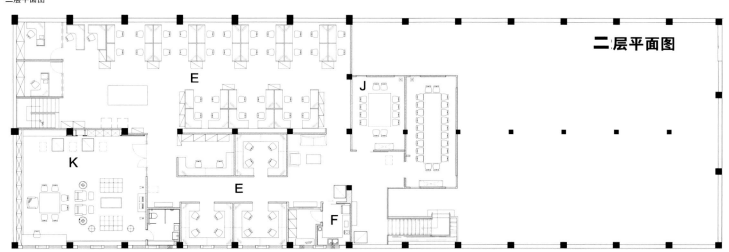

**Section**
剖面

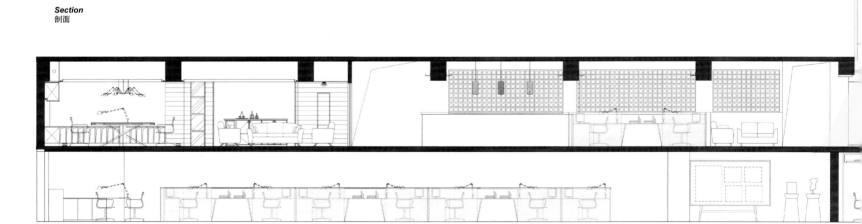

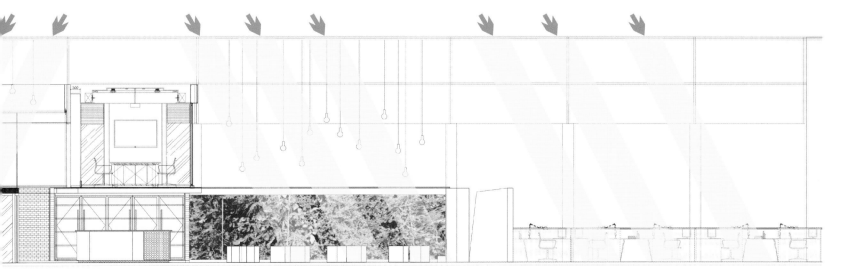

# 以"理"服人
## ——建筑理念新传承

# Convinced "concept"
## —new heritage of architecture idea

建筑伴随着人类的产生而产生，从穴地而居到高楼大厦，从茅草之屋到美轮美奂的建筑，在漫长的历史过程中经历了无数次的转变，每次转变都伴随着建筑理念的传承与创新。尤其是到了现代以后，人们对建筑提出了更高的要求，建筑的形式也跳出了几种经典的范畴，各种各样的流派和思潮，层出不穷，犹如百家争鸣，纷繁复杂。

建筑是一个美妙的东西，是人类文献中最伟大的记录，也是对时代、地域和人的最忠实的记录。建筑反映的不只是建筑本身的造型和内部的空间关系，从建筑里你能解读到建筑大师的人生观和建筑所蕴含的文化和艺术。从一个建筑看世界，你能从中领略到永恒。概括来说，目前比较流行的建筑理念主要包括以下几方面：

一是崇尚自然的建筑观。经过上百年的锤炼，建筑基本上已经褪去了它原有的本质，随着人们思想观念的进一步解放，越来越多的人开始向往那种"结庐在人境，而无车马喧。问君何能尔，心远地自偏"的原始情怀。因此，崇尚自然的建筑理念在近几年开始"蹿红"。

二是构建"活"的有机建筑。这里说的"活"的建筑，是指反映人类社会生活的真实写照的建筑。这种"活"的建筑观念，能使建筑师摆脱固有形式的束缚，注意按照使用者、地形特征、气候条件、文化背景、技术条件以及材料特征的不同情况而采取相应的对策，最终取得自然的结果，而并非是任意武断地加强固定僵死的形式。"活"的建筑可以借助于建筑结构的可塑性和连续性进而去实现其整体性。

三是表现出建筑真正的"本性"。著名建筑师赖特曾说"每一种建筑材料有自己的语言……每一座建筑都有自己的故事……"追求建筑艺术性的前提是必须先尊重建筑的"本性"，即该建筑存在的理由。

四是现下最为流行的观念——绿色、生态的可持续建筑观。绿色建筑不仅仅将价值取向指向生态与科学，亦指向了审美化的生命情感。正如德国诗人荷尔德林在他的诗歌所写的那样"诗意地安居"，而这简短的五个字在经历了上百年的历史洗礼后，却发展成为今天都市人群追求居住场所的最高境界。

Building is accompanied by the generation of human, from the caves to the high-rise buildings, and from the thatched houses to the magnificent buildings, which has undergone numerous changes in the long course of history and every shift is accompanied by the architectural concept of inheritance and innovation. Especially to the modern, people put forward higher requirements for the construction, and the forms of the buildings have also out of the scope of several classic category, so a variety of genres and ideas emerge endlessly, like contending, which are numerous and complicated.

Architecture is a wonderful thing, which is not only the greatest record in the human literature, but also is the most faithful record for the age, geography and people. Architecture not only reflects the relationship between the shape of the building itself and the interior space. You can interpret the architects' views of life and the cultures and arts implied in the architectures from buildings to buildings. And you are able to experience the eternity from one building to see the world. Generally speaking, the more popular architectural concept mainly includes the following aspects:

The first is the concept of building respects for nature. After hundreds of years of polish, buildings basically have faded its original nature, and with the further liberation of people's ideas, more and more people began to dream of the primitive feelings of "I built my hut in peopled world, Yet near me there sounds no noise of horse or coach, Would you know how that is possible? Secluded heart makes secluded place". Therefore, the concept respecting for nature has been "leaping up to red" in recent years.

The second is to build a "live" organic architecture. That "live" architecture is a true portrayal reflecting the human social life. This "live" architecture concept, enable the architects to get rid of the bondage of inherent forms, thus they will pay attention to taking relevant countermeasures in accordance with different circumstances for users, topography, climatic conditions, cultural background, technical conditions and material characteristics, and ultimately achieve natural result, rather than arbitrarily strengthen the fixed and dead forms. The "live" architecture can by means of the plasticity and continuity of the building structure to achieve its integrity.

The third is to show the real "nature" of the architecture. The famous architect Wright once said, "every building material has its own language ......every building has its own story......", so the pursuit of architectural artistry premises that you must respect the "nature" of the architecture, namely the reason existing in the construction.

The fourth is the most popular idea at this moment—green and ecological concept of sustainable architecture. Green building not only makes value orientation to the ecology and science, but also points to the aesthetic life emotions. Just as the German poet Holderlin wrote in his poetry "poetic settlement", while the short five words experienced hundreds of years' baptism, has been developed into the highest state of place to live for today's urban people to pursue.

# 韩国昌原国立庆尚大学医院
## Changwon Gyeongsang National University Hospital

设计单位：BAUM Architects
项目地址：韩国昌原
竣工年份：2015
占地面积：79 743m²
建筑面积：12 261m²
总建筑面积：105 580m²
设计团队：李贞兔　元亨竣　金相烨　金乔恩
　　　　　郑容圭　孙南一　孙基永　孙静娥
　　　　　李昶勋　李根映　白贤镐　崔在赫
　　　　　秦达瑞　韩知延　高一熙

Designed by: BAUM Architects
Location: Gyeongnam, the Republic of Korea
Completion: 2015
Site Area: 79,743m²
Building Area: 12,261m²
Gross Floor Area: 105,580m²
Project Team: Jungmyon F. Lee, Hyungjoon Won, Sangyeop Kim, Joeun Kim,
Namil Son, Kiyoung Son, Jungah Son, Changhun Lee,
Keunyoung Lee, Hyunho Baek, Youngkyu Jeong, Jaehyuck Choi,
Dalrae Jin, Zeeyeon Han, Ilhee Ko

**项目概况**

庆尚大学医院计划在昌原市成立一所综合专业医疗机构,来满足庆尚南道省中部人们的就医需要,旨在发展成为21世纪最先进的医院,强化医疗质量和服务,以引领韩国医疗的未来。

**理念与规划**

该建筑以"治愈浪潮"的理念为病人提供最新的医疗设备模范、以人为本的医疗环境以及和谐共存的自然与人文气氛。

建筑物底层部的大堂"医院街",环绕在自然景色之中,为病人及访客带来生态友好的环境。另外,它还成为一条城市街道,可进行各种各样的公共活动和休息等。

**立面与内部空间**

建筑每间病房设计都摒弃了巨大的六面体结构,而是采用平缓的弧线形外观,使得整座建筑都向大自然和城市敞开,为病人、家属以及访客营造舒适的休息区域。建筑虽构思为单一综合体,但其立面却具有多种形态,在功能运行相互独立的基础上,将每个空间都有机、合理地连接起来。

内部空间设计也打破了常规房间的庞大结构模式,从而形成创造性的空间,为病人提供安全、有效的医疗服务。

弧线形外观
街道大堂

Curvy Shape
Street Lobby

## Profile

Changwon Gyeongsang National University Hospital plans to establish a specialized medical institution in Changwon city to respond to essential medical needs of the peoples in mid-Gyeongsangnam Province, and aims to grow and develop as the world's foremost state-of-the-art hospital in the 21st century. It endeavors to enhance the quality of medical and patient services to lead the way for the Korean medical future.

## Concept and Planning

The construction has been designed under the concept of "healing wave" that provides a new paradigm for medical facilities, in which a human-oriented environment and air of harmonizing nature and culture are provided to patients. Hospital Street, which is a lobby in the lower part of the building, actively embraces outside nature for patients and visitors of hospital to enjoy an eco-friendly environment within the space. As a result, it realized an urban street with a diverse events, rest and space for community activities.

## Façade and Internal Space

Each ward rejects monolithic hexahedral form and pursues gradual curvy shape to demonstrate that the construction has been planned to be open to nature and the city, providing pleasant resting spaces for patients, families and visitors all together. Changwon Gyeongsang National University Hospital is composed as a single complex but also as a façade with a variety of formation. Furthermore, each space is planned to be organically connected but to have independence to perform clear functions as well.

Interior spaces also break with conventions for a monolithic structure of required rooms and design creative spaces to provide safe and efficient medical services for patients by materializing medical plans.

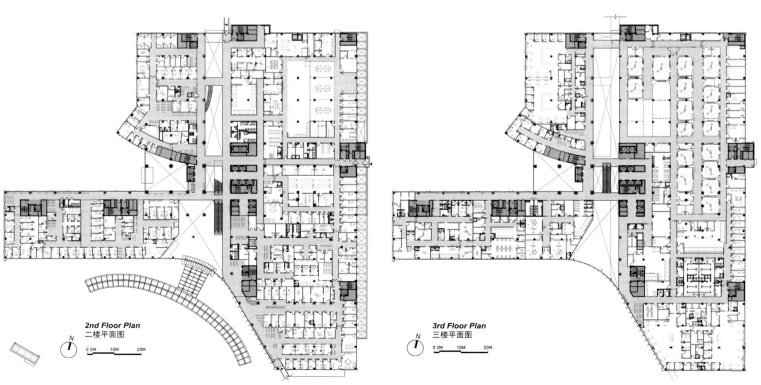

*2nd Floor Plan* 二楼平面图

*3rd Floor Plan* 三楼平面图

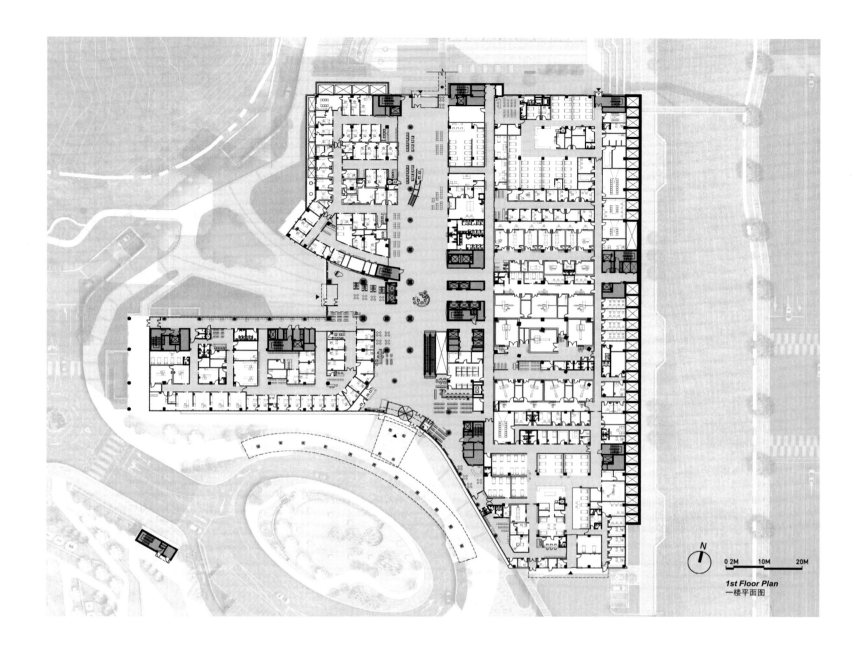

**1st Floor Plan**
一楼平面图

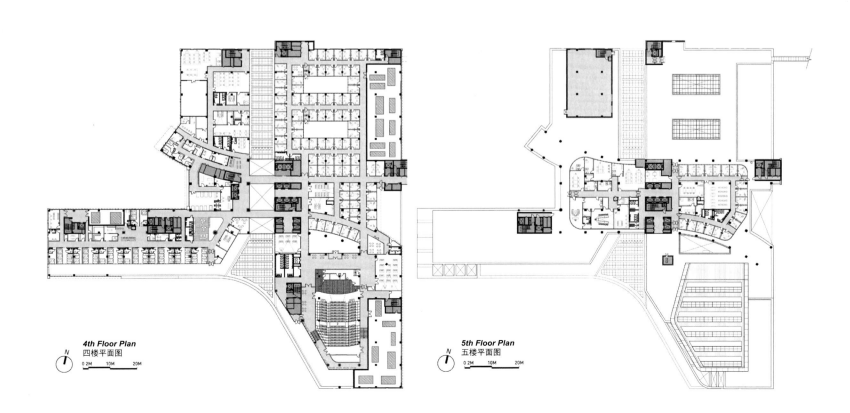

**4th Floor Plan**
四楼平面图

**5th Floor Plan**
五楼平面图

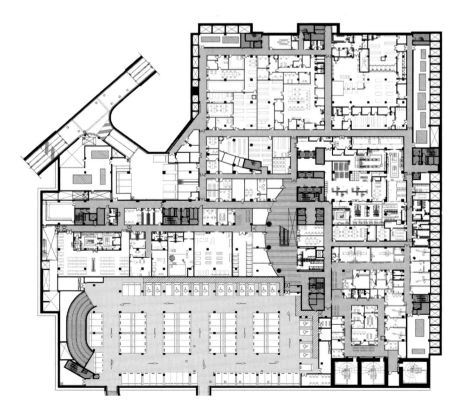

**B1 Floor Plan**
地下一楼平面图

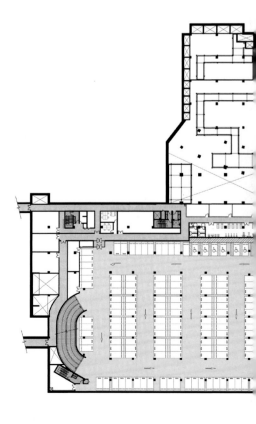

**B2 Floor Plan**
地下二楼平面图

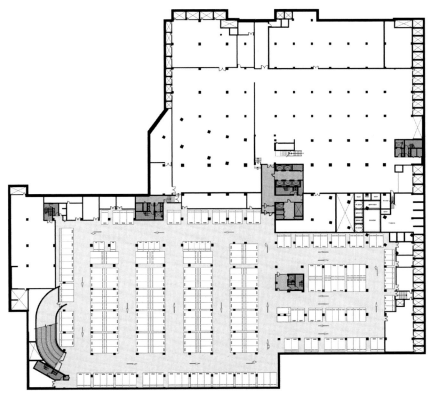

**B3 Floor Plan**
地下三楼平面图

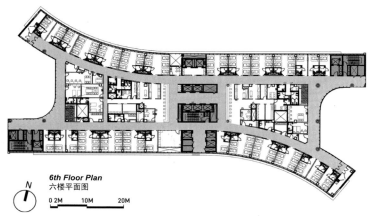

**6th Floor Plan**
六楼平面图

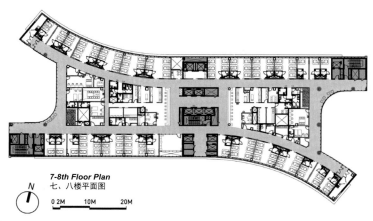

**7-8th Floor Plan**
七、八楼平面图

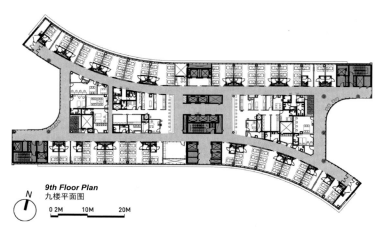

**9th Floor Plan**
九楼平面图

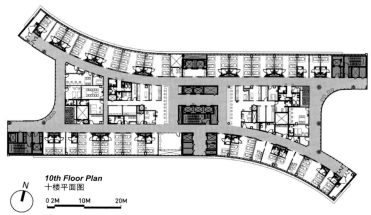

**10th Floor Plan**
十楼平面图

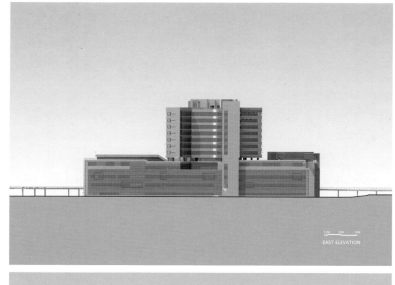
EAST ELEVATION

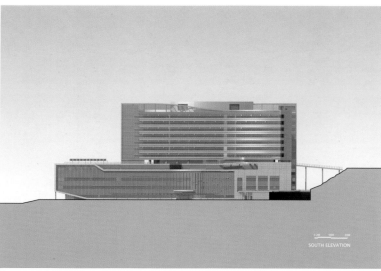
SOUTH ELEVATION

WEST ELEVATION

NORTH ELEVATION

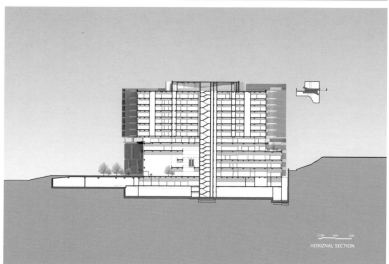
HORIZNAL SECTION

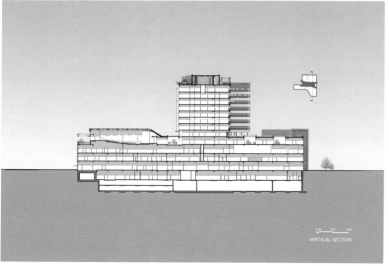
VERTICAL SECTION

# 绿谷
## Green Valley

设计单位：schmidt hammer lassen architects
合作单位：华东建筑设计研究院
委托方：上海世博建设开发有限公司
项目地址：中国，上海，云台路，绿谷
建筑面积：50 000m²

Designed by: schmidt hammer lassen architects
Cooperation Unit: East China Architecture and Design Institute
Client: Shanghai EXPO Construction Development Co. Ltd.
Location: Green Valley, Yuntai Road, Shanghai, China
Building Area: 50,000m²

## 项目概况

新绿谷项目坐落于2010上海世博会原展区，毗邻中国馆，由丹麦SHL建筑师事务所（schmidt hammer lassen architects）、华东建筑设计研究院和上海世博建设开发有限公司共同打造。该项目旨在塑造一个多样化、充满活力和包容性的社区，建成以后将成为上海乃至整个华东地区的一个绿色的、可持续的地标。

## 建筑设计

由于绿谷是在原世博展馆的基础上建造的，因此，它具备强大而成熟的基础设施——绿地公园、行人通道、文化景点以及现代办公设施等，坚固而扎实的基础条件为其成为上海新中心城市发展项目提供了有利的保障。此外，绿谷的设计还结合了可持续发展的理念，场址中央的开放空间由绿色植物、水以及柔和的景观贯穿而成，它们就如同绿谷的山脊。这个开放空间又将地块隔成两个同样大小的区域，每一边均有两栋主建筑。建筑群庞大贯通的结构是为了打造新城市生活而精心设计的。这些建筑将被用作现代办公设施，用材高端，具有灵活性，同时兼顾环境因素与低运营成本。

兼容性
可持续性
标志性

Compatibility
Sustainability
Iconic

## Project Overview

It starts construction of the new Green Valley project on the site of the former 2010 Shanghai Expo by schmidt hammer lassen architects, East China Architecture and Design Institute, and Shanghai Expo Construction Development Company together.

The project aims to building a community, giving people a diverse, vibrant and inclusive feeling. It will be a green, sustainable landmark for the city and for the entire region. The Green Valley development is expected to be completed in 2015.

## Project Design

The Green Valley project will mark the heart of the new permanent development on the site. The design expresses openness and accessibility, with a strong identity. A central open space composed of greenery, water and a soft landscape runs through the middle of the site. It functions as the spine of the Green Valley. This open space splits the site equally into two, with two major buildings located on each side. The buildings are designed to offer modern office facilities with a high standard of finish, flexibility, consideration of environmental issues, and low operating costs. The green hanging gardens inside the open atriums will be visible from the surrounding areas, and the people working in the buildings will be offered a great view to the greenery and city beyond. In addition, the design has a large, connected structure, tightly choreographed to set a new scene for urban life. It will act as a guiding element in the development of the entire area.

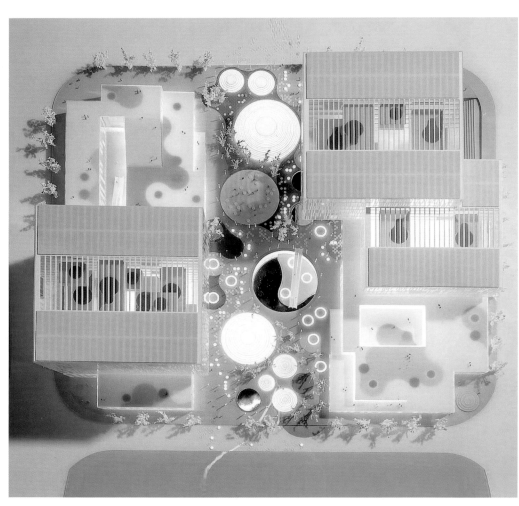

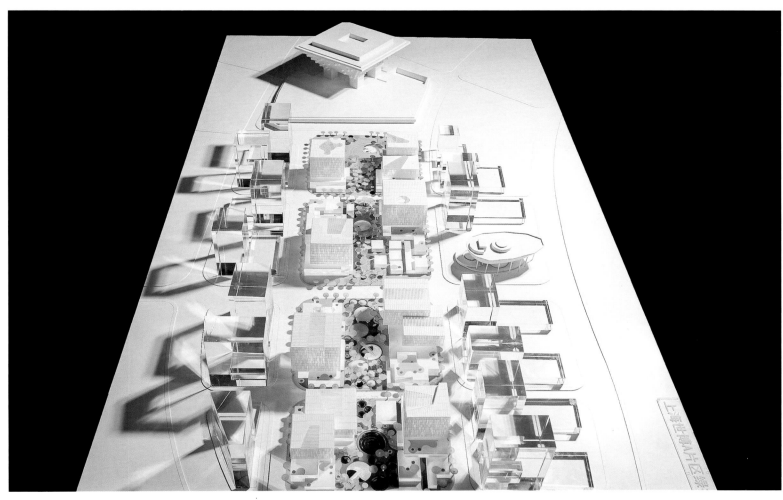

# 韩国大邱图书馆
## Daegu Gosan Public Library

设计单位：SURE Architecture 绿舍都会
开发商：韩国大邱图书馆
项目地址：韩国大邱
建筑面积：约 3 625m²

Client: South Korea Daegu Gosan Public Library
Designed by: SURE Architecture
Location: Daegu, Korea
Building Area: About 3,625m²

"突触式"的信息交换
"神经元式"的连接系统

"Synaptic" Type of Information Exchange
"Neuron" Type of Connection System

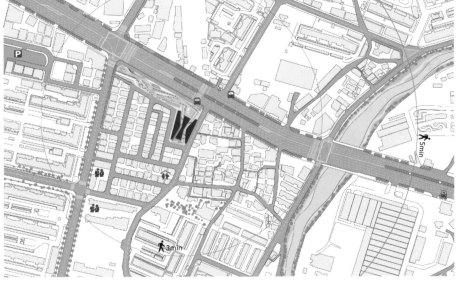

## 项目概况

信息交换作为学习的一种方式是使得人类发展和建立关系的基础。神经元每刻在我们的身体里这样做。它们通过突触进行信息交换。突触是一个过程,让整个神经系统互联,为刺激和处理信息作出回应。在这个思路的基础上,韩国大邱图书馆被设计为连接整个程序空间,考虑该地块北部的绿化带。

## 建筑设计

该项目旨在生成一个组织,其中在公园和建筑之间有连续性。因此,在该项目的设计中,设计师们通过引入绿化带,从而间接引导人们进入图书馆,振兴居委会,并建立新的运动坐标轴,使公民获得更多的学习和求知的机会。

该建筑是由三个体量组成:侧面两个和由连接器连接的中心一个。侧面的两个体量在侧边更高,它们是从一楼而上。因此,它们侧面与在中间的体量相接,并创建一个立面朝向邻近的街道。中间的体量没有接触到地面,使得公园作为一个公共空间深入地块。与中间由玻璃制成的体量相比,两侧的体量有更紧凑的外形。它们有一个像素化的立面合成物,有一个从底部缩减到顶部的紧密坡度。周边百叶窗的设计强调垂直梯度和与公园水平的连接。

建筑物内的用途和空间的位置也响应更多的公众到更多的私人的梯度及更多的噪音到更安静的梯度。中间的体量充当中间连接器,并且更公开了。

此外,它的意图是该建筑为了用户互动和交换信息,其本身可能是一个学习的元素。在建筑中设计有LED屏,其能量来自太阳能收益及用于绿化带植物灌溉的屋顶收集雨水。此外,还有一种路面发电用于公园的照明。因此,图书馆用户不知不觉地变得熟悉回收和节能。

## Project Overview

The exchange of information as a way of learning is the base of knowledge which makes humans develope and establish relationships. Neurons are doing it in our bodies at every moment. They exchange information through synapse. The synapse is a process that lets the entire nervous system be interconnected to respond to stimulus and process information. From the basis of this idea, the project has been designed to link the whole program spaces considering the green belt in the north of the plot.

## Building Design

The proposed project aims to generate a tissue in which there is continuity between the park and the building. Thus, in the design of the project, the tissue through the green belt designed by the architects indirectly leads people to enter into the library, revitalize the neighborhood and establishing new axes of motion to bring citizens learning and knowledge.

The building is composed by three volumes: two in the laterals and one in the center connected by linkers. The volumes in the laterals are taller in the side and they are up from the ground floor. Thus, they flank the volume in the middle and create a façade facing to the adjacent streets. The volume in the center is not touching the ground level to let the park to continue into the plot as a public space. The volumes of the sides have a more compact appearance in comparison to the intermediate volume, which is made of glass. They have a pixelated façade composition and there is a gradient of compactness decreasing from bottom to top. The design of perimeter louvers emphasizes the gradient vertically and the connection to the park horizontally.

The location of uses and spaces within the building also responds to a gradient of more public to more private and of more noise to quieter. The volume in the middle acts as an intermediate connector and is more public. Moreover, it is intended that the building can be itself an element of learning for the users to interact and exchange information. There are LED screens designed in the building whose energy comes from solar gains and the roofs collect rainwater for irrigation of the plants in the green belt. Furthermore, there is a kind of pavement power generator for lighting in the park. Thus, the library users become familiar with recycling and energy saving unconsciously.

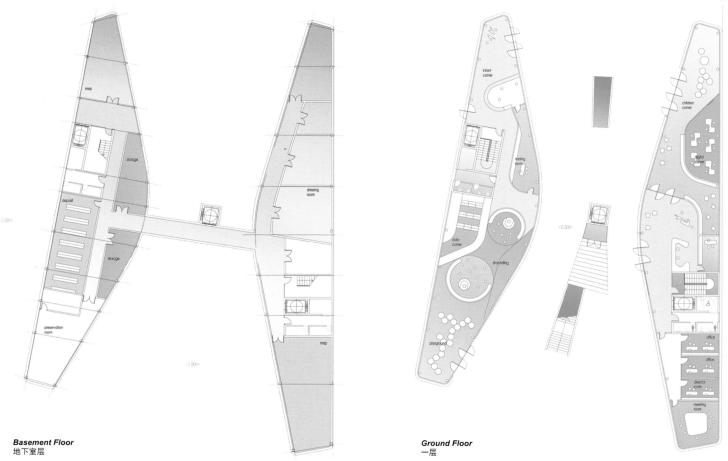

**Basement Floor**
地下室层

**Ground Floor**
一层

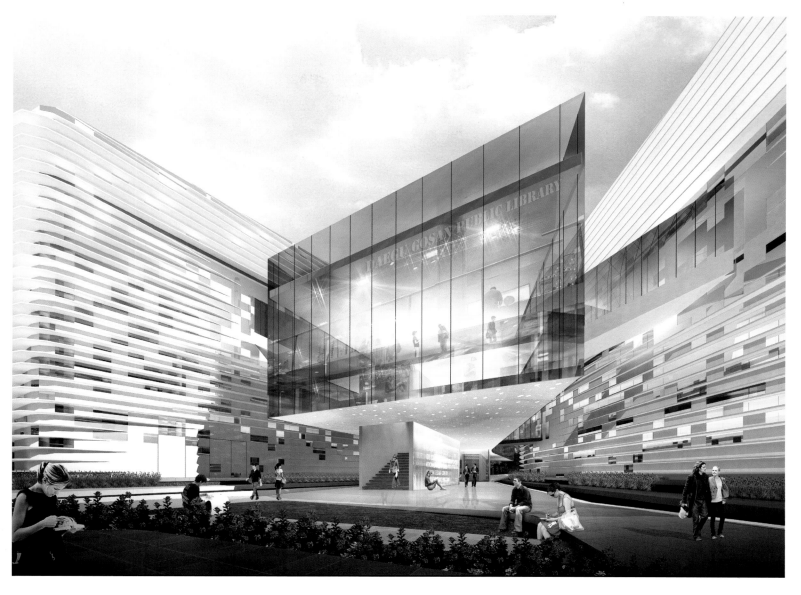

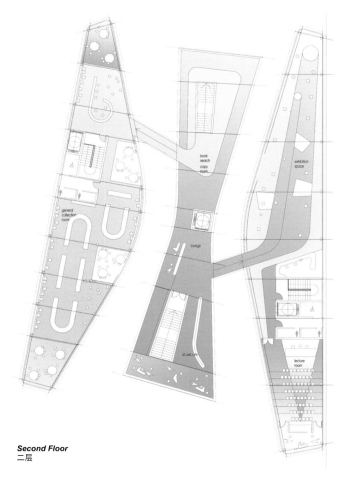

**Second Floor**
二层

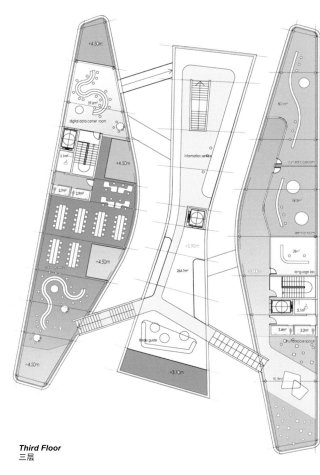

**Third Floor**
三层

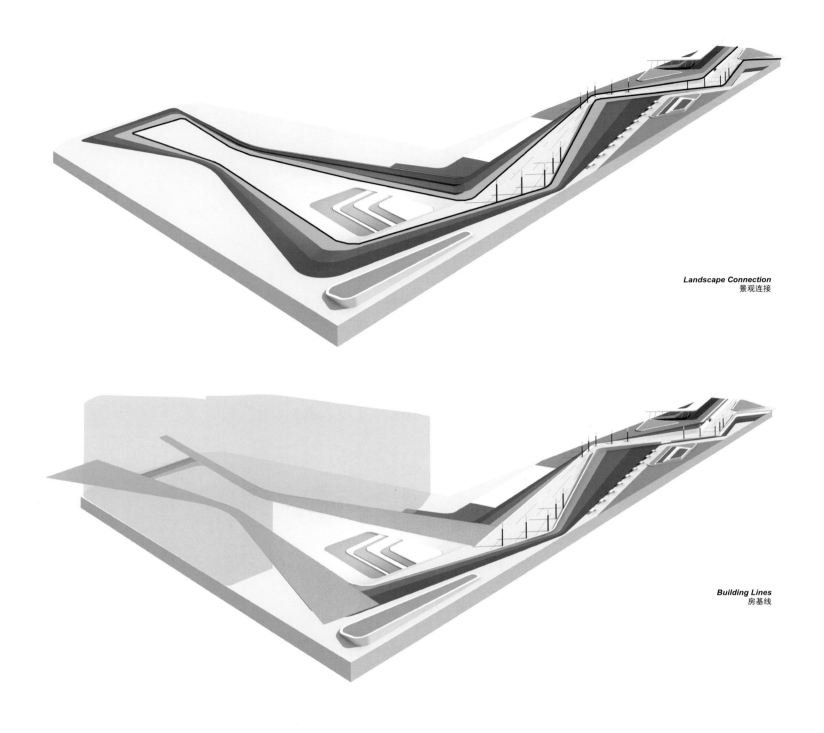

*Landscape Connection*
景观连接

*Building Lines*
房基线

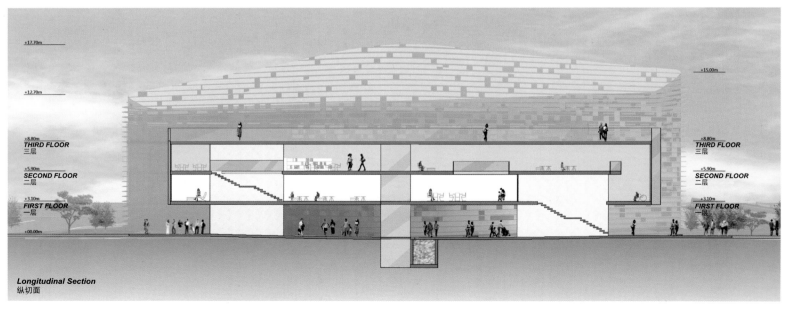

*Longitudinal Section*
纵切面

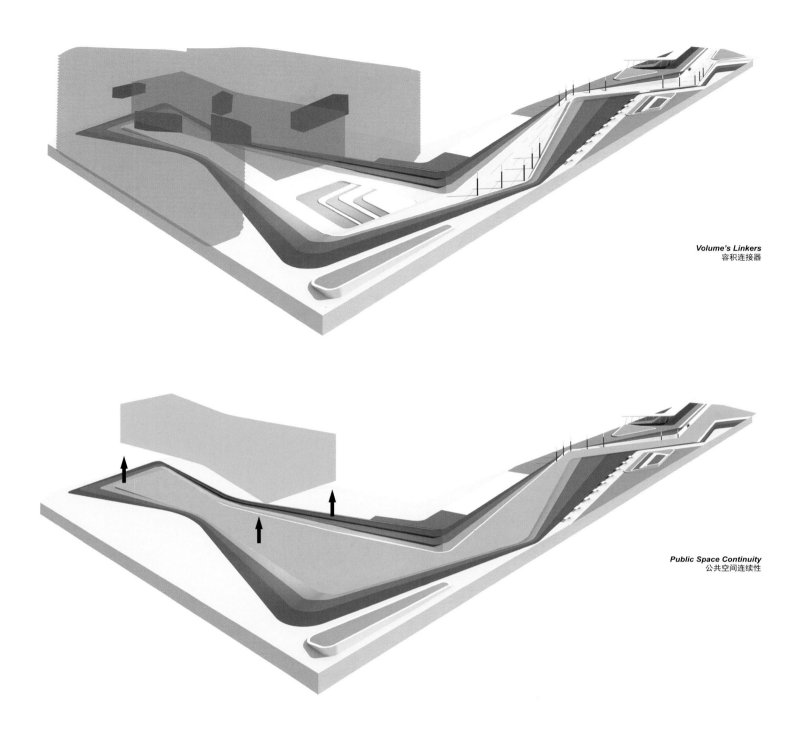

**Volume's Linkers**
容积连接器

**Public Space Continuity**
公共空间连续性

*South Elevation*
南立面

Compact Structure
Contrast Façade
Low Energy Consumption

紧凑式结构
立面的对比
低能耗

# 因斯布鲁克办公楼
## Office Building in Innsbruck

开发商：安瑞斯有限公司
项目地址：奥地利因斯布鲁克
使用面积：8 300m²(地面 5 900m²，地下 2 400m²)
总建筑面积：9 450m²(地面 6 900m²，地下 2 550m²)
摄影师：Thilo Härdtlein　München

*Client: ATRIUM amras GmbH*
*Location: Grabenweg 58, Innsbruck*
*Usable Area: 8.300 m² ( Above Ground 5.900 m², Underground 2.400 m² )*
*Gross Floor Area: 9.450 m² (Above Ground 6.900 m², Underground 2.550 m²)*
*Photographer: Thilo Härdtlein, München*

**建筑设计**

紧凑式结构——一个充满阳光的中庭。该大楼有一个几乎是正方形的平面图，由发展的布局产生，并且被设计为一个中庭类型。通过屋顶的内部区域进入内部，其延伸到多个楼层。类似看台的楼梯从一楼入口大堂到达一楼，其中小活动和会议的多功能区在这里。中庭周围办公空间的组织是提供不同尺寸的出租单元的布局。沿着中庭的互动空间，屋顶露台为该建筑的员工提供了另一种半专用的会议点，在顶层的大露台也注重该位置提供给周围群山和安布拉斯城堡至南面的风景。

立面是对比的相互作用。立面是由预制混凝土和每一个单独的楼层的高度构建的。这些创建了一个非均匀网格，由两个不同尺寸窗开口交替形成不同的图案组成。窗户是一个复合设计，超薄型材配置，中间为防风防晒的保护系统。每层楼立面元素的轻微偏移创造了光明和黑暗的相互作用，标志着每层的边缘，用户由此可能看见各层地板。光明和黑暗的对比度是被表面的粗糙度差异加剧的。粗糙和亚光砂岩彩色混凝土表面和光滑反射玻璃元素对比。此外，大楼的外墙壳是该结构的一个独特的部分，它成为该项目一个企业形象和图标。

精艺技术——低能。透明的最佳组合和封闭墙的外面，高品质的外墙外保温玻璃提供余热回收，使该建筑物的能效等级参考 HWB 为 A+ 级，13.6 kWh/m²a。该办公室的窗户都是具有最佳的建筑物理性能的复合材料设计。结构特性允许精益空调制冷设计。

**Building Design**

Compact structure—an atrium flooded with light. The building has an almost square ground plan, resulting from the layout of the development, and was designed as a type of atrium. Access to the interior is through the roofed inner area, which extends over multiple floors. Stairs that resemble bleachers lead from the entrance foyer on the ground floor up to the first floor, where a multipurpose area for small events and meetings is located. The office spaces located around the atrium are organised to provide variable layouts of differently sized rental units. Alongside the interaction space of the atrium, a roof terrace provides another semiprivate meeting point for people employed within the building. The large terrace on the top floor also emphasises the views that the site provides of the surrounding mountains and Ambras Castle to the south.

The façade is constructed from precast concrete components, each the height of an individual storey. These create a non-uniform grid composed of two differently sized window openings that alternate to form a varied pattern. The windows are a composite design with slim profiles and sun protection systems fitted in the middle for protection from the wind. The slight offset of the façade elements at each storey creates interplay of light and dark, marks the edges of each level and makes it possible to read the floors. The contrast of light and dark is intensified by differences in the roughness of the surfaces. The rough and matt sandstone-color concrete surfaces are contrasted with smooth reflective glass elements. The exterior shell of the building is such a distinctive part of the structure that it becomes a corporate identity and an icon for the project.

Lean technology—low energy. The optimum mix of transparent and closed exterior wall surfaces, high-quality exterior insulation and glazing provides waste heat recovery that gives the building an energy efficiency class of A+, with HWBref of 13.6 $KWH/m^2 a$. The office windows are a composite design that has optimum building physics performance. The structural properties allow a lean HVACR design.

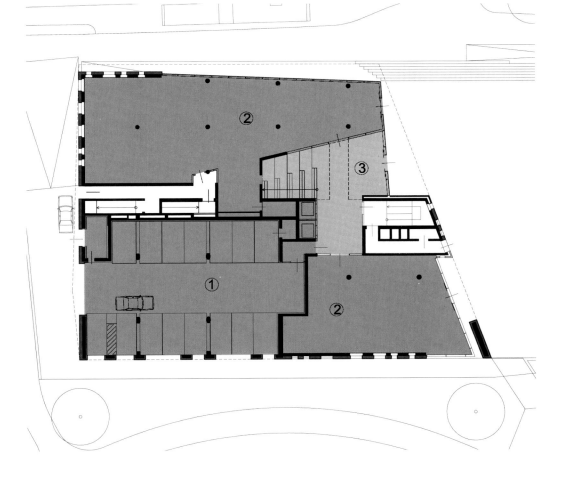

**Level Plan Ground Floor**
底层平面图

1. **CARPORT**
 1. 车位
2. **RETAIL**
 2. 零售
3. **ENTRANCE**
 3. 入口

0　5　10　15　20

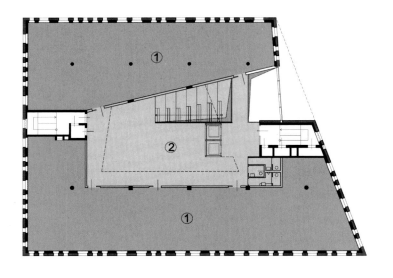

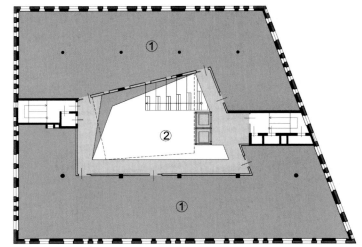

*Level Plan 1st Floor*
一层平面图

1. **OFFICE**
1. 办公室
2. **ATRIUM**
2. 中庭

*Level Plan 2nd Floor*
二层平面图

1. **OFFICE**
1. 办公室
2. **ATRIUM**
2. 中庭

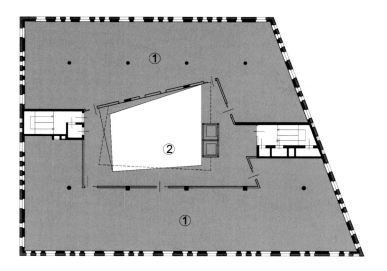

**Level Plan 3rd Floor**
三层平面图

**1. OFFICE**
1. 办公室
**2. ATRIUM**
2. 中庭

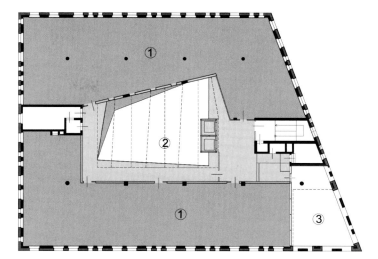

**Level Plan 4th Floor**
四层平面图

**1. OFFICE**
1. 办公室
**2. ATRIUM**
2. 中庭
**3. TERRACE**
3. 露台

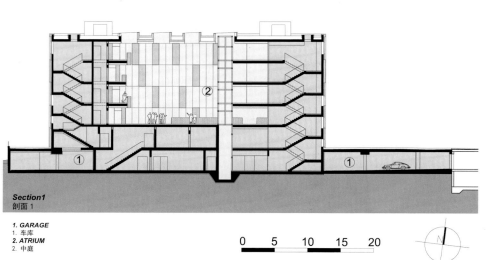

**Section1**
剖面 1

1. *GARAGE*
   1. 车库
2. *ATRIUM*
   2. 中庭

0  5  10  15  20

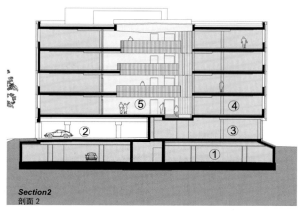

**Section2**
剖面 2

1. *GARAGE*  3. *RETAIL*  5. *ATRIUM*
   1. 车库       3. 零售       5. 中庭
2. *CARPORT* 4. *OFFICE*
   2. 车位       4. 办公室

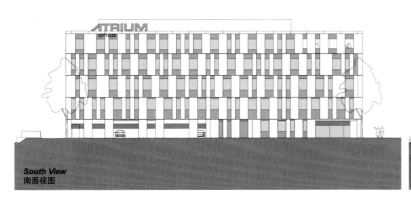

**South View**
南面视图

**North View**
北面视图

**West View**
西面视图

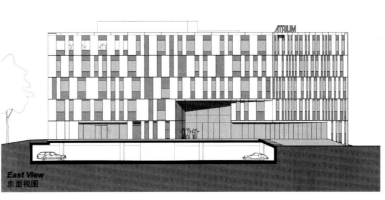

**East View**
东面视图

ATR_Atrium amras _ Fertigteile Fassade - Prefabricated concrete elements facade

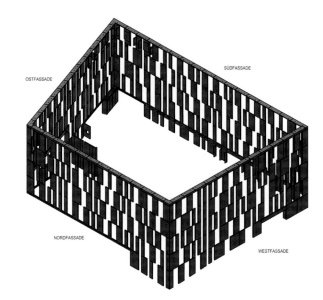
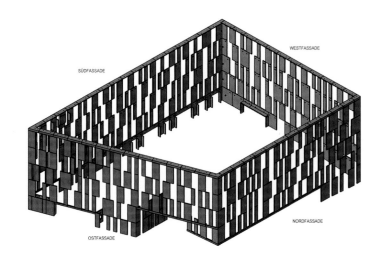
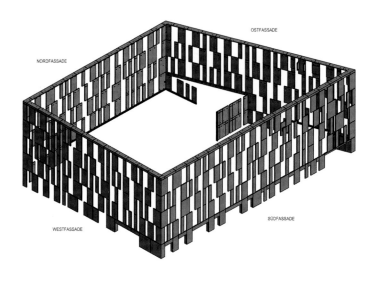
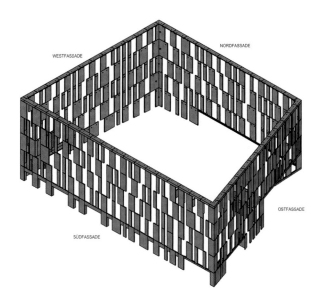
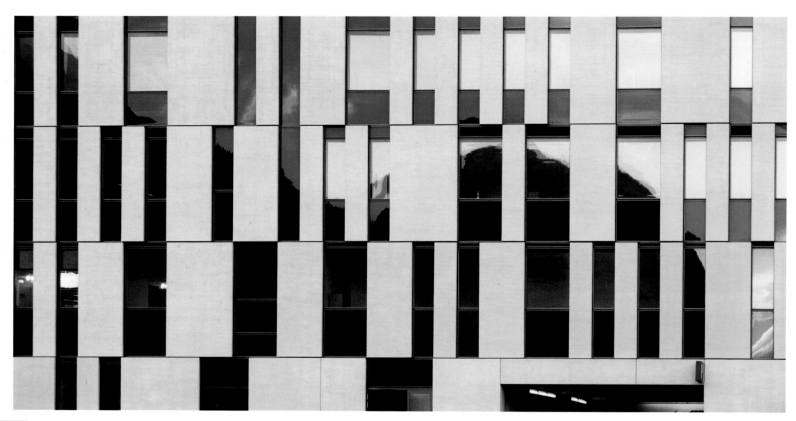

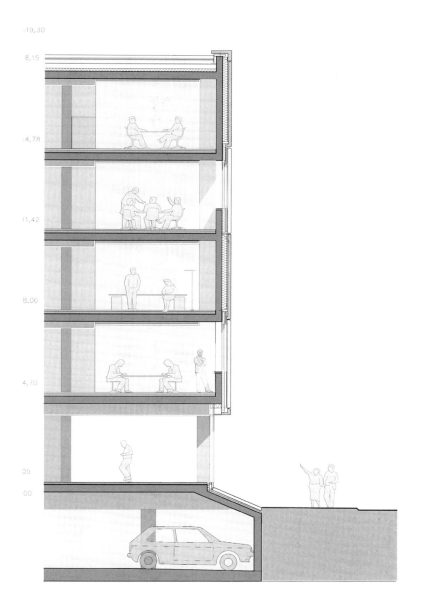

DETAILSCHNITT FASSADE

ANSICHT FASSADE

# 苏州中吴红玺
## Suzhou Zhongwu Hongxi

设计单位：HIC 翰创设计
开发商：江苏中吴置业有限公司
项目地址：苏州高新区浒关镇

**Designed by: HIC Design**
**Client: Jiangsu Zhongwu Properties Limited**
**Location: High-tech Zone, Huguan Town, Suzhou**

**项目概况**

本项目位于素有"江南要冲地、吴中活码头"之称的江苏省苏州高新区浒关镇，全镇河道纵横、公路四通八达。中国交通大动脉沪宁铁路、沪宁高速公路、312国道、京杭大运河均穿越镇区；一小时内均可直抵东西北向的上海虹桥（浦东）国际机场、光福机场和无锡硕放机场，独特的区位优势、便捷的水陆空交通条件，形成了发达的人流、物流、财流和信息流。

建筑设计结合高端的市场定位，以创造法式生活体验为主题，各类生活和公共活动配套设施完善，从高端餐饮到养生SPA，健身娱乐一应俱全，增强小区高端气氛的同时，也为未来客户的生活情趣平添了一份色彩。

**建筑设计**

本设计充分利用创造法式园林这一主题，在引入轴线对称布局的同时结合内部架空层的绿化渗透，引入自然清新的气息。同时中轴对称的规划逻辑，也使得景观自然而然、分级有序。在中心开阔视野的法式景观轴线统帅下，清新宜人的绿化带、富有情趣的小区组团绿地，以及进入每个单元的半室外庭院分合有序，浑然天成。在立面处理上借鉴了法式建筑的手法，用不同的材质组合与色彩对比，形成端庄、优雅、精致的立面造型。

## Project Overview

The project is located in High-tech Zone, Huguan Town, Suzhou, Jiangsu which known as "the Jiangnan Crossroads, the Wuzhong Live Dock", the rivers of the town are vertical and horizontal and highways are extended in all directions. Chinese traffic arteries—Shanghai-Nanjing Railway, Shanghai-Nanjing Expressway, 312 National Road and Beijing-Hangzhou Grand Canal are through the town; it can be straight to the orientation with east, west and north of Shanghai Hongqiao (Pudong) International Airport, Guangfu Airport and Wuxi Shuofang Airportwithin one hour. The unique geographical advantages and convenient transportation conditions form a well-developed flow, logistics, financial and information flows.

Architectural design combined with high-end market positioning to create the theme of a French-type life experience; all kinds of life activities and public facilities are perfect; from high-end catering to the health SPA, the fitness and entertainment is readily available, which is to enhance the high-end residential atmosphere, also added to a color for the future customers' interest in life.

## Building Design

This design makes full use of this theme creation of French gardens, it introduces the axis symmetry layout, while combines with internal empty space greening penetration, bringing natural and fresh breath. Meanwhile, axial symmetry planning logic also makes landscape naturally, hierarchical and orderly. In the French commander of the center axis of the open landscape vision, the fresh and pleasant green belt, green groups with rich appeal and each inside unit of an outdoor courtyard orderly merge-joint, totally like nature itself. It borrowed from French architecture method on the façade processing and used different material combinations and colors contrast to form a dignified, elegant and refined façade style.

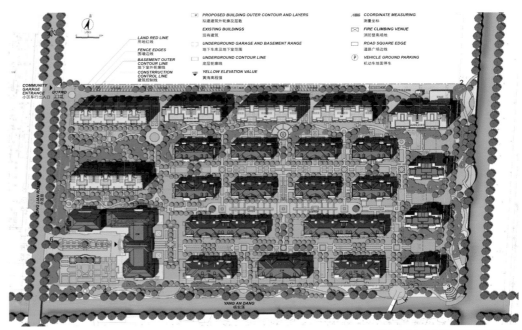

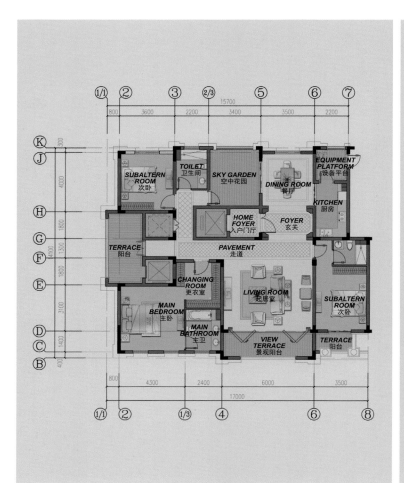
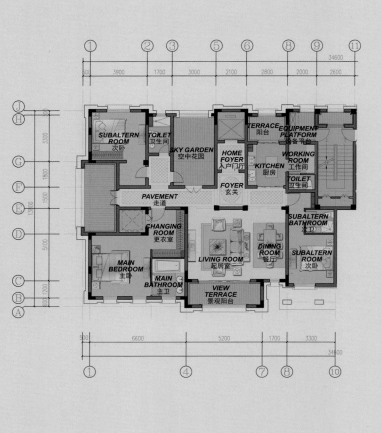

# 意大利曼图亚多功能综合体
## Multi-purpose Complex DUNANT

设计单位：Studio Rodighiero Associates
建筑师：Arch. Massimo Rodighiero
开发商：Rudiana Immobiliare
项目地址：意大利曼图亚
项目面积：10 000m²

*Client: Rudiana Immobiliare*
*Architect: Arch. Massimo Rodighiero*
*Designed by: Studio Rodighiero Associates*
*Location: Mantua, Italy*
*Building Area: 10,000m²*

# Cloak-like Roof
# Respecting the Nature
# High Energy Efficiency

# "斗篷"屋顶
# 尊重自然
# 高效节能

## 项目概况

在风景优美的 Morenic Hill，距离加尔达湖几公里的地方，就是该创意多功能中心。这一个位于城市郊区的重要结构需要在建筑上和结构上与周围的景观相协调，项目包括一个酒店、一个会议中心、一个大型餐厅和一个购物综合体。

## 功能分区

新的多功能中心包括3栋大楼，分别位于中央三角形庭院的三侧。第一栋大楼包含了杜南酒店，其舒适的客房分布在4个楼层，每间房间都提供了 Moorenic Hills 无与伦比的视野，使每位客人都能享受到整个中心的丘陵景观。

第二栋大楼包含了底层的一个 900m² 的购物综合体，可以为邻近区域的人们服务，大楼内的一个大型中央庭院使其也可作为一个聚会场地、一个休闲场地。而会议中心位于第三栋大楼，直接与酒店相连。

## 建筑设计

这一壮观的建筑经过精心的设计，使建筑元素与周围景观完美融合在一起。酒店、会议中心和餐厅拥有"斗篷"状的屋顶，其圆滑、流线型的形态让人不禁联想到周围山体的柔软线条。

建筑立面应用了现代工业方法和混凝土预制板，采用的面板结构饰以浅浮雕，创造性地实现了建筑与众不同的视觉标识。

另外，整个建筑的设计还旨在实现高能效，光伏太阳能板隐藏在酒店的屋顶，在不影响综合体美感的同时，还能为建筑提供能源和热水。

## Profile
Among the scenic Morenic Hills, a few kilometers from Garda Lake, is the innovative Multi-functional Center. This important structure on the outskirts of the city has required architectural and structural solutions in harmony with the landscape. The complex has in fact a hotel, a congress center, a large restaurant and a shopping complex.

## Function Division
The new Multipurpose Center consists of three buildings that are located on the sides of a central triangular courtyard. The first building houses the Hotel Dunant, with his comfortable guest rooms which are spread over 4 floors. Every room offers unparalleled views of the Moorenic Hills allowing guests of the hotel to enjoy the hilly landscape in which the entire center is set.

## Architectural Design
Through careful design, the architectural elements blend in with the landscape. The roof, which looks like a cloak, is gently placed over the hotel, the conference center and the restaurant, creating a sleek and streamlined reminiscent of the soft lines of the surrounding hills.

The structure is then completely built by the application of a modern industrial approach, the concrete prefabricated panels. The bas-relief designed by the architect camouflages the structural joints of panels used, resulting in a innovative building with a character that is immediately recognizable, unique and distinctive.

Additionally, the entire structure is designed to achieve high energy efficiency; photovoltaic solar panels were hidden on the top of the roof in the hotel, which are not influence the complex beauty, and also provide energy and hot water for the building.

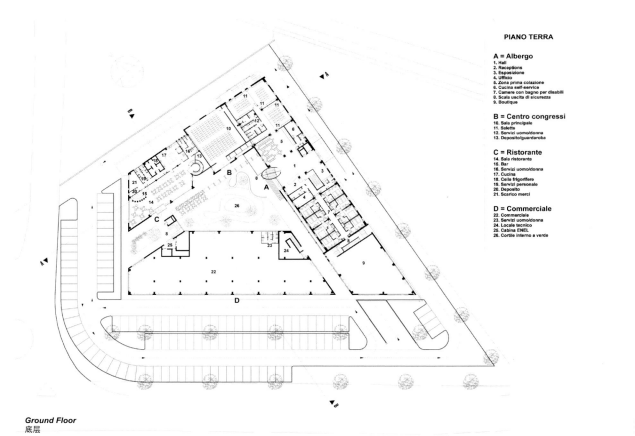

**PIANO TERRA**

**A = Albergo**
1. Hall
2. Receptions
3. Esposizione
4. Ufficio
5. Zona prima colazione
6. Cucina self-service
7. Camere con bagno per disabili
8. Scala uscita di sicurezza
9. Boutique

**B = Centro congressi**
10. Sala principale
11. Salette
12. Servizi uomo/donna
13. Deposito/guardaroba

**C = Ristorante**
14. Sala ristorante
15. Bar
16. Servizi uomo/donna
17. Cucina
18. Cella frigorifera
19. Servizi personale
20. Deposito
21. Scarico merci

**D = Commerciale**
22. Commerciale
23. Servizi uomo/donna
24. Locale tecnico
25. Cabina ENEL
26. Cortile interno a verde

*Ground Floor*
底层

*Basement*
地下室

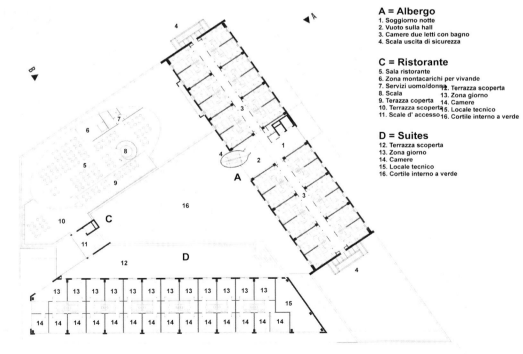

## PIANO INTERRATO

1. Parcheggio interrato
2. Locale tecnico
3. Locale filtro
4. Vasca d'acqua
5. Ingresso parcheggio
6. Uscita parcheggio
7. Collegamento al piano terra

## PIANO PRIMO

### A = Albergo
1. Soggiorno notte
2. Vuoto sulla hall
3. Camere due letti con bagno
4. Scala uscita di sicurezza

### C = Ristorante
5. Sala ristorante
6. Zona montacarichi per vivande
7. Servizi uomo/donna
8. Scala
9. Terrazza coperta
10. Terrazza scoperta
11. Scale d' accesso
12. Terrazza scoperta
13. Zona giorno
14. Camere
15. Locale tecnico
16. Cortile interno a verde

### D = Suites
12. Terrazza scoperta
13. Zona giorno
14. Camere
15. Locale tecnico
16. Cortile interno a verde

*First Floor*
一楼

# 北京中国国家美术馆
# National Art Museum of China

| | |
|---|---|
| 设计单位：Office for Metropolitan Architecture | Designed by: Office for Metropolitan Architecture |
| 开发商：中国国家美术馆 | Client: National Art Museum of China |
| 项目地址：中国北京市 | Location: Beijing, China |
| 建筑面积：30 000m² | Building Area: 30,000m² |

设计团队：Alessandro De Santis, Jing Chen, Midori Hasuike, Martin Hejl, Jinman Jo, Anu Leinonen, Jue Qiu, Adrienne Lau, Pietro Pagliaro, Yu Wang, Ippolito Pestellini, Yanfei Shui, James Westcott, Espen Vatn, Junjie Yan, Nurdan Yakup, Haohao Zhu, Dongmei Yao, Kostya Miroshnychenko

Project Team: Alessandro De Santis, Jing Chen, Midori Hasuike, Martin Hejl, Jinman Jo, Anu Leinonen, Jue Qiu, Adrienne Lau, Pietro Pagliaro, Ippolito Pestellini, Yanfei Shui, Espen Vatn, Yu Wang, James Westcott, Junjie Yan, Nurdan Yakup, Haohao Zhu, Dongmei Yao, Kostya Miroshnychenko

**项目概况**

在过去的二十年里，博物馆建筑发生了日新月异的变化。它们的规模将不仅仅是一座大楼，更是一座微缩的城市。而中国国家美术馆就是世界上第一座这种新典范的博物馆，是第一座被构思为小型城市的博物馆。

**规划设计**

项目以"城市"这一概念为核心进行规划，混合着"官方"或基层的分区，它的中心和边缘包含可规划为中国和国际区域、现代和历史区域、商业和政府区域等，就像一座城市一样，区域之间可以重新定义、修复甚至替换。

将美术馆规划为一座城市并不是说它不能提供博物馆的本质功能，和所有城市一样，其单独的部分可能很小，但是能提供博物馆所特有的多种服务。

**建筑设计**

主基座建筑是一些传统的正交博物馆空间，还有更现代化的、形态更自由的建筑，仿佛正在讲述或探索着中国的艺术故事。建筑基于一个五角星造型，主要流通线路从外围的多个入口点到达中心区，"五角星"连接到"灯笼"，这也是该美术馆的立体象征。红色的立面表皮，完美地展现了"中国红"这一传统的色彩元素。

单一的切线将"灯笼"与鸟巢联系起来，与城市的错综复杂性相比，"灯笼"结构的六层主楼提供了宽敞、开放的空间，以至于建筑并未与展览会的布局相冲突。建筑的内部组织也很合理，其灵活的表皮从金属框架中伸展出来，使它看起来就像一个谜。

"灯笼"
"五角星"
微缩城市

"Lantern"
"Five-pointed Star"
Small City

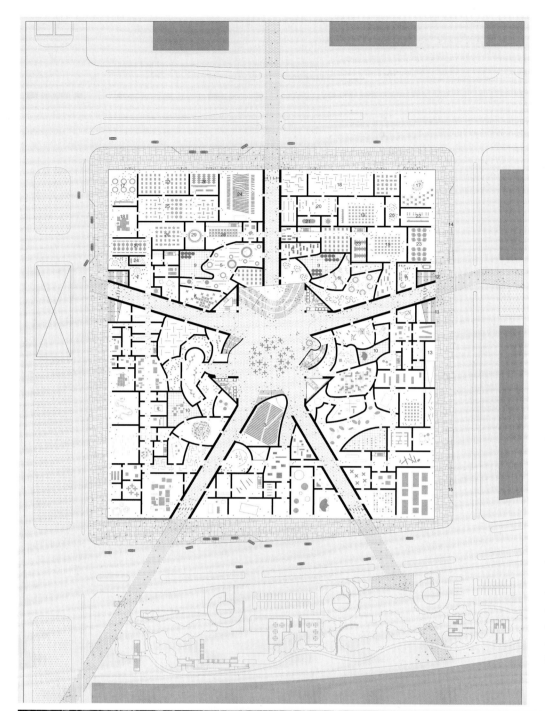

## Profile

In the past two decades, our museums have become larger and larger; they have now reached a scale at which they can no longer be understood as large buildings, but only as small cities. NAMOC can be the first museum in the world based on this new paradigm, the first museum conceived as a small city.

## Planning Design

The project is planned around the concept of "city", which can mix sectors, "official" and grassroots. It could have a centre and a periphery, a Chinese and an international district, modern and historical areas, commercial and "government" neighborhoods. Like a city, areas can be redefined, renovated, or even replaced, without compromising the whole.

To plan NAMOC as a city does not mean that it cannot offer the intimacy that remains the essence of the museum experience. Like any city, its individual parts can be small, humane, but it will offer a degree of variety that will be unique for a single museum.

## Architectural Design

The architecture of the main plinth offers a range of classical, orthogonal museum spaces, to more contemporary, freer forms. The story of Chinese art can be told, or discovered. The main circulation of the city is based on a five-pointed star that leads from the multiple entry points on the periphery to the centre. Here, the star connects to the "lantern", a multistory stack of platforms, wrapped in a red skin, on which temporary exhibitions and events are arranged with the smooth efficiency of a convention centre.

A single tangential axis relates the Lantern to the Bird's Nest. In contrast to the intricacy of the city, the six main floors of the Lantern offer wide open spaces, so that the architecture does not interfere with the organization of the exhibitions or events. Although its internal organization is rational, the elastic skin stretched around the metallic frame makes it look like a mystery.

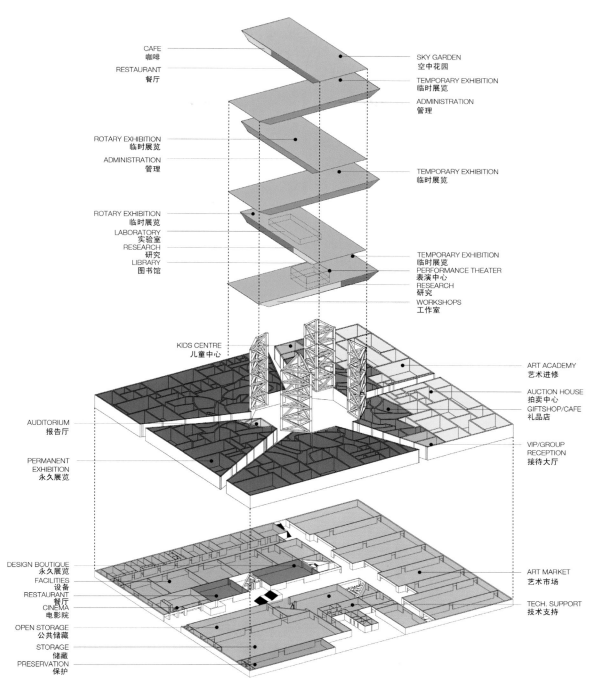
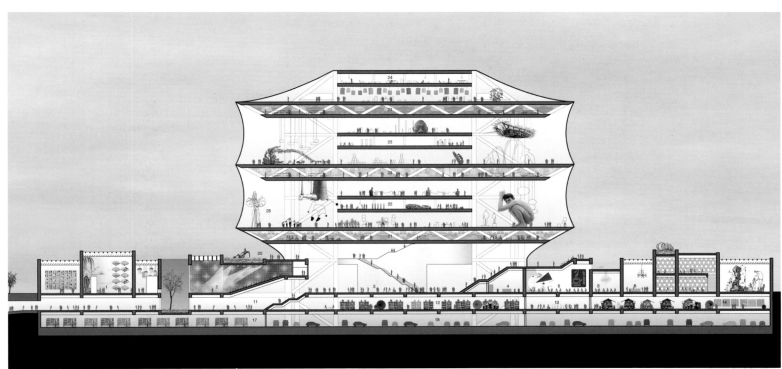

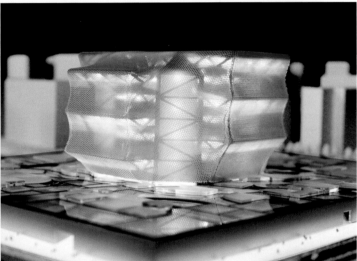

# 内蒙自治区鄂尔多斯基督教堂
## Ordos Protestant Church

设计单位：三磊设计
开发商：鄂尔多斯市规划局
项目地址：内蒙古自治区鄂尔多斯市
建筑面积：3 000m²
设计团队：张华　周厚陶

Designed by: Sunlay Design
Client: Ordos Planning Bureau
Location: Ordos, Inner Mongolia
Building Area: 3, 000m²
Design Team: Zhang Hua, Hotao Chow

"和平鸽"
混凝土

"Dove of Peace"
Concrete

**项目概况**

该基地教堂坐落在鄂尔多斯的一座小山上，其舒展柔美的建筑形态代表着飞翔、轻盈、优雅和平静，和谐地融入周边环境之中。

**设计灵感**

建筑以《圣经》故事里的"和平鸽"为意象，通过重新诠释和平鸽用喙衔回橄榄枝的故事，赋予了教堂新的寓意和诗意。另外，其山地地形也成为设计灵感来源之一，设计师首先从地形作为设计出发点，综合气候、环境等因素，努力营造出独特的教堂景观。

**建筑设计**

该教堂是混凝土结构，外立面为白色涂料表面，流线的外形顺应了地块周边的道路走向。建筑内外空间关系是通过对光线和阴影的描写，来凸显出教堂更为深层次的空间。建筑设计既传承了传统的教堂文化，又融入了新的设计元素。柔美的建筑线条加上微妙的光线处理，创造出全新的供人们祈祷以及庆祝的空间，使其成为一座现代感十足而又不失传统文化底蕴的建筑。

**Profile**

The site is located on top of a hill of Ordos. The stretching and mellow architectural form represents flying, lightness, elegance and peace, harmoniously blending in with the surrounding environment.

**Design Inspiration**

The scheme takes image of "Dove of Peace" in Bible, and gives new metaphor and poetry to the church by re-interpreting a contemporary and abstract silhouette of the bird caring a branch of Olive in its beak. The project also takes its inspiration from the topography of the hilly land. With combination of climate and environment, the designers manage to create unique church landscape.

**Architectural Design**

The church is a concrete structure with white crepi finish for its façades. Its dynamic shape follows the adjacent curved road that crosses the site. The dialogue between the outside and the inside space is emphasized by the play of shadows and light that creates complexity and depth in the reading of the space.

The design inherits the traditional church culture and also integrates new elements. The mellow lines and subtle light create new spaces for people to pray and celebrate, generating a building with both modernity and traditional culture.

**Section AA**
剖面 AA

**Section BB**
剖面 BB

*Dove Concept*
鸽子概念

*East View*
东部视角

*South View*
南部视角

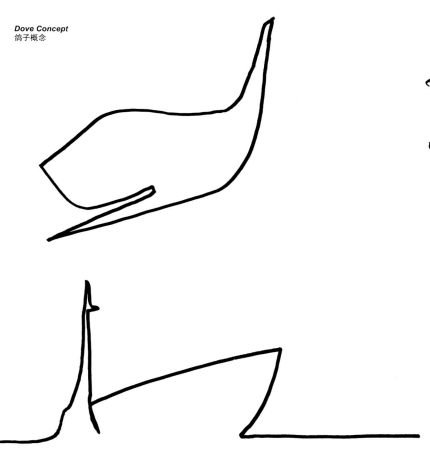

**Section**
剖面

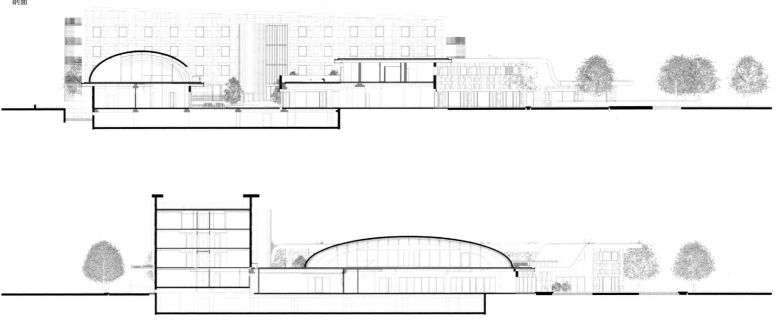

# 德国慕尼黑宝马办公大楼
## BMW Office Building

设计单位：plajer & franz studio
开发商：BMW Group
项目地址：德国慕尼黑
建筑面积：15 650m²

*Designed by: plajer & franz studio*
*Client: BMW Group*
*Location: Munich, Germany*
*Building Area: 15,650m²*

### 项目概况
宝马公司拥有巨大的室外场地，而这座办公大楼则正是坐落于其中的轻巧干净的"透明工厂"，是宝马汽车设计师们工作的地方。

### 建筑设计
为保证汽车设计行业的独创性、排他性和私密性，办公大楼的设计必须与外界形成一种相对的隔离，尤其是临街一侧。瑞士制造的巨大玻璃幕墙，使它的立面效果非常壮观，就像一个巨大的拥有透明屋顶的"玻璃盒子"。

建筑最外侧设计有较高的景观墙，层层穿插的设计充分保证了汽车设计过程的保密性。软质的花园设计与严格、宽敞的玻璃立面形成鲜明对比，为设计师们创造出一个安静、放松的创作环境。

### 设计理念
汽车的外形日新月异，建筑形式的更新却需要相当长的时间。Plajer & Franz 建造这座大楼贯彻始终的理念就是：不过分设计，不矫揉造作，尽量自然地塑造出富有人情味的空间。

为此，设计选择质朴实用的材料，如：黑色大理石地砖、实木地板，甚至是最普通的混凝土表面材料，从而满足不同工作类别对地面的要求，使大楼使用者感到便捷高效且贴心温暖。

### Profile
This clear and crisp building with its large outside areas in the middle of the Bavarian capital is an office building for the BMW designers.

### Architectural Design
To ensure creativity, exclusiveness and secret of car design, the building is relatively isolated from the outside, especially the façade to the street front. The isolating glass components built by a Swiss company make the façade quite splendid, like a giant "glass box".

The outside courtyard is surrounded by high landscaping walls to guarantee secrets of car design process. Soft and round garden design is in striking contrast with the rigid but open glass façade, providing tranquil and relaxed working environment for the designers.

### Design Concept
The appearance of car changes quickly, while building form stays for a long time. Plajer & Franz sticks to the concept: not over design, not fake and try to naturally create human space.

Therefore, simple and useful materials are selected for the building design, such as dark marble floor tile, acacia wood floor and even the most common concrete surfaces that meet the designers' needs for roughness and make the users gain convenience, efficiency and warmth.

"玻璃盒子"
景观墙
"人情味"空间

"Glass Box"
Landscaping Walls
Human Space

*Ground Floor*
底层

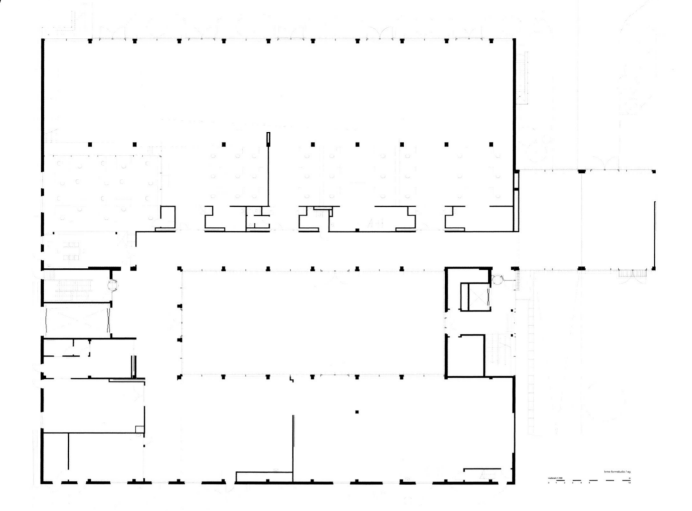

*Upper Floor*
上层

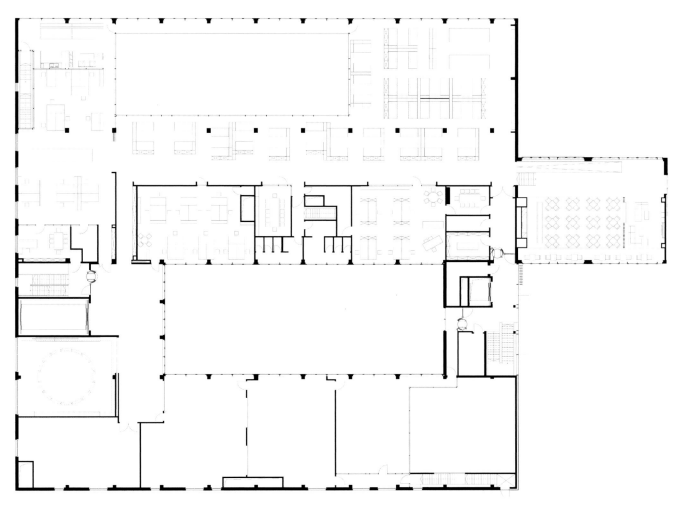

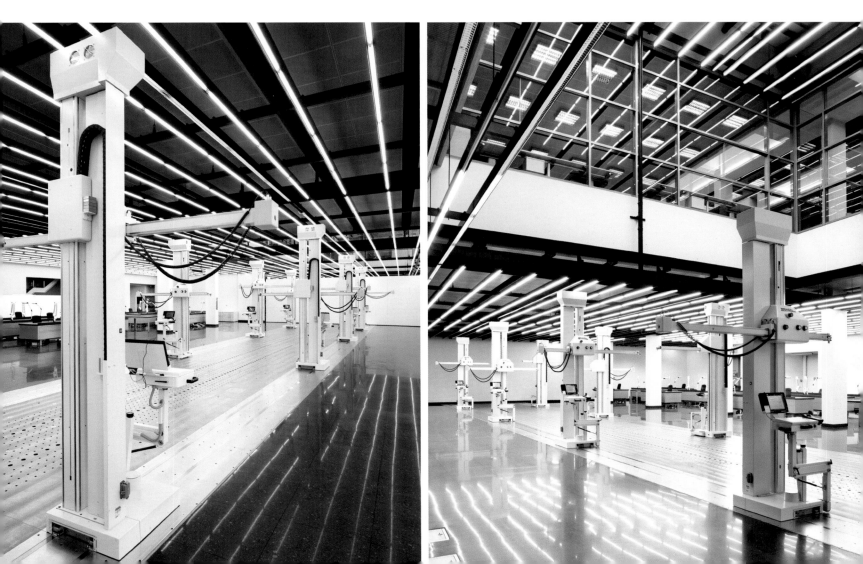

# 美国加利福尼亚州湖畔工作室
## Lakeside Studio

设计单位：Mark Dziewulski Architect
项目地址：美国加利福尼亚
建筑面积：111.48m²
奖项：美国建筑师协会奖
　　　太平洋海岸建筑师奖

*Designed by: Mark Dziewulski Architect*
*Location: California, U.S.*
*Building Area: 111.48m²*
*Awards: American Institute of Architects Design Award*
*Pacific Coast Builders Award*

**项目概况**

这座位于湖畔的工作室是一座集办公室、艺术室、灵活起居室以及画廊于一体的建筑综合体。建筑最大化地融入到周边湖景环境中，为业主营造出一个安静、庇护的工作和生活空间。

**建筑设计**

建筑雕刻式的弯曲造型使其能有效利用优美的河景，延伸的悬臂式楼层使居住者立于湖面之上，通过透明的玻璃楼板能更好地俯瞰湖水。

水面上骨骼般的金属楼梯伫立在湖畔的石柱底座上，不间断的玻璃墙使房间向景观开放，并模糊了室内外的界线。长长的屋檐将空间遮蔽得恰到好处，阳光经过水面反射，在天花板上形成不断变换、跳动的光斑，宛如一幅梦幻的意象画。

天窗的设计使日光深深地照进室内，使得屋顶的造型显得格外清晰。外部屋顶上的一个圆孔进一步加强了屋顶的渗透力，而设有透明玻璃面板的弯曲墙壁则形成视觉上的百叶窗，为雕塑提供明亮的背景。

**建筑与环境**

自然环境和建筑的关系是本设计的重点，该建筑完美地回应了自然环境的动感，通过人造与自然的对比，与环境融为一体。建筑泰然自若地伫立在湖面上，形成的优美倒影和建筑与景观的关系进一步加强了其雕塑般的品质。

这一开阔、灵动的浮动式结构在自然环境中形成一处明亮、高雅的形态，成为自然中最平静而精致的风景。

弯曲墙壁
浮动式结构
悬臂式楼层

Curved Wall
Floating Structure
Cantilevered Floor

## Profile

The lakeside studio is a complex with a combination of office, art studio, flexible living space and gallery. It integrates into the surrounding environment, creating tranquil and sheltered living/working space.

## Architectural Design

The sculptural form of the structure curves and twists to take advantage of spectacular river views. The stretching cantilevered floor carriers the observer well over the lake, an experience which is reinforced by views down to the water through areas of transparent glass floor.

Projecting over the water, the skeletal metal stair comes to land at a massive stone plinth set into the bank. Uninterrupted glass walls open the room up to the landscape and the boundaries between exterior and interior are blurred. The space is shaded by the dramatically extended roof overhang. Sunlight reflecting off the water dapples the ceiling with dancing patterns, constantly changing with the progress of day.

Skylights bring light deep within the interior and articulate the roof form. Continuing the sequence of roof penetrations, an oculus in the exterior roof allows a slowly moving circle of sunlight to animate the floor. A curved wall of angled translucent and transparent glass panels creates a optical louver that directs views towards the river and provides an illuminated backdrop for the sculpture.

## Architecture and Environment

The relationship between natural environment and architecture is the key of the design. The building perfectly responds to the dynamics of the natural context. The form is integrated with this context and yet at the same time makes use of the contrast between the man-made and the natural. It appears poised over the lake and uses its reflected image and its relationship to the landscape to reinforce its sculpture-like quality.

The openness and apparent agility of the floating structure also alludes to the lightness and elegance of forms found in nature. The structure becomes the most tranquil and delicate scenery in nature.

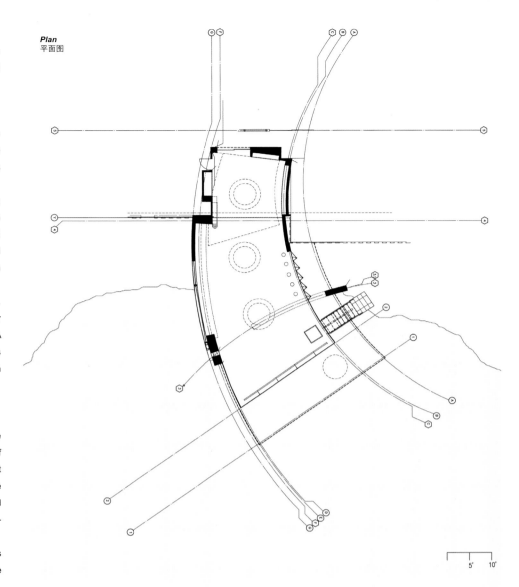

*Plan*
平面图

# 荷兰费嫩达尔 Panorama 社区中心
Panorama

| | |
|---|---|
| 设计单位：Architectenbureau Marlies Rohmer | Designed by: Architectenbureau Marlies Rohmer |
| 开发商：费嫩达尔市政府 | Client: Municipality Veenendaal |
| 项目地址：荷兰费嫩达尔 | Location: Veenendaal, Netherlands |
| 设计团队：Marlies Rohmer  Fabian van den Bosch  Floris Hund  Kate Griffin  Thomas van Nus  Begoña Masiá | Project Team: Marlies Rohmer, Fabian van den Bosch, Floris Hund, Kate Griffin, Thomas van Nus, Begoña Masiá |
| 摄影：Daria Scagliola  Marlies Rohmer | Photography: Daria Scagliola, Marlies Rohmer |

# 可持续设计
# "迷彩"立面
# 斜面屋顶

# Sustainable Design
# "Camouflage" Façade
# Sloping Roof

**项目概况**

Panorama 项目位于一片绿地之中,其全方位圆形场馆造型显得格外引人注目。这是一个对所有年龄段人群开放的社区中心,融合大约 60 种文化,并设有体育馆。

**建筑设计**

建筑圆形的外观将 Panorama 独立式的特征表现得尤为突出,就像一座置身于 20 世纪 60 年代公寓楼社区中的巨型雕塑。包围着建筑的这片草地半径达 30~40cm,使得整个场馆似乎轻微凹陷于地下,又像一个伫立于草丛中的人。这片草坪也成为有遮盖物的休闲场所。

依据可持续性设计理念,场馆还设计有绿色屋顶,屋顶的植被可起到蓄热器的作用,可缓冲雨水且吸收空气中的微粒。稍高的体育馆体量设有很大的朝南的斜面屋顶覆盖层——530m² 的太阳能板和 10m² 的太阳能集热器。一个非常大的外悬式雨棚形成全方位的阳台,并为室内遮挡夏季的烈日,雨棚和太阳能板形成的斜面也共同构成这座建筑的基本形态。

◀ 立面——这栋圆形建筑通过全玻璃立面向周边开放，显得格外动人，并形成室内外空间的融会贯通。玻璃后面依稀可见弧形的内墙，立面上的全方位着色是由当地居民设计，黑白条纹的鲜明对比使人想起军舰使用的迷彩图案，而内部鲜艳的图案则代表附近区域 60 种不同的民族文化。

Façade—the round building opens itself up invitingly to its surroundings through fully glazed façade. The result is a fluid interpenetration of exterior and interior spaces. Behind the glazing, a curved inner wall is visible. Painted entirely with designs by local residents, it forms a colorful inner lining, which recall the dazzle painting formerly used to camouflage warships and represent all the sixty or so minority cultures in the surrounding district.

**laag 0**

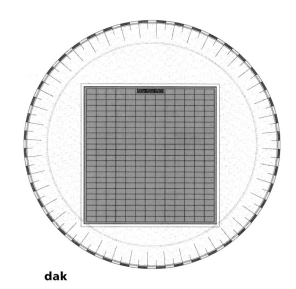

**dak**

1 Balie
2 Gymzaal
3 Toestelberging
4 Kleedkamer
5 Computerlokaal
6 Kookcursussen / vergaderruimte
7 Activiteitenruimte jongeren
8 Deelbare grote activiteitenruimte
9 Huiskamer
10 Kantoor medewerkers

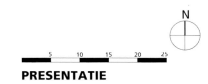

**PRESENTATIE**

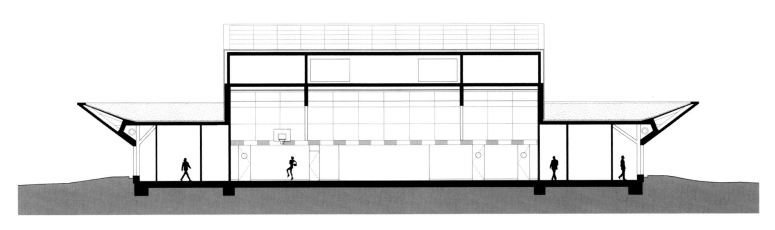

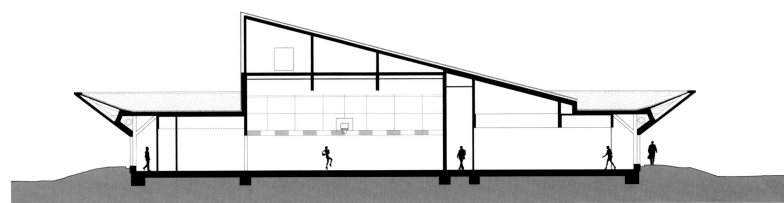

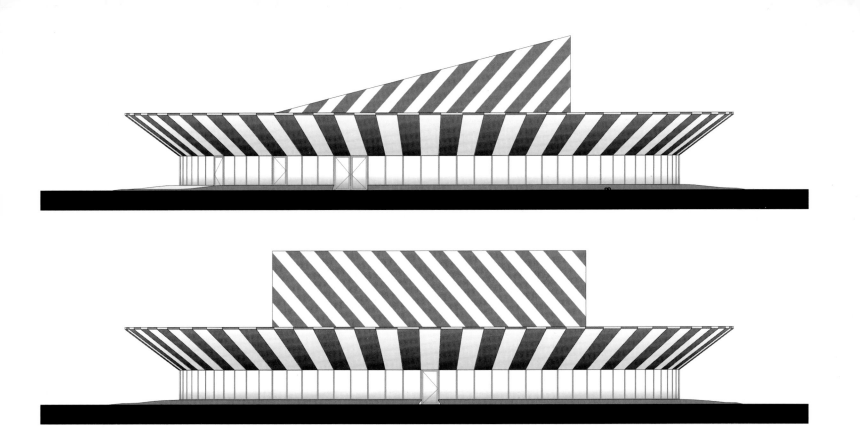

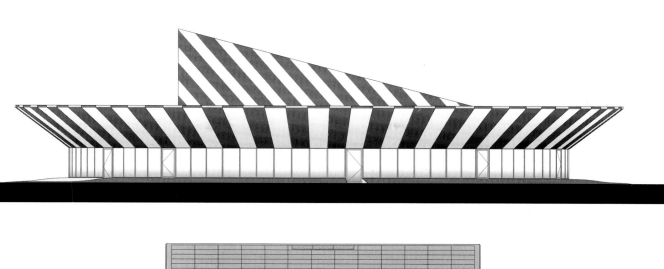
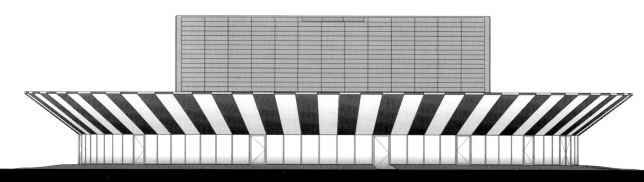

## Profile
The Panorama project is sited in a green space and takes the form of an inviting, omnidirectional, circular pavilion. This community center is open to people of all ages of 60 cultures with gym in Engelenburg.

## Architectural Design
The building's round shape places maximum emphasis on its freestanding character, like a large sculpture amid the neighboring 1960s blocks of flats. The lawn that encircles the building rises outwards by 30cm to 40cm. This makes the pavilion seem slightly sunken into the ground, like someone standing with their feet buried in undergrowth. The grass slope also serves as a sheltered leisure area.

According to the concept of sustainability, a green roof is designed on the pavilion. The rooftop vegetation acts as a heat accumulator, buffering rainwater and absorbing particulates. The slightly taller gym volume has a large, south-facing sloping roof clad with 530m$^2$ of solar panels (electricity) and 10m$^2$ of solar collectors (hot water). A very large overhanging canopy forms an all-round veranda and shades the interior from the hottest summer sun.

## Stappenstrategie

De 'low-tech' benadering is bij ons gebouw Tric Trac een resultante van onderstaande stappenstrategie. Hierbij gaan wij er van uit dat de te nemen maatregelen in onderstaande volgorde genomen worden.

### Stap 1 - Beperking energievraag:

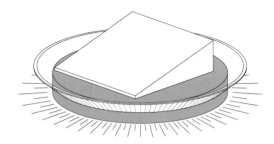

**1. Voorkomen transmissie en infiltratieverlies**
- Het gebouw is compact (rond) en heeft een optimale verhouding gevel- grondoppervlak.
- Hoge isolatie van dichte geveldelen (Rc=5.0 m²K/W).
- Hoge isolatie van gevelopeningen door toepassing van drie dubbel glas en hoogwaardig geïsoleerde kozijnen (u-raam = 1.2).
- Hoge luchtdichting van aansluitdetails en driedubbele kierdichting van kozijnen.

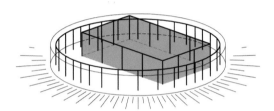

**2. Generieke hoofdstructuur**
- Lange levensduur door toepassing van een kolommenstructuur die het gebouw tot in lengte van dagen flexibiliteit geeft.
- Houtskeletbouw - cradle to cradle

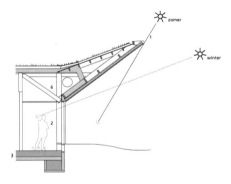

**3. Voorkom oververhitting 'intelligente gevel'**
- De luifel houdt zomerzon buiten en laat winterzon binnen.
- De daglichttoetreding wordt verbeterd door een glas met een hoge LTA toe te passen terwijl het glas oververhitting beperkt door een lage ZTA.
- Thermische massa: massieve vloer en dak zonder verlaagde plafonds i.v.m. warmteaccumulatie.
- Kierstand van ramen in verblijfsruimte maken natuurlijke ventilatie per ruimte mogelijk.

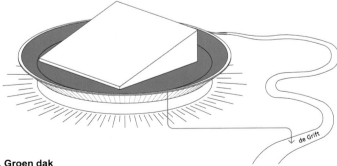

**4. Groen dak**
- Het groene vegetatie dak heeft een – warmte accumulerende massa van 150 kg/m², en fungeert als extra isolatie.
- Buffert hemelwater en stof en loost overtollig water op "de Grift"

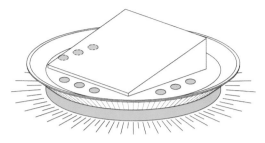

**5. Daglicht (reductie kunstlicht)**
- De rondlopende pui tussen openlopend maaiveld en overhangende luifel zorgt ervoor dat rondom licht naar binnen stroomt.
- De daglichttoetreding van het gebouw wordt geoptimaliseerd middels toepassing van solartubes in locaties buiten de daglichtzone.
- Verlichting wordt geschakeld door daglichtafhankelijke regeling en aanwezigheidsdetectie.

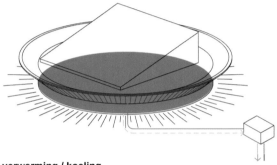

**6. verwarming / koeling**
- Vloerverwarming / vloerkoeling die individueel per ruimte regelbaar is, uitgevoerd in een laag-temperatuursysteem in combinatie met een 'omkeerbare' warmtepompinstallatie.

### Stap 2 - Hergebruik Restenergie

**7. Warmteterugwinning**
- Warmteterugwinning uit ventilatielucht.
- Warmteterugwinning uit douchewater.

### Stap 3 - Duurzame energie, benut de natuur

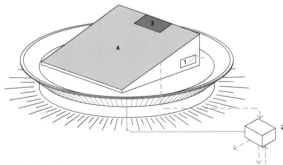

**8. Natuurlijke bronnen**
- Hybride ventilatiesysteem in de gymzaal. Afhankelijk van de gemeten $CO_2$ concentratie en de buitentemperatuur worden, kleppen in de zijgevel van de gymzaal open gestuurd. Dit verlaagt de benodigde ventilatie energie en verhoogt de functionaliteit en flexibiliteit van het gebouw door de veel grotere ventilatie debietten (1).
- De warmtebron voor de warmtepomp is een WKO (warmtekoude opslag) (2).

**9. Duurzame energie**
- Voor warm tapwater worden zonneboilers toegepast (3).
- Er worden PV-cellen toegepast (ca. 540m²). Daar waar elke andere maatregel energie bespaart, wekken PV-cellen energie op. (4)

"村庄"概念
弧形结构
"舰艇"

Concept of "Village"
Curved Shape
"Vessel"

# 英国 Birnbeck 岛 "村庄"
## Birnbeck Island Village

设计单位：安东尼奥·卡迪罗
开发商：RIBA Competitions
项目地址：英国萨默塞特
建筑面积：12 000m²

*Designed by: Antonino Cardillo Architect*
*Client: RIBA Competitions*
*Location: Somerset, England*
*Building Area: 12,000m²*

**项目概况**

该项目以"村庄"为概念,每一部分似乎都有变化,并在建筑群中展现出新的含义,彼此产生共鸣,让人仿佛置身于一个留下生命和时间轨迹的古老村庄之中。

**建筑设计**

该建筑主体依托着原有的码头和建筑,开发出新的道路系统,建筑就像是一艘巨大的舰艇,弧形的结构之中包含着不同的构成元素,建造出一个新的地下室取代旧的混凝土平台。

地下室设在低于码头 5m 的地方,从 Birnbeck 岛主入口旁的斜坡可通往这里。大量灵活的空间将岛上所有的建筑物都连接起来,舰艇的设计与原有的地面标志物形成对照,而底部的混凝土和顶部的板材这两种不同的材料将建筑表面区分开来。

设计师综合了多种传统建筑布局来填补空旷的空间,包括塔楼、悬桁、入口等。在每个狭窄的立面顶端,透过巨大的窗户都可观看到壮观的海景,形成全景效果,就像许多灯塔一样。

另外,主体部分与外部空间相连接,创造出丰富的层次感。

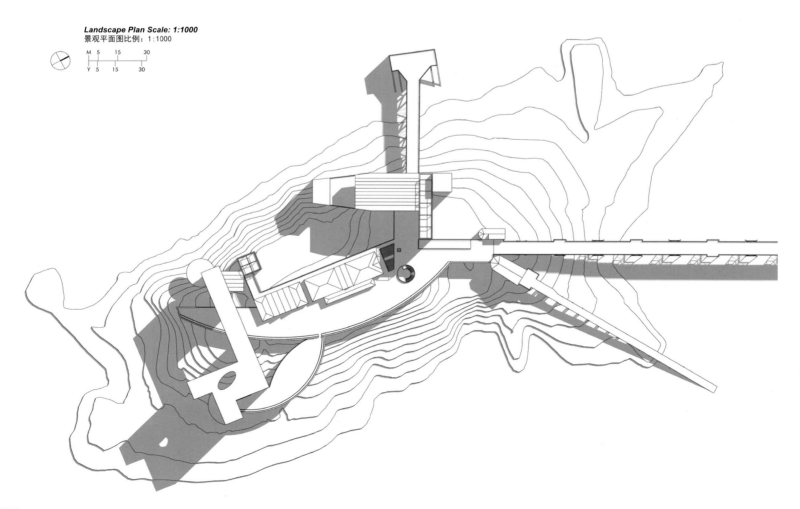

**Landscape Plan Scale: 1:1000**
景观平面图比例：1:1000

1. MAIN OLD PIER
1. 主要的老码头
2. INFO POINT
2. 信息点
3. RNLI STATION
3. 英国皇家救生艇协会站
4. BASEMENT BUILDING ENTRY (RAMP)
4. 地下室建筑入口（坡道）
5. BASEMENT BUILDING ENTRY (STAIR)
5. 地下室建筑入口（楼梯）
6. CLOCK TOWER
6. 钟塔
7. OFFICE TOWER ENTRY
7. 办公大楼入口
8. GALLERY RECEPTION
8. 画廊接待处
9. THEATRE FOYER
9. 剧院大厅
10. STALLS
10. 正厅前座
11. STAGE
11. 舞台
12. LANDING STAGE
12. 栈桥
13. PUBS/BARS
13. 小酒馆／酒吧
14. SHOPS
14. 商店
15. BASEMENT BUILDING ROOF
15. 地下室建筑屋顶
16. HOTEL RECEPTION
16. 酒店前台
17. HALL
17. 门厅
18. OFFICE
18. 办公室
19. PANORAMIC LOUNGE BAR
19. 全景沙发吧
20. SPA + PANORAMIC RESTAURANT + DISCOTHEQUE ENTRY
20. 水疗 + 全景餐厅 + 迪斯科舞厅入口
21. VOID TO AUDITORIUM
21. 空地至礼堂
22. RESIDENCE RECEPTION(ALSO LINKED TO PANORAMIC RESTAURANT)
22. 住宅接待处（连接全景餐厅）

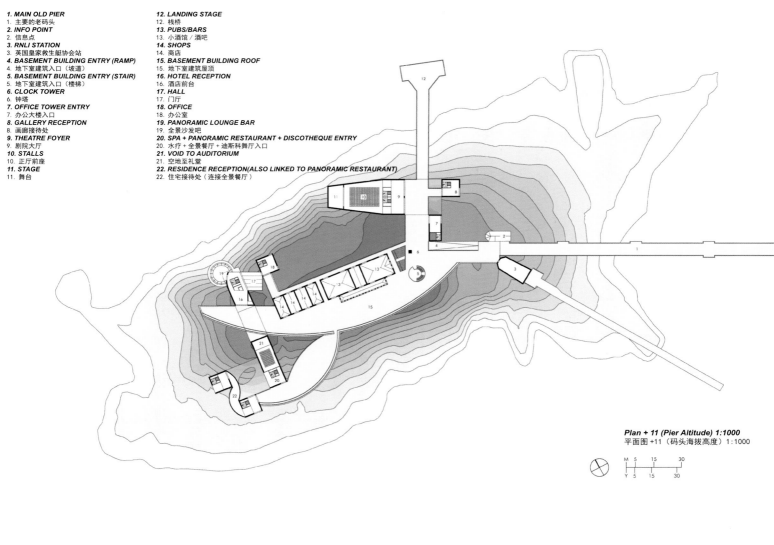

**Plan + 11 (Pier Altitude) 1:1000**
平面图 +11（码头海拔高度）1:1000

1. BASEMENT BUILDING ENYRY (RAMP)
1. 地下室建筑入口（坡道）
2. BASEMENT BUILDING ENTRY (STAIR)
2. 地下室建筑入口（楼梯）
3. SERVICE PARKING
3. 停车服务
4. GALLERY STORAGE
4. 画廊存储室
5. GALLERY ENTRY
5. 画廊入口
6. PLANT ROOM
6. 机房
7. THEATRE STORAGE
7. 剧院存储
8. RNLI STATION EXPANSION
8. 英国皇家救生艇协会站扩展区
9. MUSEUM RECEPTION
9. 博物馆接待处
10. EXHIBITION SPACES
10. 展览空间
11. LOUNGE BAR
11. 沙发吧
12. BAR STORAGE
12. 酒储
13. HOTEL STORAGE
13. 酒店存储
14. OFFICE
14. 办公室
15. BOOKSHOP
15. 书店
16. SPA + PANORAMIC RESTAURANT + DISCOTHEQUE ENTRY
16. 水疗 + 全景餐厅 + 迪斯科舞厅入口
17. AUDITORIUM
17. 礼堂
18. RESIDENCE ENTRY
18. 住宅入口

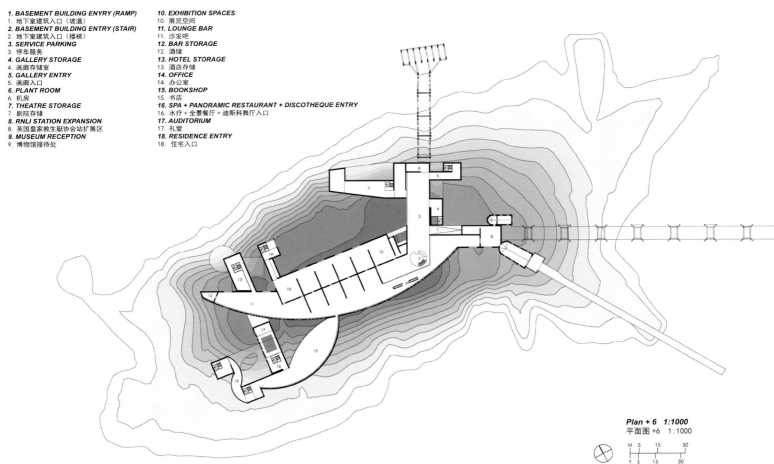

**Plan + 6 1:1000**
平面图 +6 1:1000

## Profile

Taking the concept of "village", the building reveals new meanings in the complex, with each episode appearing transfigured. Each part resonates in another constructing a stratified reality, as in an ancient village where life and time leave tracks on the ground.

## Architectural Design

This architectonic complex is born out of the pre-existing piers and buildings developing into a system of pathways. Like a big naval vessel, a curved shape embraces all the diverse elements of the composition, substituting an old concrete platform with the new basement building.

Constructed five meters beneath the main pier, it is accessible from a ramp near the main gate of Birnbeck Island. Its sequence of flexible spaces links all the buildings on the island from below. The vessel design counterpoints the new tall buildings, in plan strategically oriented following the double orthogonal reference system created through pre-existing signs. Two different materials divide these building surfaces: concrete at the bottom and planking at the top.

The method of occupying the void synthesizes diverse traditional architectural layouts (tower, linear, cantilever and city gate). In each of their narrow frontages, at the top, a big window marks the seascape creating a panorama like multitude of lighthouses. Moreover, these primary volumes are written through several unconventional signs that communicate to the outside the different spatial situation of the interiors, according to an "urban" poetry that makes a complex stratification of meaning.

*Cavalier Perspective 1:400*
散点透视图 1:400

**Elevation & Sections 1:1000**
立面 & 剖面 1:1000

# Maat-Pita Gaudham 寺庙综合楼
## Maat-Pita Gaudham Temple Complex

| | |
|---|---|
| 建筑师：VSGS ARCHITECTS AND PLANNERS | Architect: Vsgs Architects and Planners |
| 开发商：Maat-Pita Gaudham Trust | Client: Maat-Pita Gaudham Trust |
| 项目地址：印度，旁遮普，莫哈里，Khalawar 村 | Location: Village Khalawar, Mohali, Punjab, India |
| 建筑面积：30 000m² | Building Area: 30,000m² |

宗教信仰
"曼陀罗"
可持续发展 Religion "Mandala"
Sustainable
Development

**项目概况**

该项目总体规划体现了"曼陀罗"（印度教和佛教精神和仪式的象征，代表着宇宙），并且用了一种类似于"灵性如何保持人类大脑和身体的平衡"的方式，使其在功能和设计之间倾向平衡。在某种程度上，该计划模型是以一种理想层次，实现了建筑功能和可用空间之间的平衡，这有助于牛群、游客和校园居民之间的无限循环。

**建筑设计**

该设计理念来源于当地历史、传统、哲学和地区地形的因素。如玻璃和钢铁等现代材料的引入，赋予了传统的印度寺庙建筑的新定义，并使得大量的自然光线进入"garbha-griha"（印度教寺庙的小熄灭的神殿和至圣所内部的梵文）的内部空间。此外，像石头和木头这样的传统材料，与保持丰富的视觉解剖新的词汇的使用，保留了圣地的神圣性。除此之外，该项目还兼有一个可持续发展的使命，这一使命将这一地区的环境、经济和社会文化的可持续发展的理念浓缩到该项目的建设过程中来。

该项目在选址上就兼顾了可持续发展的理念——它将尽可能地利用选址内的各种可能的自然资源，如雨水收集、太阳能、沼气能源等。校园沼气能源就是利用牛群产生的废料而产生；选址内种植的药材和水果将被用来治疗特定疾病；养殖的牛群生产的牛奶和天然肥料将有助于创造额外收入，进而维持项目内的各项活动的开展与持续进行。

该方案的另一个优势就是，它将建筑功能扩展到了选址以外，它将在最大范围内传播有关有机农业实践的教育和认识，其目的在于通过村到村的传播，实现天然肥料代替化学肥料的传统耕作方式。此外，该选址同样提供产品测试实验室、研讨厅、人力资源和免费图书馆所需的各种基础设施，力图大力提升当地农业科技水平的同时，实现农业的可持续发展。

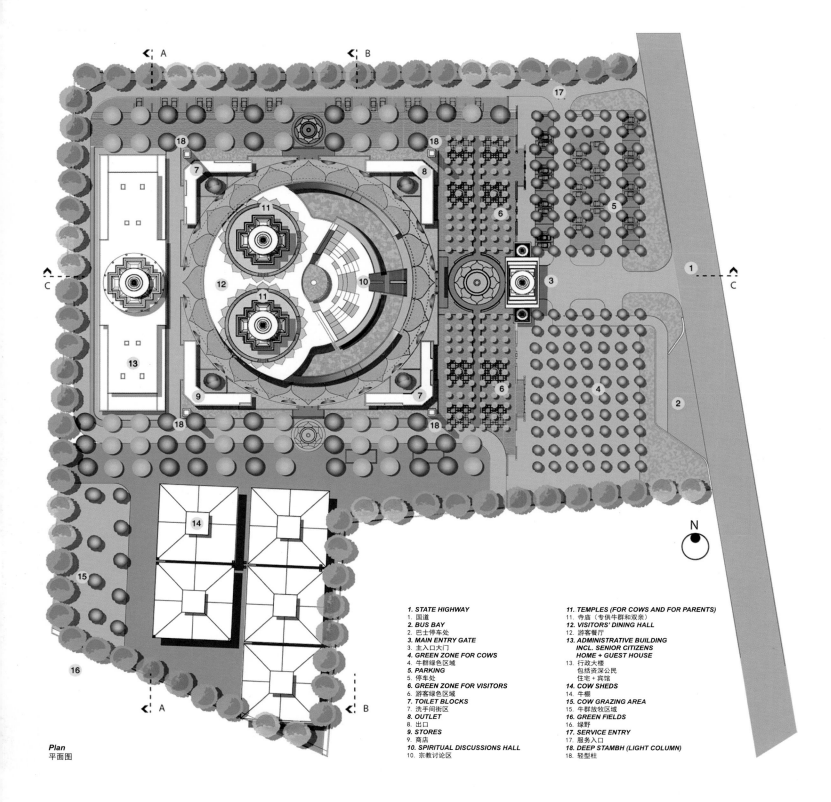

**Plan**
平面图

1. STATE HIGHWAY
1. 国道
2. BUS BAY
2. 巴士停车处
3. MAIN ENTRY GATE
3. 主入口大门
4. GREEN ZONE FOR COWS
4. 牛群绿色区域
5. PARKING
5. 停车处
6. GREEN ZONE FOR VISITORS
6. 游客绿色区域
7. TOILET BLOCKS
7. 洗手间街区
8. OUTLET
8. 出口
9. STORES
9. 商店
10. SPIRITUAL DISCUSSIONS HALL
10. 宗教讨论区
11. TEMPLES (FOR COWS AND FOR PARENTS)
11. 寺庙（专供牛群和双亲）
12. VISITORS' DINING HALL
12. 游客餐厅
13. ADMINISTRATIVE BUILDING INCL. SENIOR CITIZENS HOME + GUEST HOUSE
13. 行政大楼 包括资深公民 住宅＋宾馆
14. COW SHEDS
14. 牛棚
15. COW GRAZING AREA
15. 牛群放牧区域
16. GREEN FIELDS
16. 绿野
17. SERVICE ENTRY
17. 服务入口
18. DEEP STAMBH (LIGHT COLUMN)
18. 轻型柱

## Project Overview

The Master Plan reflects the 'mandala' (a spiritual and ritual symbol in Hinduism and Buddhism, representing the Universe) and enables a radial balance between the functions and their design, in a way similar to how spirituality provides balance to the human mind and body. The program matrix has been laid on the site in a way that an ideal hierarchy is achieved between the functions and usable spaces, which aids the circulation between the cows, visitors and on-campus residents.

## Building Design

The design derives elements from history, traditions, philosophy and topography of the region. Introduction of modern materials like glass and steel give a new definition to the traditional Indian temple architecture and enhances the internal spaces with abundant access to natural light inside the 'garbha-griha' (a small unlit shrine of a Hindu temple and a Sanskrit word for the interior of the sanctum sanctorum). Moreover, use of conventional materials like stone and wood with a new vocabulary maintains the rich visual anatomy that retains the sanctity of a sacred place. Apart from this, there is a sustainable mission attached to the program that will work directly and indirectly, towards environmental as well as economical and socio-cultural enrichment of the region.

The site is self-sustainable using all possible natural resources of producing and consuming energy, like rain water harvesting, solar energy and bio-gas energy. Biogas energy for the campus will be produced by the waste from cows housed within the site. Medicinal herbs and fruits are planned throughout the site for treating specific diseases in addition to producing fodder for cows' onsite using organic farming methods. Production of milk and natural fertilizers from the cow waste will help generate income to sustain activities within the program.

Another advantage of the program is that it extends outside the site and aims at spreading education and awareness about organic farming practices from village-to-village. This will be achieved in the beginning by adopting a farmer, then a village, to orient the conventional farming practices towards natural fertilizers instead of chemical fertilizers. The site acts as the infrastructure needed for the same by providing product testing labs, seminar halls, human resources and a free library. It tries best to enhance the local agricultural science and technology level, as well as achieve the sustainable development of the agriculture.

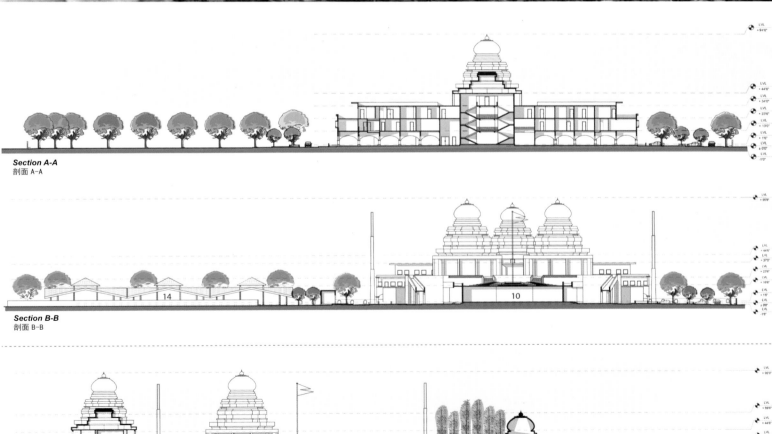

**Section A-A**
剖面 A-A

**Section B-B**
剖面 B-B

**Section C-C**
剖面 C-C

# FACING TO THE FUTURE
## —ARCHITECTURE TREND

# 面向未来
## ——建筑趋势2015

### 下册

龙志伟　编著
Edited by: Long Zhiwei

**栋梁之"材"**——建筑材料新运用
Shaped "material"—new application of architecture material

**奇"形"巧状**——建筑外形新趋势
Special "appearance" and attract structure—new trend of architecture shape

**一"技"之长**——建筑技术新发展
Excellent "technique"—new development of architecture technology

广西师范大学出版社
·桂林·

# 目录 Contents

## 8 栋梁之"材"
### ——建筑材料新运用
### Shaped "material"
### —new application of architecture material

| | | |
|---|---|---|
| **10** | 德国耶拿多功能重建大楼 | Reconstruction Multifunctional Building in Jena |
| **20** | 墨西哥比亚埃尔莫萨Liverpool Altabrisa百货商店 | Liverpool Altabrisa Department Store |
| **24** | 美国圣塔莫尼卡市民中心停车场 | Santa Monica Civic Center Parking Structure |
| **28** | 法国贝休恩水上活动中心 | Aqua Center, Bethune |
| **36** | 鹿特丹巴讷费尔德的巴士站 | Bus-stop in Barneveld Noord |
| **42** | 丹麦Egtved天然气压缩工厂 | Gas Compressor Station Egtved |
| **48** | 墨西哥库埃纳瓦卡La Estancia教堂 | La Estancia Chapel |
| **56** | 英国伯恩利"彩虹门" | Shell Lace Structure |

## 62 奇"形"巧状
### ——建筑外形新趋势

**Special "appearance" and attractive structure—new trend of architecture shape**

**64** 杭州新天地商务中心项目
Hangzhou Xintiandi Business Center Project

**68** 法国伊西莱穆利诺Galeo大楼
Galeo

**72** 华尔街巨蛋 —— 太阳能清真寺
The Wall Dome – Solar Powered Mosque

**76** 金州新区医疗中心
Jinzhou New Area Medical Center

**80** 中央洛杉矶公立高中视觉和表演艺术的洛杉矶联合校区
Central Los Angeles Public High School for the Visual and Performing Arts of the Los Angeles Unified School District

**86** 哥伦比亚波哥大EAN大学E1 Nogal 校区教学楼
Classroom Building of El Nogal Campus, Universidad EAN

**92** 比利时科特赖克Sint-Amand学院住宅塔楼
Housing Tower of Sint-Amand College, Kortrijk

**100** CUBE BIOMETRIC CENTER "魔方"
Cube Biometric Center

**108** 丹麦艾尔西诺文化中心
The Culture Yard

**120** 瑞典斯德哥尔摩朋友竞技场
Friends Arena

| | | |
|---|---|---|
| **124** | 比利时布鲁塞尔托儿所和福利办公室<br>A Childcare Center and Welfare Office, C.P.A.S. | |
| **132** | 德国文登Drehmo公司新大楼<br>New Building for Drehmo, Wenden | |
| **142** | 荷兰阿尔梅勒宠物农场<br>Petting Farm | |
| **150** | 美国加利福尼亚州F65中转村<br>F65 Center Transit Village | |
| **156** | 荷兰Zwaluwen Utrecht 1911<br>Zwaluwen Utrecht 1911 | |
| **164** | 韩国京畿道"岛轩"<br>Island House | |
| **174** | 湖北十堰大美盛城会所营销中心<br>Marketing Center of Dameisheng City Club | |
| **180** | 辽宁葫芦岛比基尼广场<br>Huludao Bikini Plaza | |
| **184** | 世界休闲体育大会体育场馆工程<br>World Conference Leisure Sports Stadium Engineering | |
| **194** | 荷兰特克塞尔博物馆<br>Netherlands, Texel Museum | |
| **202** | V Lesu销售办事处<br>Sales Office "V Lesu" | |
| **206** | 瑞典哈拉斯树上酒店<br>Tree Hotel in Harads | |
| **212** | 格鲁吉亚第比利斯和平之桥<br>The Bridge of Peace | |

## 218 一"技"之长
### ——建筑技术新发展
## Excellent "technique"
### —new development of architecture technology

**220** 加拿大蒙特利尔Bota Bota水疗中心
Bota Bota, Spa-sur-l'eau

**234** 上海帝斯曼上海总部
DSM Headquarters, Shanghai

**240** 澳大利亚悉尼哈雷·戴维森澳大利亚总部
New Headquarters for Harley Davidson, Australia

**246** 西班牙拉斯帕尔马斯El Lasso社区中心
El Lasso Community Center

**254** 英国伦敦设计研究实验室十周年纪念亭
[C]space DRL10 Pavilion

**260** 上海JNBY
JNBY

# 栋梁之"材"
## ——建筑材料新运用

# Shaped "material"
## —new application of architecture material

新型建筑材料是在传统建筑材料的基础上产生的新一代建筑材料,主要包括新型墙体材料、保温隔热材料、防水密封材料和装饰装修材料。新型建筑材料及其制品工业是建立在技术进步、环境保护和资源综合利用的基础上的新兴产业。一般来说,新型建筑材料具有复合化、多功能化、节能化、绿色化、轻质高强化和工业生产化等特点。

在本章节中主要介绍了以下几种新型建筑材料:

**涂层铝板**——铝板是指用纯铝或者铝合金材料通过压力加工制成的厚度均匀的矩形材料,主要用作照明灯饰、建筑外观、家具橱柜、天花板和墙面等。而铝板幕墙则是经过铬化处理后,再用氟碳喷涂加工制成的。氟碳涂层具有卓越的抗腐蚀性和耐候性,能抗酸雨、烟雾和各种空气污染物,耐冷热性能极好,能抵御强烈紫外线的照射,具有长期不褪色、不粉化、使用寿命长等特点。

**耐候钢板**——耐候钢板是指耐大气腐蚀的钢板,该钢种的耐候性是普碳钢的2~8倍,涂装性为普碳钢的1.5~10倍,其钢板表面是一层自然锈蚀的锈红色,展示出原始材料的粗糙肌理,给建筑带来不同一般的外立面效果,并且还具有重大的经济意义,符合当今高效、长寿、节能、环保等"绿色"观念和国家政策导向。

**U形玻璃**——U形玻璃又称槽形玻璃,是一种新颖的建筑型材玻璃。因截面呈U形,使之比普通平板玻璃具有更高的机械强度,并具有理想的透光性、较好的隔音性以及保温隔热性等,既能节省大量的金属材料,又便于施工,正日益广泛地适用于建筑的内外墙、屋面和门窗,起到独特的装饰效果。

**聚碳酸酯面板**——聚碳酸酯(简称PC)是分子链中含有碳酸酯基的高分子聚合物,根据酯基的结构可分为脂肪族、芳香族、脂肪族-芳香族等多种类型。聚碳酸酯无色透明,耐热,抗冲击,阻燃BI级,在普通使用温度内都有良好的机械性能。其主要分为防静电PC、导电PC、加纤防火PC、抗紫外线耐候PC、食品级PC、抗化学性PC。聚碳酸酯的应用开发是向高复合、高功能、专用化、系列化方向发展,应用于光盘、汽车、办公设备、箱体、包装、医药、照明、薄膜等多种产品制造领域。

New building materials is a new generation of building materials produced on the basis of traditional building materials, mainly including the new wall materials, insulation materials, sealing materials and decoration materials. New building materials and its products industry are the emerging industries built based on the technological progress, environmental protection and comprehensive utilization of resources. In general, it has features with a compound, multi-function, energy conservation, green, lightweight high strengthen and industrial production.

In this chapter, it mainly introduces the following new building materials:

**Coating aluminum plate**—aluminum plate refers to the thickness uniform rectangular material which is made of pure aluminum or aluminum alloy materials formed by pressure and processing, mainly using for lighting, architectural appearance, furniture, cabinets, ceiling and metope, etc. The aluminum curtain wall is after dealing with the chromium, made of fluorine carbon spraying processing again. Fluorocarbon coating has excellent corrosion resistance and weather resistance, which can resistant to acid rain, smog and air pollutants with excellent resistance to thermal performance, and can withstand the strong ultraviolet radiation with features of not fading and not pulverization in a long-term, as well as long service life.

**Weather-proof steel plate**—it refers to the steel plate for the atmospheric corrosion resistance, which the weather-ability is as 2-8 times as carbon steel, and the painting is as 1.5-10 times as carbon steel with a layer of natural corrosion rust red of the steel surface, demonstrating raw materials rough texture to bring different façades effect for the buildings, and also having great economic significance, so it is in line with today's high efficiency, long life, energy saving, environmental protection and other "green" ideas and national policy guidance.

**U-shaped glass**—it is also known as trough type glass and is a novel construction profiles' glass. Because of the U-shaped cross section, the mechanical strength is higher than the ordinary flat glass, and has the desired light transmission, good sound insulation and thermal insulation, etc., which both can save a lot of metal materials and convenient for construction, are increasingly widely using for the interior and exterior walls of the buildings, the roofs and windows to play a unique decorative effect.

**Polycarbonate panels**—Polycarbonate (PC) is the high-molecular polymer that containing carbonic acid ester in a polymer chain. According to the ester structure, it can be divided into aliphatic, aromatic, aliphatic-aromatic and other types. PC is colorless and transparent, heat resistant, shock resistant, flame retardant grade BI, and is used under the normal temperature with good mechanical properties. It is mainly divided into anti-static PC, conductive PC, fire and defibrillators PC, anti-ultraviolet PC, food-grade PC, chemical resistance PC. The application and development of the polycarbonate is oriented to be highly complex, high function, specialty, standardization, applying to a variety of product manufactures such as discs, automobiles, office equipment, boxes, packaging, medicine, lighting, film and so on.

# 德国耶拿多功能重建大楼
## Reconstruction Multifunctional Building in Jena

| | |
|---|---|
| 设计单位：wurm + wurm architekten ingenieure | *Designed by: wurm + wurm architekten ingenieure* |
| 开发商：Christoplan GmbH | *Client: Christoplan GmbH* |
| 项目地址：德国耶拿 | *Location: Jena, Germany* |
| 建筑面积：3 400m² | *Building Area: 3,400m²* |
| 摄影：Ester Havlova | *Photography: Ester Havlova* |

Façade Bands
Coating Aluminum Plates

条带立面
涂层铝板

**项目概况**

该项目的任务是将一栋空置的预制混凝土类型的厂房改造成一栋现代化的灵活办公楼和工作间建筑,以满足当下的现代需求。

**建筑设计**

为了形成一个拥有 8 个独立功能单元的结构,获得更多的所必需的可用面积,该建筑在高高的天穹上用独立式承重钢筋混凝土结构另设了一个天花板,使其在两层楼上的中心轴形成一个公共门厅和流通区,并设有两个户外入口可通往街道和庭院。

除了这些空间设计区域外,还有两个可供步行的中庭,减少了现有大楼的高度,并为办公空间和门厅带来自然光照。

另外,建筑原有的宽阔混凝土夹层板已移除,现有的大厅结构得到加固。窗户立面由幕布般的立面条带覆盖,分别由可折叠的、高度和厚度各不相同的涂层铝板构成,使得办公空间灵活分布,营造出现代舒适的高品质办公环境。

## Profile

The task in this project is to remodel an empty production facility, which is built in type design of precast concrete, into a modern, flexible office and workshop building, which meets modern demands.

## Architectural Design

In order to create a structure that allows holding eight independent functional units, and obtain the required usable area, an additional floor ceiling is installed in the hall high ceilings with an independent, load-bearing reinforced concrete structure. The center axis of the total of 9 fields is formed as a common foyer and circulation zone over two levels and with two outside entrances, streets and courtyard positioned.

Besides this spatially expressive designed area in the center axis two walkable atriums are arranged, which reduce the high depth of the existing building and supply the office space and the foyer itself with light.

Additionally, the wall lining large concrete sandwich-panels have been removed, and the existing hall structure is upgraded statically. The small-scale distribution of the new window façade, owed the flexible divisibility of the office space is covered with curtain-like façade bands, consisting of individually folded coated aluminum plates, that vary in height and depth.

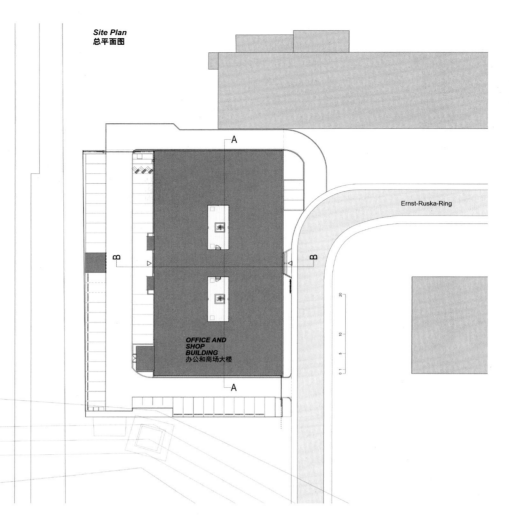

**Ground Plan**
底层平面图

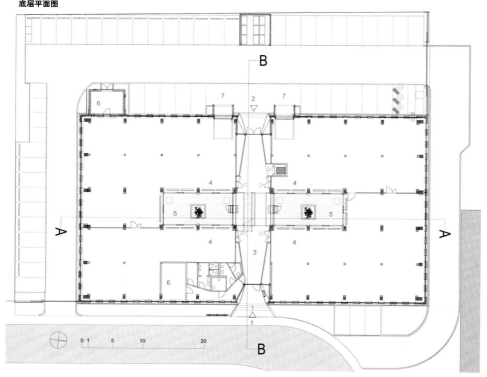

▲ 铝板——铝板是指用纯铝或者铝合金材料通过压力加工制成的厚度均匀的矩形材料，可用于照明灯饰、建筑外观、家具橱柜、天花板、墙面等。

　　而铝板幕墙一般经过铬化处理后，再采用氟碳喷涂处理。氟碳涂层具有卓越的抗腐蚀性和耐候性，能抗酸雨、烟雾和各种空气污染物，耐冷热性能极好，能抵御强烈紫外线的照射，具有长期不褪色、不粉化、使用寿命长等特点。

Aluminum plate—aluminum plate is a rectangular material with even thickness, which is made by pressure processing of pure aluminum and aluminum alloy material. It can be used for decorative lighting, building exterior, furniture, ceiling and wall surface, etc. After chromating, aluminum curtain wall is coated with fluorocarbon. Fluorocarbon coating features excellent corrosion resistance and weather resistance, which can also resist acid rain, smog, air pollutant, strong ultraviolet rays and maintain original color, long life, without pulverization.

*Second Floor*
二楼

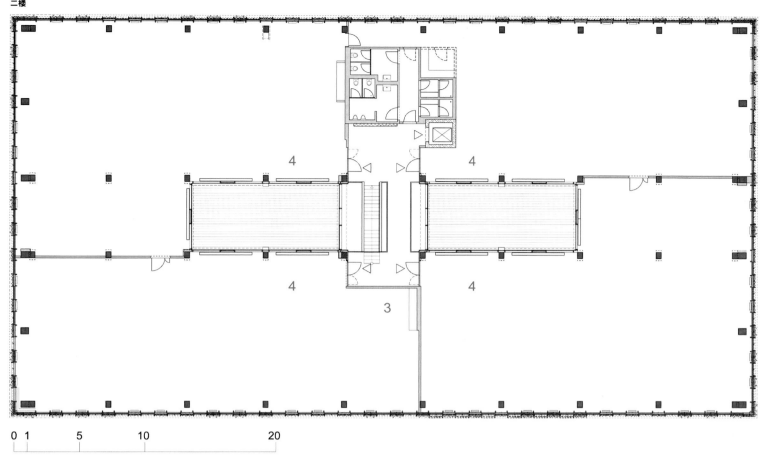

**Section AA**
剖面 AA

**Section BB**
剖面 BB

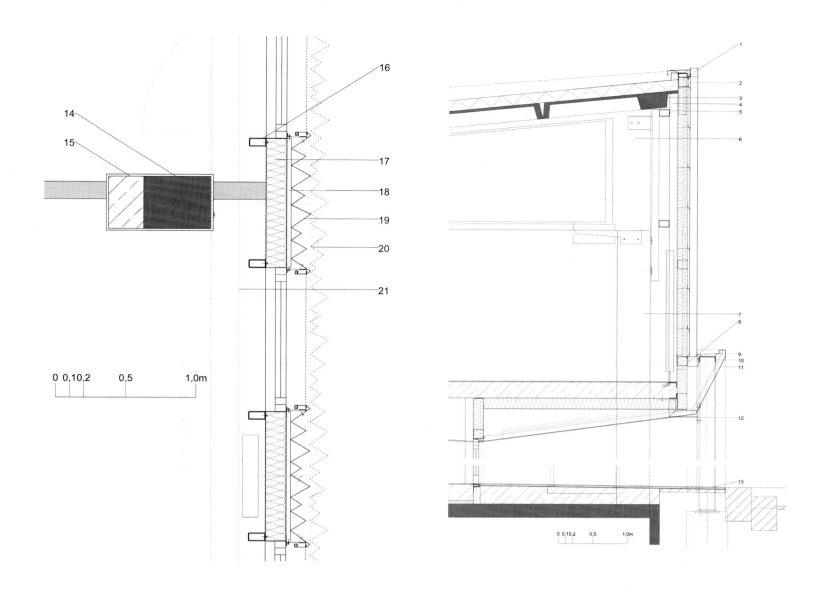

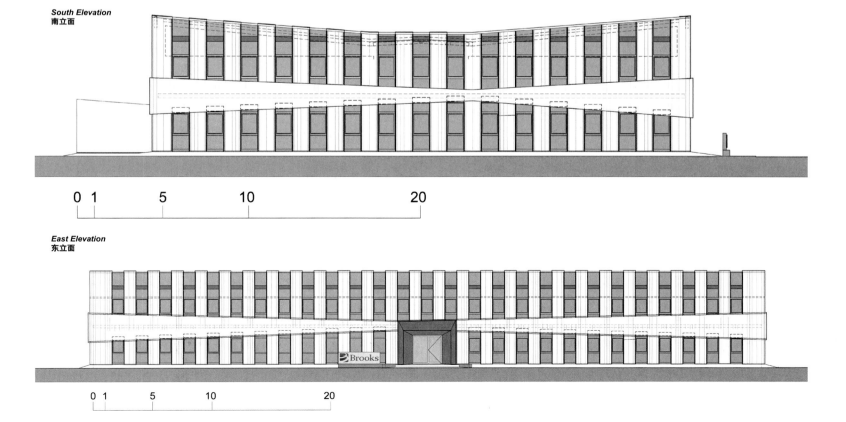

*South Elevation*
南立面

*East Elevation*
东立面

Concrete Moving Façade

混凝土"移动"立面

# 墨西哥比亚埃尔莫萨 Liverpool Altabrisa 百货商店

Liverpool Altabrisa Department Store

| | | |
|---|---|---|
| 设计单位：Iñaki Echeverria | | |
| 开发商：Liverpool 百货 | | |
| 项目地址：墨西哥比亚埃尔莫萨 | | |
| 建筑面积：8 070m² | | |
| 设计团队：Iñaki Echeverria | Ivan Parra | Josue Lee |
| Osvaldo Estrada | Jonathan Hajar | Jorge Duran |
| Roberto Fuerte | Javier Marroquin | Guillermo Lopez |
| Leonel Alcantara | Rafael Brioso | Fredy Lamadrid |
| Fernanda Tellez | Victor Garcia | Israel Meneses |
| Xochitl Zuñiga | Daniela Gonzalez | |
| 摄影：Luis Gordoa | | |

*Designed by: Iñaki Echeverria*
*Client: Liverpool*
*Location: Villahermosa, Mexico*
*Building Area: 8,070m²*
*Design Team: Iñaki Echeverria, Ivan Parra, Josue Lee, Osvaldo Estrada, Jonathan Hajar, Jorge Duran, Roberto Fuerte, Javier Marroquin, Guillermo Lopez, Leonel Alcantara, Rafael Brioso, Fredy Lamadrid, Fernanda Tellez, Victor Garcia, Israel Meneses, Xochitl Zuñiga, Daniela Gonzalez*
*Photography: Luis Gordoa*

### 项目概况

Altabrisa 购物中心广场位于城市南端一个战略性地区，是塔巴斯科州比亚埃尔莫萨市新的发展中心。该项目的挑战就是要在中心广场寻求一套简单有效的建筑系统，并且还要以现代动感的外立面为目标，为墨西哥最大的奢侈品零售商提供全新的形象。

### 建筑设计

由于塔巴斯科州属于热带气候，太阳强度和湿度都很高，因此设计选用混凝土这种耐久、抗老化的材料作为建筑外立面的主要材料。建筑采用5种不同类型的螺旋桨般的预制件，按轴线旋转180°，根据位置的不同安装在16m到20m的高度上。

在创新建造技术的帮助下，项目形成了新的形象，简单但多变的表皮形式是由不同部分组成，但又在整体中给人一种运动感，纯白的建筑形象，宛如一朵飘浮于城市间的白云，充满着梦幻色彩。不论是白天在日光之下，还是夜晚在LED人工照明下，建筑都通过这些简单的、受控制的变化创造出立面标志性的形象。

### Profile

Located in a strategic area on the south end of the city, the shopping center Plaza Altabrisa is a new development pole for Villahermosa, Tabasco. The project's challenge was to find a simple and effective construction system that would accelerate production, assembly and installation of the façade while providing a bold architecture. The aim was to design a dynamic and modern façade that would provide a new image for the largest luxury retailer in Mexico.

### Architectural Design

Given Tabasco's extreme tropical climate with severe solar incidence and humidity levels, concrete was selected as the project's design material; a material both resistant and one of extraordinary aging qualities. The result is a façade that's constructed by combining five different types of precast pieces shaped like a propeller. Each propeller rotates 180° on its axis; heights vary between 16 to 20 meters, depending on their position. With the assist of innovative building technique, a new image is created. The simple and variable skin composes of different segments, while reflecting overall movement. The white building image is like a cloud floating in the city, full of dream-like colors. Both sunlight and controlled lighting with LED create a changing image over time. These simple and controlled variations produce an iconic image for the façade.

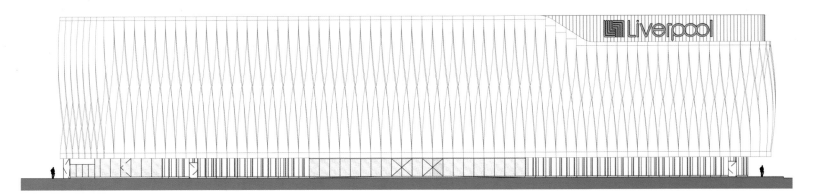
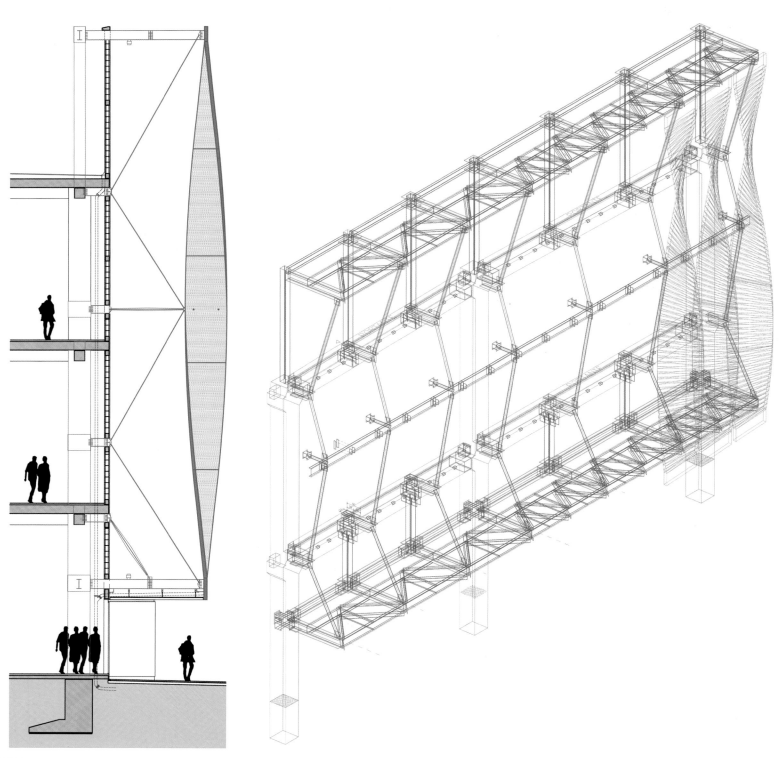

▲ 混凝土 ——混凝土是一种充满生命力的建筑材料，原料丰富，价格低廉，且生产工艺非常简单，是建筑中最受欢迎的建筑材料之一。

混凝土具有抗压强度高、耐久性好、强度等级范围宽等特点，能够适应恶劣的气候环境，因此，不仅在土木工程中广泛使用，甚至造船业、机械工业、海洋的开发、地热工程等领域，混凝土也是重要材料。

Concrete—it is a construction material full of vitality. With rich raw material, low cost and simple manufacturing technique, concrete is one of the most popular building materials.

With high compressive strength, excellent durability and wide rate range of strength, concrete can adapt to harsh climatic environment. It is not only widely used in civil engineering, even in the industries of shipbuilding, machinery, marine exploitation and geothermal engineering.

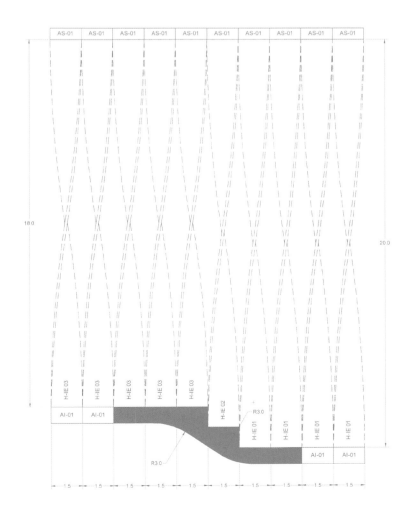
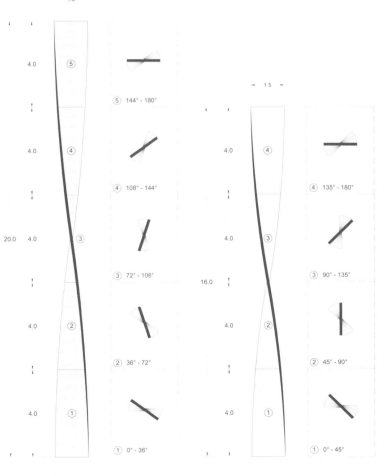

# 美国圣塔莫尼卡市民中心停车场
## Santa Monica Civic Center Parking Structure

设计单位：Moore Ruble Yudell Architects & Planners
项目地址：美国圣塔莫尼卡
摄影：John Edward Linden

Designed by: Moore Ruble Yudell Architects & Planners
Location: Santa Monica, U.S.
Photography: John Edward Linden

生态环保
光影立面
粉煤灰

Environmental Protection
Façade of Shadows and Light
Fly-ash

## 项目概况

该停车场位于城市民用房屋聚集区，不仅能提供882辆车的停车位，而且还能为社区提供多种便利设施，是一个极具特色的建筑物。它并不是传统意义上的停车场，而是一个体现城市生活的多功能动感建筑。

## 生态设计

该建筑是美国第一批获得LEED认证的停车场之一，在项目建造过程中，设计师秉持着生态环保的建筑理念，采用了大量的节能材料：屋顶光伏板不仅起到遮阳的效果，还能提供大量能源；建筑材料方面则用粉煤灰取代传统的水泥，以及支架采用了可回收的钢框架等。

## Profile

The parking structure is located within the cluster of civic building in the city, effectively providing not only 882 parking spaces, but also a wide variety of amenities to the community. This structure is conceived as much more than a traditional parking garage—rather, a functionally dynamic celebration of civic life.

## Ecological Design

Many factors contribute to the building's status as one of the first LEED™ certified parking structures in the United States. In the construction processing of the building, the architects adhere to the eco-friendly architectural concept to use plenty of energy saving material: photovoltaic panels on the roof provide shade and a significant portion of total energy needs; in the aspect of architectural materials, it uses fly-ash to replace the traditional cement, and the supports adopt the recycled-content reinforcing steel.

◀ 粉煤灰——在混凝土中掺加粉煤灰代替部分水泥或细骨料，不仅能降低成本，而且能提高混凝土的和易性、不透水、气性、抗硫酸盐性能和耐化学侵蚀性能，降低水化热，改善混凝土的耐高温性能，减轻颗粒分离和析水现象，减少混凝土的收缩和开裂以及抑制杂散电流对混凝土中钢筋的腐蚀。

Fly-ash—adding fly-ash to concrete, as the replacement of some cement and fine aggregate, can not only reduce cost, but also improve workability, waterproof, gas, sulfate resistance, chemical attack resistance, reduce hydration heat, improve resistance to elevated temperatures of concrete, relieve particle separation and bleeding, and reduce shrinkage and cracking of concrete and restrain corrosion of steel bar caused by stray current.

光影立面——建筑立面上设计有棱纹的混凝土板材，形成富有节奏的图案，使得停放的车辆在立面上也能形成丰富的光影变化。彩色玻璃形成的一系列凹处在白天不断变换色彩，宛如城市中璀璨闪烁的霓虹，格外绚烂夺目。在晚上，停车场结构也不断变化着，在灯光的照耀下，整座建筑成为一块迷人的光与色彩翩翩起舞的幕布。

Façade of Shadows and Light—the parking structure is a visually iconic presence in the Civic Center. Ribbed concrete panels are set in a shifting, rhythmic pattern on the façades, capturing a rich play of shadows while screening the presence of parked cars. A series of bays made of colored channel glass bring a lively, ever-changing quality by day, like bright neon in the city. The character of the structure changes yet again in the evening, when the glass is illuminated to become a dancing curtain of color and light.

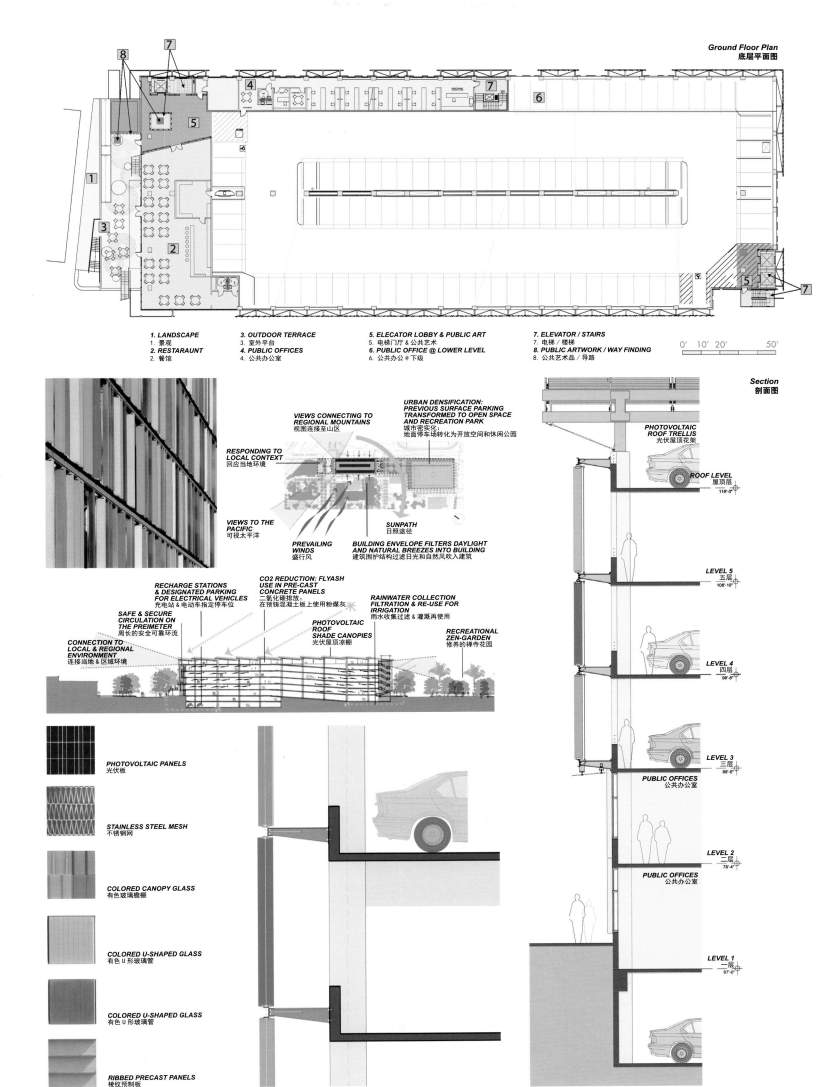

# 法国贝休恩水上活动中心
## Aqua Center, Bethune

设计单位：SAREA Alain Sarfati Architecture
项目地址：法国贝休恩
室内面积：5 000m²
室外面积：1 000m²

Designed by: SAREA Alain Sarfati Architecture
Location: Bethune, France
Inside Area: 5,000m²
Outside Area: 1,000m²

**项目概况**

贝休恩水上活动中心不管是外部还是内部都充满着惊喜，耐人寻味。复杂的造型设计保证了该公共设施的耐久性，这里不仅可以举办各种活动，还具备社交功能。

**建筑材料**

该建筑呈一座火山的形态，就像是地热水流经所产生的一处庇护站。处于沉积岩和火山岩双重地质特征中的建筑稳固又动感，在井然有序的结构中体现出开放性，是光和水的构成物。

建筑外立面充分运用染色混凝土、聚碳酸酯面板、钻孔和折叠的金属板，以及涂漆铝制复合板、彩色玻璃等材料，在这些材料的巧妙组合中形成独具创造性的建筑形态，极具视觉标识性。

入口设计由高技术的板材构成，与建筑内部的玄武岩瓷砖、悬浮声学光纤天花板以及金制阳极氧化铝板等技术材料完美结合，缤纷的室内空间使得建筑宛如一个神奇的万花筒。

多种材料
万花筒
火山庇护站

Various Materials
Kaleidoscope
A Shelter in Form of Volcano

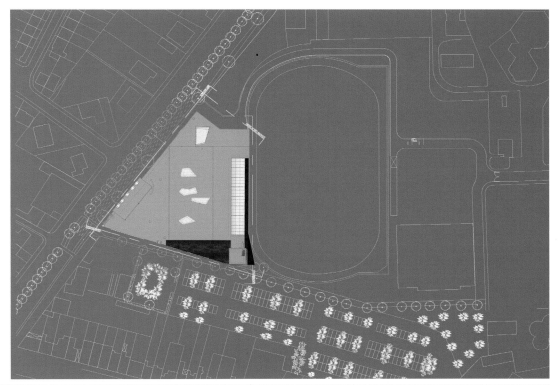

## Profile

The Aqua Center in Bethune is full of surprise and joy from the exterior development to interior exploration. Complexity and wealth of form ensure durability of the public facility in which not only a wide range of activities can be developed but where there is a desire to create a social connection through the shared modern asset that it is.

## Building Materials

In the image of a volcano, the building is like a shelter under which the telluric waters flow with their string of beneficial properties. The architecture is both stable and dynamic, comprised of water and lights. It has a double geological dimension; the first one is sedimentary and the other is volcanic. It is also in order with openness.

The façade of the building takes full use of dyed concrete, polycarbonate panels, punch and folded metal sheets, lacquered aluminium composite panels, glazing with colored film, etc. The combination of materials results in a creative architectural form and visual mark.

The entrance is presented as sedimentation of technical plates. Beach basalt shade tiles, suspended acoustic fiber cap ceiling and gold anodized & punch aluminium sheet used inside make the whole building a magical kaleidoscope.

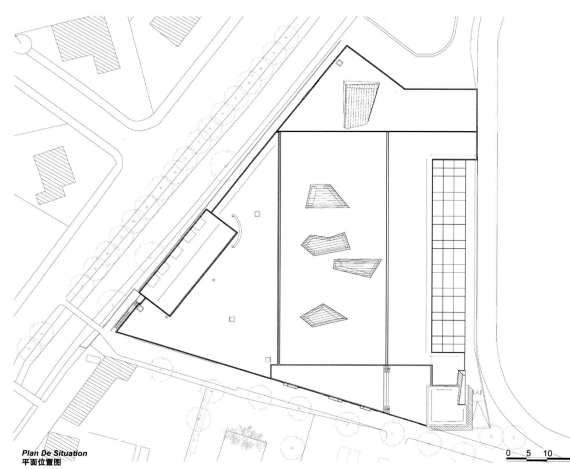

*Plan De Situation*
平面位置图

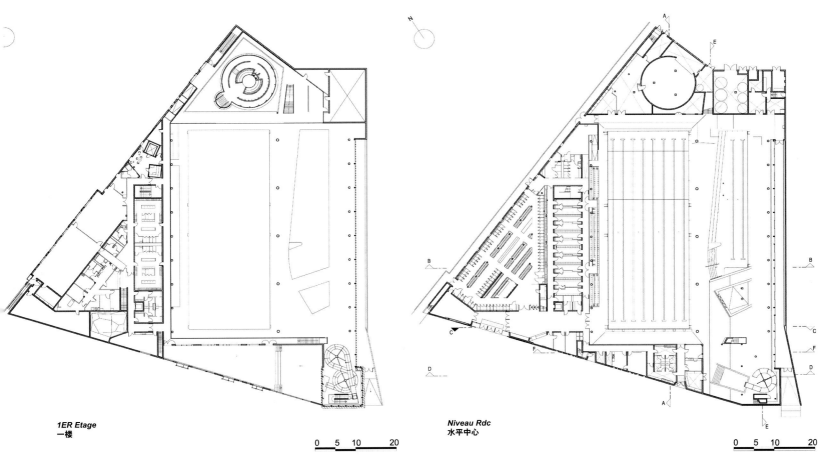

1ER Etage
一楼

Niveau Rdc
水平中心

0 5 10 20

◀ 休闲池——泳池大厅设于橄榄球场的对面，原来的泳池被休闲池所取代，共包括有戏水池、训练池、娱乐池和一个滑梯。另外，日光浴场设在楼上的正南方泳池的阳台上，朝向浴疗区域，而泳池则设在Lawe河后面，朝向体育大厦，将现有的建筑进行了最佳整合。

该片水池区域成为人们活动的中心，正如该水上活动中心的一颗明珠。

Leisure Pools—the pools' hall is kept in its current place, opposite the rugby stadium. The former pool is replaced by new leisure pools: a paddling pool, training and relaxation pool and a slide. The solarium is located due south, upstairs, on the balcony over the pools, facing the balneotherapy area. Built behind the Lawe River, the pool faces the sports block; the existing building already has optimal integration.

The pools area is the activity center for people, like a bright pearl of the Aqua Center.

**ELEVATION ENTRÉE**
立面入口

0　5　10　　20

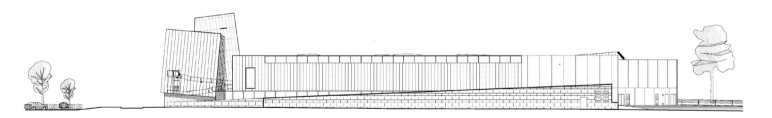

**ELEVATION EST SUR STADE**
高层舞台

0　5　10　　20

**ELEVATION COUR DE SERVICE**
高层服务阁

0　5　10　　20

# 鹿特丹巴讷费尔德的巴士站
## Bus-stop in Barneveld Noord

| | |
|---|---|
| 设计单位：NL Architects | Designed by: NL Architects |
| 开发商：ProRail | Client: ProRail |
| 项目地址：荷兰鹿特丹 | Location: Rotterdam, Netherlands |
| 建筑面积：80m² | Building Area: 80m² |
| 建筑师：Pieter Bannenberg  Walter van Dijk  Kamiel Klaasse | Architects: Pieter Bannenberg, Walter van Dijk, Kamiel Klaasse |
| 项目架构师：Gerbrand van Oostveen | Project Architect: Gerbrand van Oostveen |

人性化
安全感
集装箱组合建筑

Humanity
Security
Portfolio Container
Construction

## 项目概况

设计师 ProRail 负责荷兰的铁路网络，他在项目建设过程中发现：较小的车站通常是无人驾驶的、荒凉的，往往造成非安全性的感觉。因此，为了改善这种局面，ProRail 通过引进洗手间、无线网络、地板采暖、铁路电视等设备，旨在从功能上和美观上，对全国 20 多个车站等候区进行全面升级，打造一个富有人性化、安全感的全新候车区。

## 建筑设计

巴讷费尔德的巴士站是由四个集装箱组合建造的，这在建筑史上还是头一回。该集装箱包含空间，而且也形成空间，它们被设计师巧妙地合并在一起，共同形成一个模糊且强大的标志。设计旨在用最少的材质，提供最大的效率。

横向一字排开的三个集装箱靠底下的柱子支撑，它们共同形成一个"屋顶"，一个用于承载设备，一个用于存储，还有一个则与地面形成联合系统，通过大面积玻璃的重组，形成一个封闭但完全透明的区域，而这个区域被用作临时小卖场，为过往的旅客提供便捷的服务。第四个集装箱采用直立的方式，形成建筑塔，塔身还设计了一个时钟和风向标，不但在瞬间提升了建筑整体的立体感，而且更加完善了该建筑的外观形式，进一步丰富了该建筑作为巴士站的功能定位。除此之外，该建筑还安置了一个厕所，由一个玻璃屋顶突破。

## Project Overview

ProRail, responsible for the railway network in the Netherlands, He found that in the process of project construction these smaller stations are usually unmanned, desolate, often creating a sense of un-safety. The waiting areas of in total twenty stations throughout the country will be upgraded, both functionally and cosmetically: introduction of washrooms, Wi-Fi, floor heating, railway TV.

## Building Design

Barneveld bus stop is a combination of four container construction, this is the first time in the history of architecture. The containers contain space, but also form space. They will be combined into an explicit arrangement. Together they form an ambiguous but strong sign. Minimum effort, maximum output. Three containers lined up by pillars beneath the lateral support, which together form a "roof", a device used to carry and one for storage, then the third forms a joint system with the ground. It forms a closed region but completely transparent area through large glass of reorganization, and the region is used as a temporary small store to provide easy service for the visitors. The fourth container is flipped to an upright position. It makes an instant tower. The tower contains a clock and a wind vane. In addition, the tower holds a lavatory, 11.998mm high, topped by a glass roof.

*Ground Floor*
底层

*First Floor*
一楼

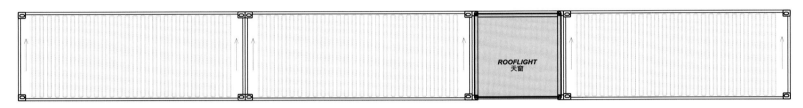

*Roof Plan*
屋顶平面图

38　面向未来——建筑趋势2015　FACING TO THE FUTURE-ARCHITECTURE TREND 2015

**Cross-Section**
横剖面

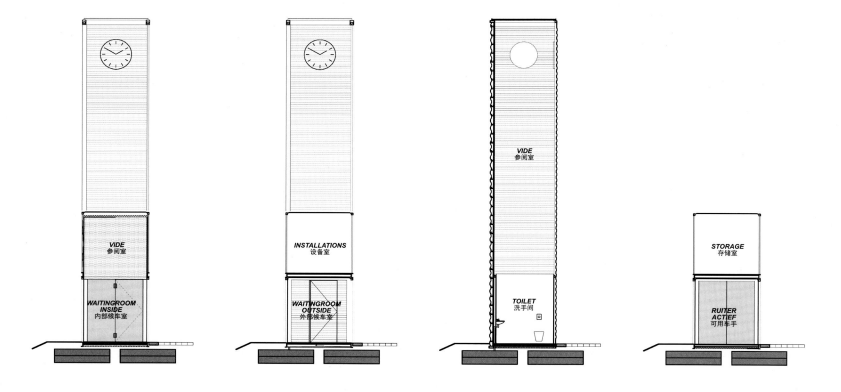

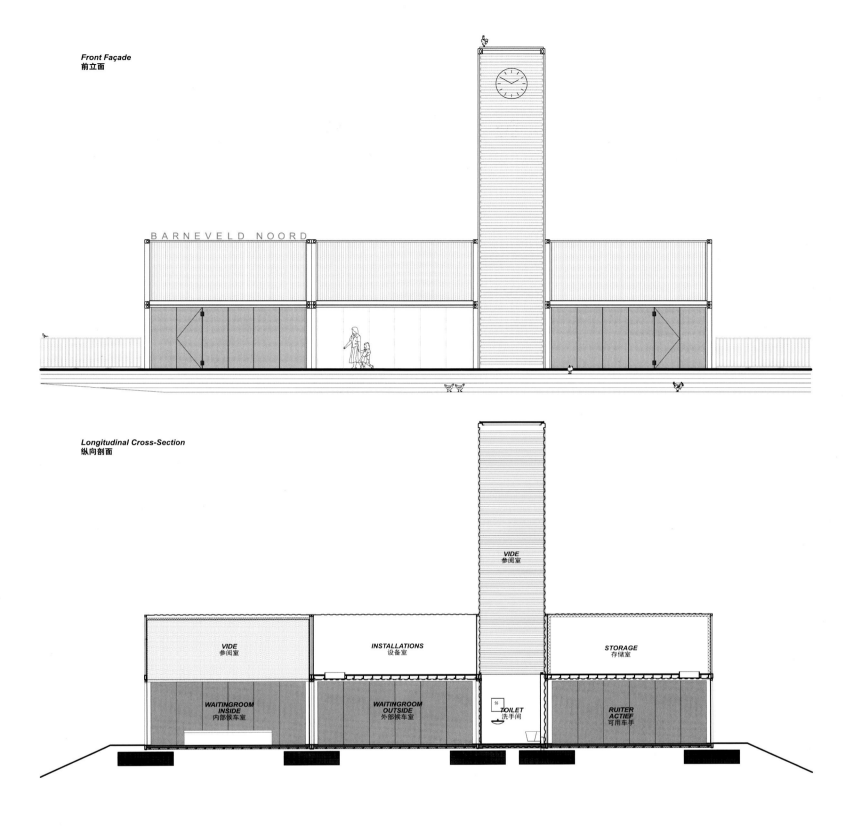

*Front Façade*
前立面

*Longitudinal Cross-Section*
纵向剖面

# 丹麦 Egtved 天然气压缩工厂
## Gas Compressor Station Egtved

设计单位：C. F. Møller Architects
开发商：Energinet.dk
项目地址：丹麦 Egtved
摄影：Jan Laursen  Julian Weyer

*Designed by: C. F. Møller Architects*
*Client: Energinet.dk*
*Location: Egtved, Denmark*
*Photography: Jan Laursen, Julian Weyer*

**项目概况**

这个新的天然气压缩工厂是从 Egtved 到德国以及欧洲大陆铺设的新"天然气高速公路"在丹麦的服务终端，将为该国提供稳定的天然气能源供应，又是德国和瑞典"天然气十字路口"的重要组成，由四个压缩机单位和服务大楼组成，叠立在开阔的自然景观之中，成为场地中的绿色堤岸。

**建筑设计**

建筑就像是盘旋在高地之上，表皮覆盖着并列设置的铁锈色耐候钢板，形成千变万化、富有生机的光影图案。这些立面材料的组合使得建筑看起来凹凸不平而又不失优雅之美。由玻璃和金属覆层的服务大楼则包括有应急发电机室和储藏室，而此服务大楼和压缩站简单而引人注目的设计又为建筑的功能性设计提供了巨大的灵活性。

**Profile**

This compressor station is the service point of the "gas motorway" from Egtved to Germany and continental Europe, which supplies stable natural gas for Denmark. It is also the central intersection of the gas pipelines from Germany to Sweden, consisting of four compressor units and service buildings, as a grassy embankment in the open landscape.

**Architectural Design**

The building appears almost to hover over the mound and is clad with rust-colored Cor-Ten steel plating. The plating is juxtaposed to create a varied and vibrant pattern of light and shadow. The combination of materials aims to make the buildings appear rugged and elegant at the same time.

The grass and iron-clad plant houses service buildings, including an emergency generator and storage rooms. The simple and striking design of the service buildings and substation also provides the opportunity for great flexibility in relation to the functional adaptation of the design process.

Grassy Embankment
Cor-ten Steel Plate

绿色堤岸
耐候钢板

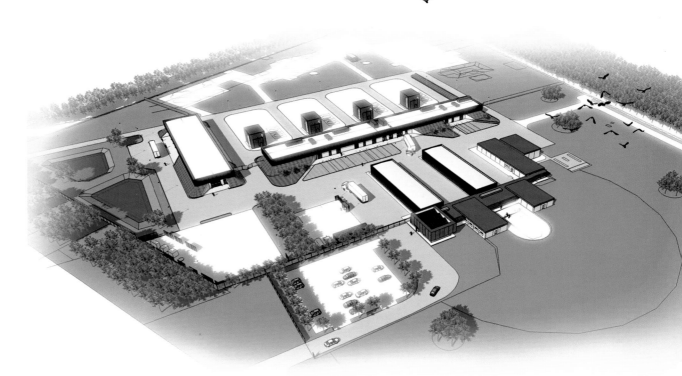

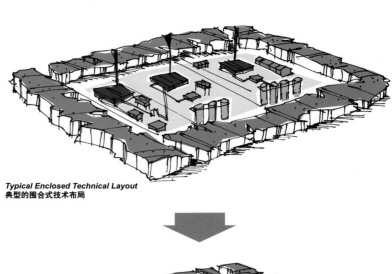

*Typical Enclosed Technical Layout*
典型的围合式技术布局

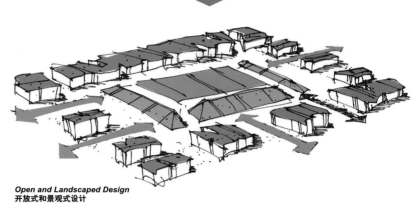

*Open and Landscaped Design*
开放式和景观式设计

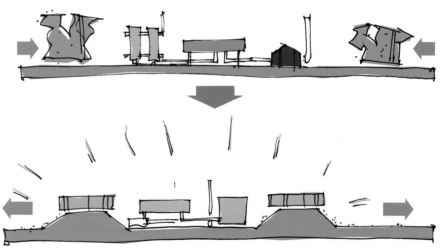

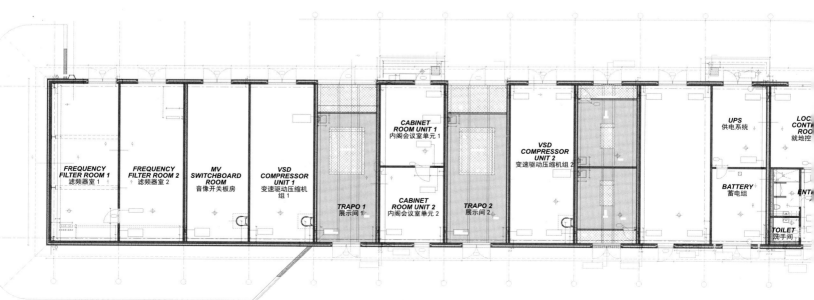

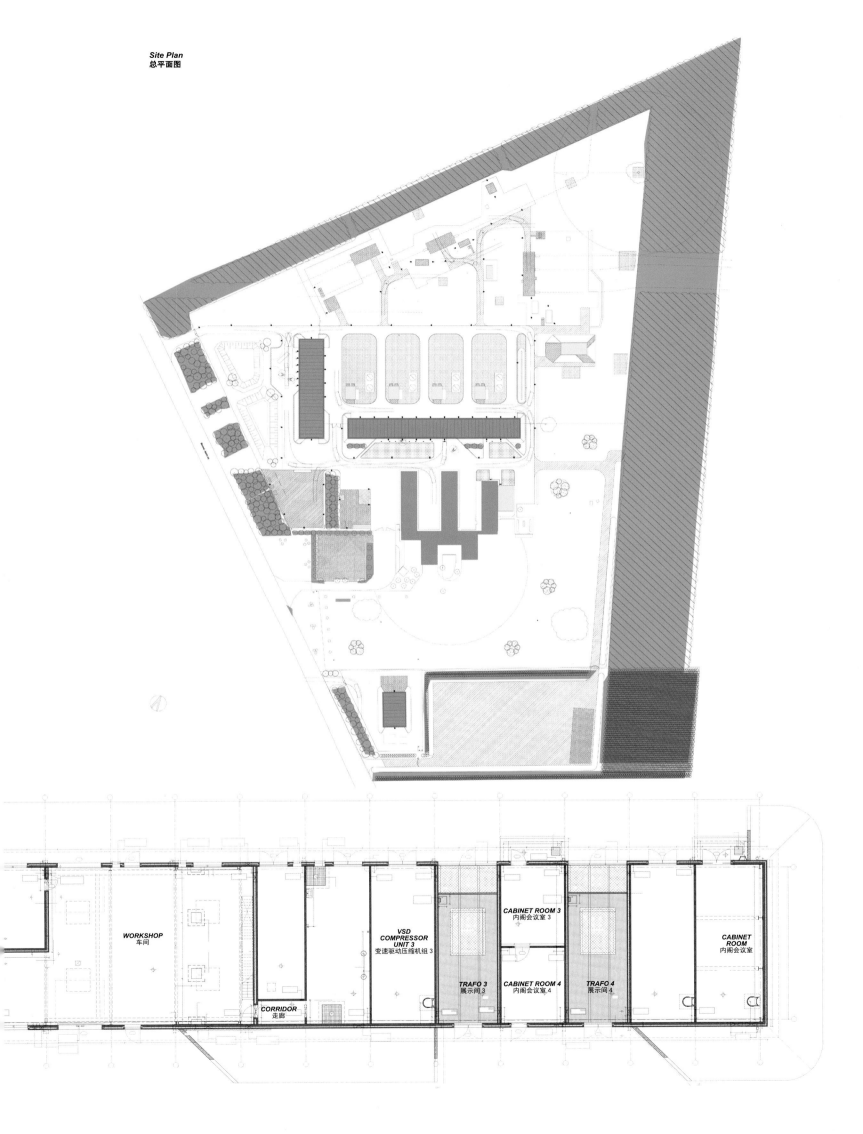

◀ 耐候钢板——耐候钢板是指耐大气腐蚀的钢板，该钢种的耐候性是普碳钢的 2～8 倍，涂装性为普碳钢的 1.5～10 倍，其钢板表面是一层自然锈蚀的锈红色，展示出原始材料的粗糙肌理，给建筑带来不同一般的外立面效果，并且还具有重大的经济意义，符合当今高效、长寿、节能、环保等"绿色"观念和国家政策导向。

Cor-Ten Steel Plate—it is a kind of weather resisting steel, whose resistance performance is twice to 8 times of plain carbon steel and coating is 1.5 to 10 times of plain carbon steel. The surface of Cor-Ten steel plate is a layer of rusty color, presenting rough texture of original materials. It provides unique façade for building and creates economic significance, conforming to the "green" concept of efficincy, longevity, energy conservation and environmental protection and policy direction of the country.

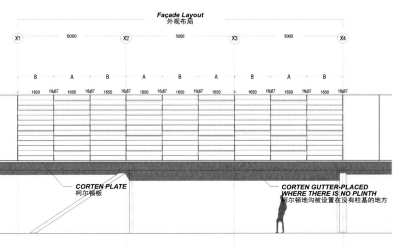

**FAÇADE LAYOUT CONSISTS OF REPEATED A + B ELEMENTS**
外观布局包含复合板 A+B 要素
**SPECIAL ADAPTIONS ARE MADE ON EACH CORNERELEMENT**
对于每个要素采取特别调整

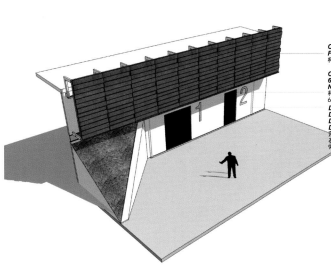

*CORTEN STEEL*
*FAÇADE ELEMENTS*
柯尔顿钢表面元素

*CORTEN GUTTER*
*60×100MM, D=3MM*
*NO INCLINATION*
柯尔顿沟
60×100 毫米, D=3 毫米无倾向

*DOWNPIPE*
*DISTANCE BETWEEN*
*DOWNPIPES IN THIS*
*DRAFT, 9600MM*
落水管
在草图上落水管之间的距离为
9600 毫米

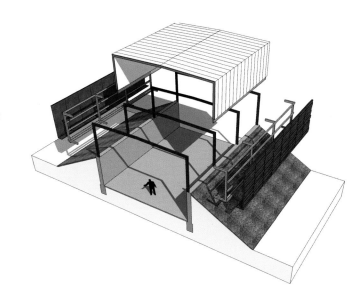

# 墨西哥库埃纳瓦卡 La Estancia 教堂
## La Estancia Chapel

设计单位：BNKR Arquitectura
合作者：Jorge Arteaga
项目地址：墨西哥库埃纳瓦卡
建筑面积：117m²
设计团队：Paola Moire　Ingrid Santoyo　Miguel Ángel Martínez
　　　　　Diego Jasso　Guillermo Bastian

*Designed by: BNKR Arquitectura*
*Collaborator: Jorge Arteaga*
*Location: Cuernavaca, Mexico*
*Building Area: 117m²*
*Project Team: Paola Moire, Ingrid Santoyo, Miguel Ángel Martínez, Diego Jasso, Guillermo Bastian*

**项目概况**

这是墨西哥某巴洛克风格花园里的婚礼小礼拜堂，依据客户的要求，这座精巧的建筑完全融入到周围优美的花园景观中，显得浪漫而迷人。

**建筑设计**

教堂的选址非常精心，位于一片茂密的植被当中，建筑不破坏场地的任何一棵树且还能获得充分的庇荫。大大的蓝花楹树为教堂形成一道天然的拱门，充分为室内遮挡了部分阳光。

为了不破坏花园景色的美感，建筑师设计了一个完全开放的玻璃教堂，与周边环境形成鲜明对比。建筑被看成是一个带有坡顶的盒子，为避免建筑在热带气候中成为一个温室，设计采用U形玻璃窗格连接在一起作为一个独立的分隔膜，使其免受日光灼射。而教堂传统的十字架元素则被巧妙地设计成立面上的十字窗户，使得整座建筑更为简单而纯净。

在优雅的花园中，这座建筑就像一颗水晶，又仿佛是天然雕饰的钻石，美好的爱情在这里得到记载与永存。

## Profile

This wedding chapel is designed in the traditional Mexican baroque colonial style. According to the client's request, the delicate architecture perfectly blends into the graceful garden landscape, quite romantic and charming.

## Architectural Design

The site for the chapel was carefully chosen within an enormous area of abundant vegetation. The location would not require the removal of any of the existing plants or trees, under large jacarandas which form a natural arch over the chapel and provide it with ample shade.

In order not to spoil the beauty of garden, the architects designed a completely open glass chapel which is in striking contrast to the surrounding environment. The chapel was conceived as a box and compressed to form a peaked roof. To avoid the chapel from becoming a greenhouse, the architects linked some U-profiled glass windows which act as a separate diaphragm to protect the building from burning of sunlight. A cross was used to create a window that looks out onto the surrounding garden, making the chapel simple and clear.

In the elegant garden, this building is like a crystal or naturally-carved diamond. Beautiful love is forever recorded here.

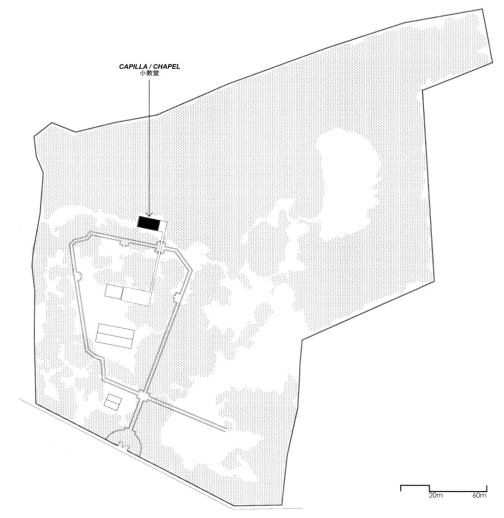

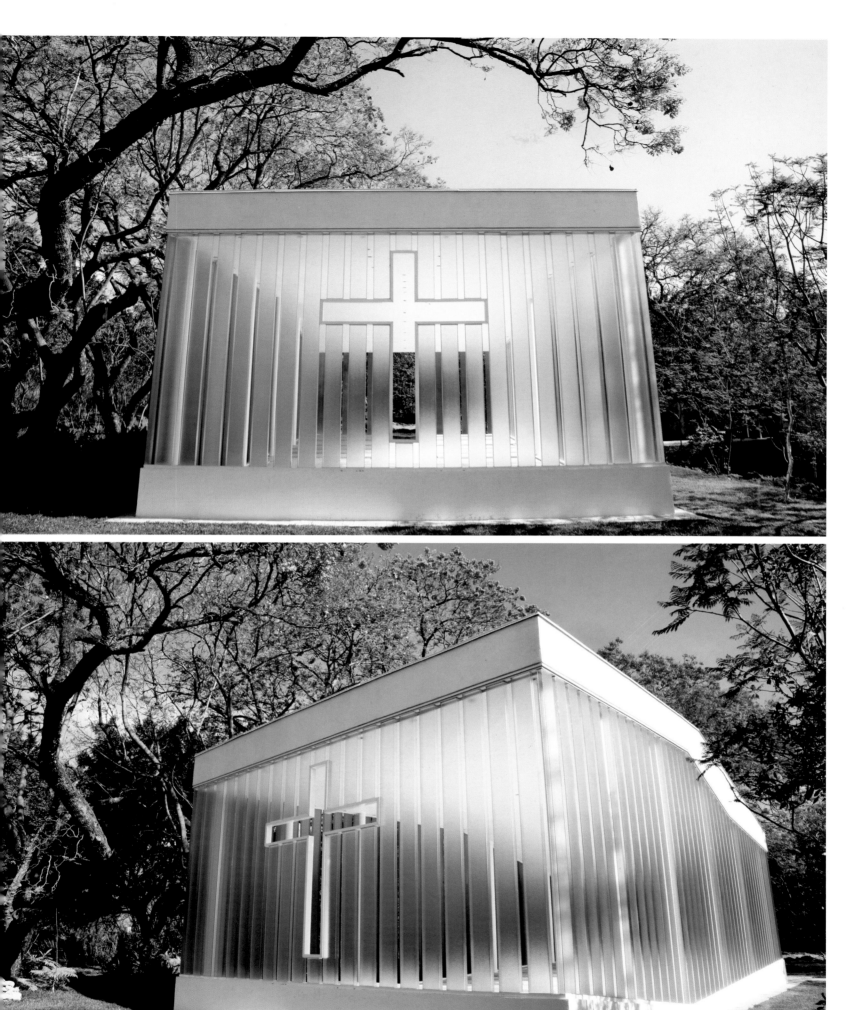

▲ U形玻璃——U形玻璃又称槽型玻璃，是一种新颖的建筑型材玻璃。因截面呈U形，使之比普通的平板玻璃具有更高的机械强度，并具有理想的透光性、较好的隔音性以及保温隔热性等，既能节省大量金属材料，又施工简便，正日益广泛地用于建筑的内外墙、隔墙、屋面及门窗等，起到独特的装饰效果。
U-profiled Glass—it (also is known as grooved glass) is an innovative building glass. Due to U shape of its section, this glass is higher in mechanical strength than common sheet glass. Additionally, it is characterized by perfect translucency, sound insulation, and heat preservation and insulation properties. Using U-profiled glass not only can reduce numerous metal materials, but also is easy to construct. It is now widely used in interior and exterior walls, partition, roofing, door and window of building, which also has unique decoration effect.

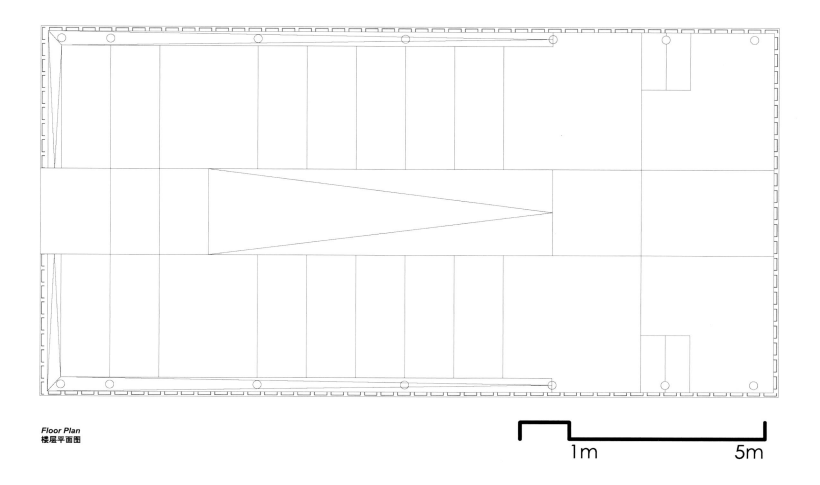

*Floor Plan*
楼层平面图

1m  5m

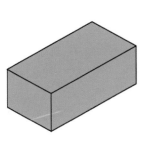
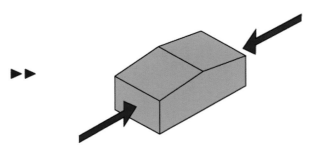
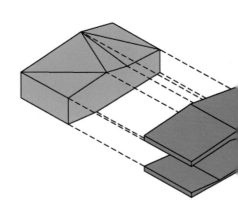

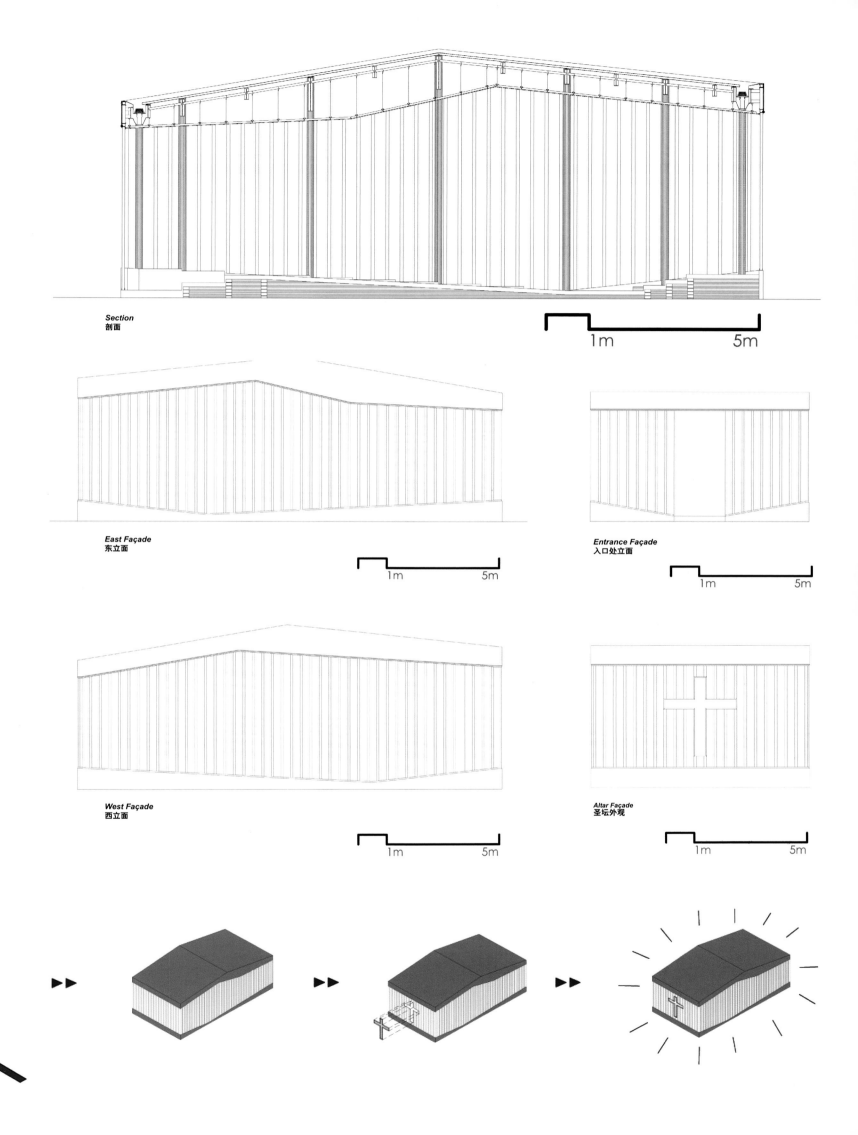

# 英国伯恩利"彩虹门"
## Shell Lace Structure

设计单位：Tonkin Liu
合作单位：ARUP
开发商：伯恩利自治市议会
项目地址：英国伯恩利
建筑面积：61m²

Designed by: Tonkin Liu
Collaborator: ARUP
Client: Burnley Borough Council
Location: Burnley, Britain
Building Area: 61m²

**项目概况**

Tonkin Liu 建筑事务所为英格兰一座河滨市镇伯恩利设计建造了一座以"彩虹门"为意象的独特亭子，作为工业革命的受益者，伯恩利这座新兴工业城市曾因为充沛的降雨来补充河水而大力发展工业。在这种概念基础上，设计师们通过"彩虹门"建筑元素向当地的经济历史致敬。

**建筑设计**

这座满是空隙的钢体以曲线勾勒出交通联系的形象，其立体造型是由平坦的激光切割的 3mm 厚的钢板构造而成，雨滴的敲打声在金属上回响，当雨水汇集成束，从亭子上的凹槽流下时，嵌在孔隙中的透镜将会捕捉阳光并将其打散为彩虹光谱，并投影到地面上，配合贝壳曲面起皱的纹理，就像是一首赞美工业革命的小诗。

另外，设计师运用一种取形于自然并且通过对材料的最优化来构成轻量化且具有自我支持性的被称为"外壳花边"的结构系统。通过整合数字制造技术和建模技术，以及加强海洋贝壳的分层几何层次，使这个建筑兼具曲线感和波纹感，而这也减少了非承重层的材料使用，令建筑在拥有活力外表的情况下也具有高效的建造结构。

### Profile
Tonkin Liu designed the unique pavilion taking the image of "rain bow gate" in riverside town Burnley in England. As a beneficiary of the Industrial Revolution, the emerging industrial city underwent great development due to its ample rainfall complementing river water. On this basis, the designer wanted to make use of the element of "rain bow gate" to respond to economic history of this town.

### Architectural Design
The perforated steel structure forms a gate that enhances traffic connection. The strong three-dimensional form is constructed from flat, laser-cut, 3mm steel sheets. The sound of rain reverberates on the thin drum-like steel roof, collecting rainwater and channeling it into the ground. The glass prism inserted in perforations captures light, and then the light becomes rain bow spectrum with shadows on the ground. The image of spectrum, with the wrinkling texture of curved shell, is like a poem praising the Industrial Revolution.

Additionally, the designer uses a technique called "Shell Lace Structure", which is taken shape from nature and is light and self-supporting by optimizing materials. Through integrating digital fabrication tools and digital modeling, as well as reinforcing geometric layer of sea shells, the architecture features sense of curve and ripple, which also reduces materials of non-bearing stratum. Besides vital appearance, the pavilion is characterized by efficient construction.

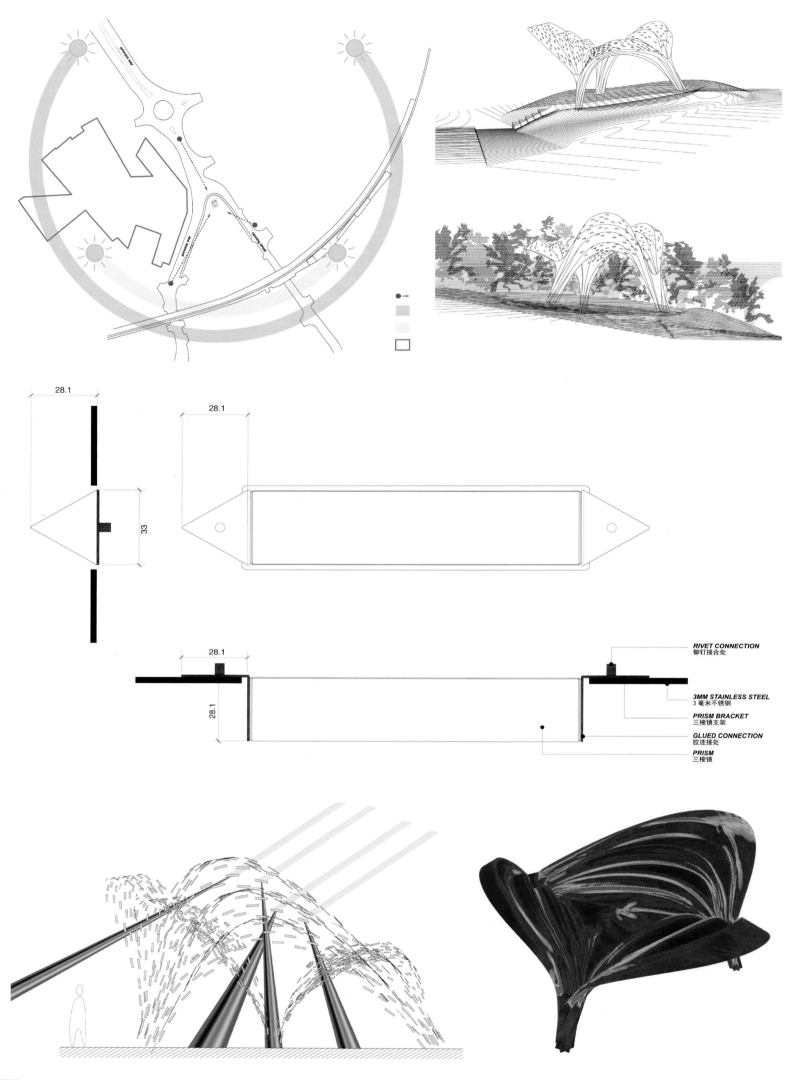

# 奇"形"巧状
## ——建筑外形新趋势

# Special "appearance" and attractive structure
## —new trend of architecture shape

随着人们对居住舒适度、居住文化品位要求的不断提高，人们对建筑外观造型、风格取向的理解及要求，也进一步加深和加强，从而形成了建筑设计风格多元化、新形式辈出的局面。

建筑外观是依附于承重结构建造的，可以看作附加结构体系上的覆盖物，但外观与结构有紧密的联系，外观不能超越结构而独立存在。最初阶段建筑外观在功能性设计时，更多的是突出建筑外观的色彩变化和形态变化，这在建筑外观的功能设计上的重点是通过颜色变化和建筑外形结构比例变化来体现。随后随着人们审美和建筑品质结合层次的逐渐深入，人们对建筑外观的需求从建筑的视觉享受开始向文化层面过渡，因此具有不同文化风格的建筑外观开始出现，这些建筑外观带有强烈的文化韵味，主要表现在以下几个方面：

**1. 欧陆风格。**"粉红色外墙，白色线条，通花栏杆，外飘窗台，绿色玻璃窗"，这种所谓欧陆风格的建筑类型，主要以粘贴古希腊古罗马艺术符号为特征，反映在建筑外形上，较多地出现山花尖顶、饰花柱式、宝瓶或通花栏杆、石膏线脚饰窗等处理，具有强烈的装饰效果，在色彩上多以沉闷的暗粉色及灰色线脚相结合。另外，这一类建筑继承了古典三段式的综合表象特征，结合裙楼、标准层及顶层、女儿墙加以不同的装饰处理。

**2. 新古典主义风格。**新古典主义风格的建筑外观吸取了类似"欧陆风格"的一些元素处理手法，但加以简化或局部适用，配以大面积墙及玻璃或简单线脚构架，在色彩上以大面积线色为主，装饰味相对简化，追求一种轻松、清新、典雅的气氛，可算是"后欧陆式"，较之前者则又进一步理性。

**3. 现代主义风格。**现代风格的作品大都以体现时代特征为主，没有过分的装饰，一切从功能出发，讲究造型比例适度、空间结构图明确美观，强调外观的明快、简洁。体现了现代生活快节奏、简约和实用，但又富有朝气的生活气息。

**4. 异域风格。**这类建筑的特点是将国外建筑式"原版移植"过来，植入了现代生活理念，同时又带有其种种异域情调空间。

**5. 普通风格。**这类建筑很难就其建筑外观在风格上下定义，它们的出现大概与商品房开发所处的经济发展阶段、环境或开发商的认识水平、审美能力和开发实力有关。建筑形象平淡，建筑外立面朴素，无过多的装饰，外墙面的材料义务细致考虑，显得普通化。

**6. 主题风格。**主题型楼盘是房地产策划的产物，本世纪伊始流行一时。这种楼盘以策划为主导，构造楼盘的开发主题和营销主题，规划设计依此为依据展开。

As the constant improvement of people's requirements on living comfort and culture taste, the understanding and requirements from people on the buildings' shape modeling and style orientation are also been further deepen and strengthen, thus forming a situation with diverse architecture styles and new forms.

Architectural appearance is built attaching to the bearing structure which can be seen as additional coverings on the structure system, but the appearance is closely with the structure and is not independently exist beyond the structure. The architectural appearance at the first stage of functional design is more highlighted the changes on colors and forms, which is importantly reflected by the color changes and the proportion changes of the building shape structure. Then as gradually deepen to the aesthetic and level of architecture quality, the demands for architectural appearance from people begin to transiting from the visual enjoy to the culture aspects, so the architectural appearances in different cultural styles gradually appear, which with a strong cultural flavor mainly manifested in the following aspects:

**1. European style.** "Pink walls, white lines, flowered railings, flared windowsills, green glass windows", what the so-called European-style architecture types, are mainly featured by pasting the ancient Greek Roman art symbol to reflect on the architectural appearance, and more represent a spire with mountain flowers, flowers and pillar decorations, vase or flower-railings, plaster moldings window and other decorative treatment, which have a strong decorative effect, and mostly combine them with dull dark pink and gray moldings on colors, in addition, this type of architectural inherited classical three-stage appearance feature, combining with podium, standard floor and top floor as well as parapet to deal with different decorations.

**2. Neoclassical style.** The building façade with neoclassical style draws some elements' handling techniques which are similar to "European style". They are to simplify or partial application together with a large area of walls and glass or a simple moldings frame as well as a large area of stitching color as the main, so the decorative taste is relative simplicity to the pursue a relaxed, fresh, elegant atmosphere, which can be regarded as "continental type" which is further rational compared with the former.

**3. Modernistic style.** Most of the contemporary-style works are based on reflecting the era characteristics, starting from functions without excessive decoration, which are paid attention to the moderate modeling proportion, clear and beautiful spatial structure, crisp and concise appearance. So it reflects the fast simple and practical pace of modern life, but riches with vigorous life atmosphere.

**4. Exotic style.** The feature of such architectures is to make the foreign architecture style "originally transplant" over, implanting the modern life idea, while with its various exotic spaces.

**5. Ordinary style.** Such architecture is difficult to define the style according to its architectural appearance. Their emergence probably relates with the level of awareness about their appearance and the economic development stage and environment the commercial housing development faced, or the developers' level of awareness, aesthetic ability and development capabilities. Plain architectural image and simple building façades, without too much decoration, and careful consideration with outer wall materials appear to be popularization.

**6. Theme style.** The theme-type buildings are the product of the real estate planning, and it is a fad at the beginning of this century. This building is leading by planning to create the development and marketing topics of the real estate, so the planning and design are started based on it.

# 杭州新天地商务中心项目
## Hangzhou Xintiandi Business Center Project

设计公司：上海栖城建筑规划设计有限公司
开发商：杭州汇庭投资发展有限公司
项目地址：浙江省杭州市
占地面积：10 144m²
建筑面积：59 368m²

Designed by: Shanghai GN Architectural and Planning Design Co., Ltd.
Client: Hangzhou, the Court of Investment Development Co., Ltd.
Location: Hangzhou, Zhejiang
Site Area: 10,144m²
Building Area: 59,368m²

扬帆远行的航船
二元有机拆分法

Sailing Journey of The Steamer
Binary Organic Separation Method

## 项目概况

杭州新天地商务中心项目位于下城区北部东新街道，项目东至杭宣铁路，南接规划中的长大屋路，西靠东新东路，北接石祥路。距武林广场约6km，距西湖约7.50km，交通便利。西临东新街、上塘河，北朝半山，周边自然环境优越。其中本项目所处的E地块位于新天地项目北区块，规划用地面积为10 144 m²。

## 建筑设计

本项目灵感取自水边的风帆，两栋塔楼与裙楼巧妙配合，犹如一艘正欲扬帆远行的航船；滨水立面采用体块穿插转折，与波光粼粼的河面相得益彰，和谐互融。面向城市的里面则以大尺度的门式造型把两栋塔楼相连，体现了办公大楼的气度与品质。这种二元有机拆分手法既让建筑拥有了极佳比例的优美体型，又使其与周边环境融为一体。并通过在裙房部分设置退台，并利用车道下穿，有机衔接原有河道景观，打造成空间层次丰富的纯步行滨河景观带。

## Project Overview

The project of Hangzhou Xintiandi Business Center is located in the Dongxin Street, North of the City, and the east is to Hang Xuan Railway, the south to the planning Changdawu Road, the west to east of the Dongxin Road and the north to Shixiang Road. It is about 6 kilometers far away from Wulin Square, and approximately 7.50 kilometers from West Lake, so the transportation is convenient. The West is neat to Dongxing Street and Shangtang River, and the North faces mid-level, so the surrounding natural environment is superior. Where the plot E the project located is located in the North block of the Plaza project, with planning site area of 10,144 square meters.

## Building Design

The project is inspired by the water's edge of sailing, and two towers and podium cleverly coordinate, like a ship which is to sail the journey; waterfront façade uses blocks with interspersed transition, with sparkling river mutual harmony melting with each other. The inside places facing the city with large-scale modeling of the door connect the two towers, reflecting the tolerance and quality of the office buildings. This binary organic separation method both makes the building have a beautiful body shape with an excellent proportion, but also makes it blend with the surrounding environment. And through the skirt portion, it sets back sets, and uses the lane beneath for organic convergence of the original river landscape, creating pure walking riverside landscape zone with rich layers.

# 法国伊西莱穆利诺 Galeo 大楼
## Galeo

设计师：Christian de Portzamparc
开发商：Bouygues Immobilier
项目地址：法国伊西莱穆利诺
建筑面积：24 000m²

Designer: Christian de Portzamparc
Client: Bouygues Immobilier
Location: Issy-les-Moulineaux, France
Building Area: 24,000m²

## 项目概况

这一项目由沿着一条街道的三栋大楼构成，其中的白色玻璃大楼被称为 Galeo 大楼，是法国房地产开发商 Bouygues Immobilier 的总部大楼，也是从巴黎进入伊西莱穆利诺的一个入口标识。

## 建筑设计

建筑形态是根据卵石的形态而来，以独特的建筑外表皮为特色，使得整座建筑看起来与城市尺度完美结合。建筑体量的几何形态是由"双层表皮"玻璃鳞片构成，令建筑宛如一颗巨大璀璨的水晶。这一结构具有实际用途，它不仅可以使建筑楼层免受周围环境温度波动的影响，同时也顾及到了大型窗体的设置，使得办公室的内部气氛变得柔和又明亮。

在夜间，这颗巨大"水晶"就像是一盏"灯"，与其他两座采用深色混凝土板材墙壁的建筑形成鲜明的反差，显得异常绚烂夺目。另外，整个建筑还凭借其高标准的环保性能，被授予了 HEQ 认证。

## Profile

This project is composed of three buildings along a street. The building of plain glass called Galeo, is the head office of France's property developer Bouygues Immobilier and denotes the entrance to the town of Issy-les-Moulineaux from Paris.

## Architectural Design

It is characterized by an envelope which fits into the urban size according to a shape of polite pebble. The geometry of this volume is constituted by a "double skin" glass scales which plays a practical role, insulating the office floors from fluctuations in environmental temperature and allowing for wide windows. The interior atmosphere of the offices is soft and light. The building looks like a giant bright crystal.

At night, this crystal becomes a "lamp" visible from the highway, offering a contrast with the other two office buildings of dark concrete panel walls. The entire project is accredited HEQ® for its high standard of environmental performance.

**Double Skin Glass Scales "Crystal"**

双层表皮 玻璃鳞片 "水晶"

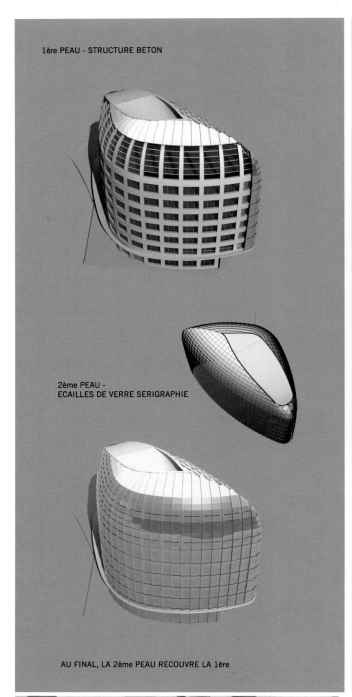

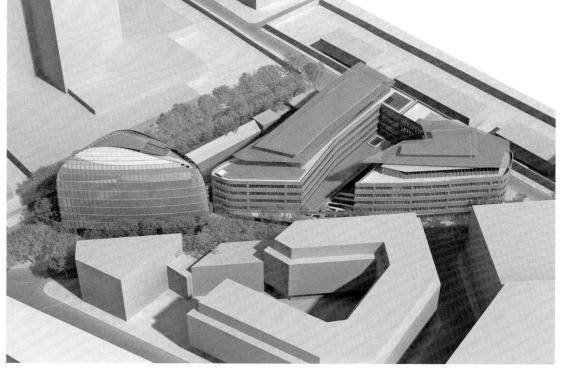
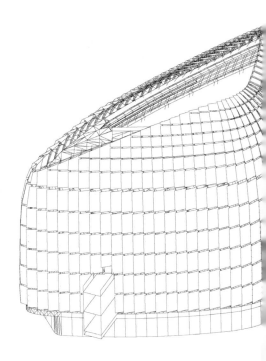

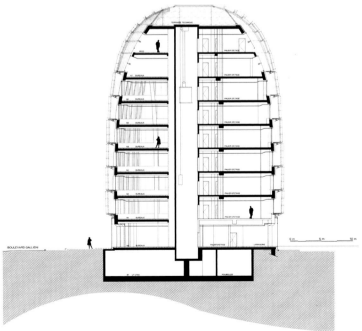
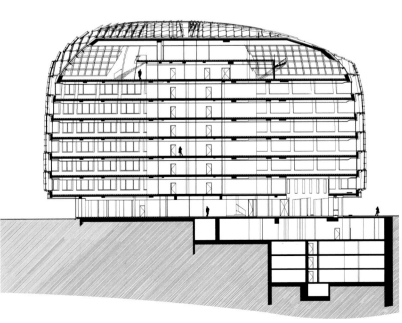

# 华尔街巨蛋 —— 太阳能清真寺
## The Wall Dome – Solar Powered Mosque

设计单位：Paolo Venturella Architecture
开发商：Mapletree Investments Pte Ltd's
项目地址：普里什蒂纳
设计团队：Paolo Venturella　Angelo Balducci
　　　　　Luca Ponsi　　　 Paolo Gaeta
建筑师：Paolo Venturella Architecture
摄影：Wemage

Designed by: Paolo Venturella Architecture
Client: Mapletree Investments Pte Ltd's
Location: Pristina
Design Team: Paolo Venturella, Angelo Balducci,
Luca Ponsi, Paolo Gaeta
Architect: Paolo Venturella
Photography: Wemage

**项目概况**

该项目目标是为普里什蒂纳城市创建一个不朽的和标志性的建筑。

它是一个单体建筑，在城市的南部，成为Dardania附近的城市支点。建筑的想法是合并清真寺的两个主要因素："穆斯林朝觐方向墙"和"巨蛋"，第一个表示祈祷的方向，第二个创造人们感觉在同一个社区的巨大的空间里。从几何的角度来说，清真寺可以被看作是从墙壁散发出来的球体，但同时作为面向麦加的社区，这代表了神圣的元素和祈祷之间的联系。新清真寺的建立将作为一个城市的图标，欢迎穆斯林信徒和其他使用者或游客，以营造全民统一的空间。

**建筑设计**

整个建筑的外壳形状被设计成一个"双层"，周围所有的祈祷大厅生成一个圆形空间。为了迎合不同时间段的祷告规模，建筑大殿分为两个独立的空间，两个空间在多个层面上进行设计，以适应周一到周四、周五及节假日的不同水平的男性和女性。外部皮肤是一系列覆盖着薄薄的薄膜光伏系统，由百叶窗制成。利用球面形状，这些太阳能电池板在保证了完美表面整天面向太阳的同时，还能优化太阳的辐射。

该项目由于其方向，还采用了被动的能量方法。朝南的穆斯林朝觐墙，充当了一个巨大的温室元素，在寒冷的时期，它会捕获热量并释放到室内。由于该项目需要四层地下停车场，因此还设计了一个地热能源系统，可用于提供加热或卫生。通道是在地下一层，辅助空间放置，其余的地下区域是教育、社会、行政和商业领域，与上面的神圣空间分开。

"巨蛋"外形 多层面结构

"Dome" Shape Multi-level Structure

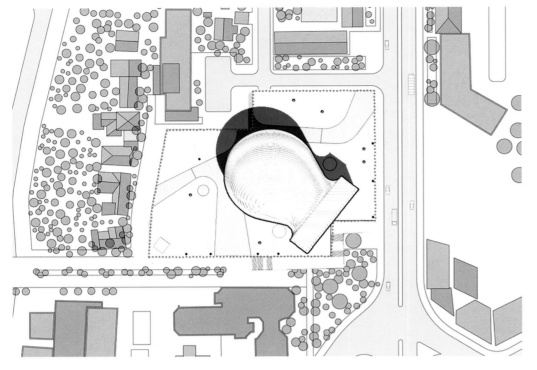

## Project Overview

The project aim is to create a monumental and iconic building for the city of Prishtina.

It is a monolithic building that becomes an urban fulcrum for the Dardania neighborhood, in the south of the city. The idea is to merge the two main elements of the mosque: the "Kiblah Wall" and the "Dome". The first indicates the direction where to pray and the second creates the huge space where people feel in the same community. From a geometric point of view the mosque can be seen as a sphere that comes out from the wall, but at the same time as the community facing the Mecca. This represents the link between the divine element and the prayers. The new Mosque establishes an urban icon that welcomes Muslims faithful and other users or tourists creating a unifying space for all.

## Building Design

The shape of the whole envelope is designed as a "double-skin" that generates a circular space all around the prayer hall. The prayer hall is divided into two separated rooms in order to organize a smaller area to pray from Monday to Thursday and a bigger one to fit more prayers during Friday and festivals. Both spaces are designed on multiple levels to accommodate men and women at different levels. The exterior skin is made by a series of louvers covered with a thin photovoltaic film system that harvest energy for the use of the mosque and the other services. Taking advantage of the spherical shape, these solar panels face the sun rays during the day from the morning to the end of the day. The curved form guarantees a surface perfectly oriented to the sun during all the day so to optimize the radiation of the sun. The project also uses a passive energetic approach thanks to its orientation. The Kibla wall, which in south oriented, works as a huge greenhouse element that captures heat and releases it to the interior when needed in cold periods. Since the project requires four levels of underground parking is also designed a geothermic energy system that provides for either heating or sanitary purposes. The access is at the underground level where ancillary spaces are placed and the entrance to the Mosque is at the ground floor. In the rest of the underground areas are educational, social, administrative and commercial areas, separated from the sacred space above.

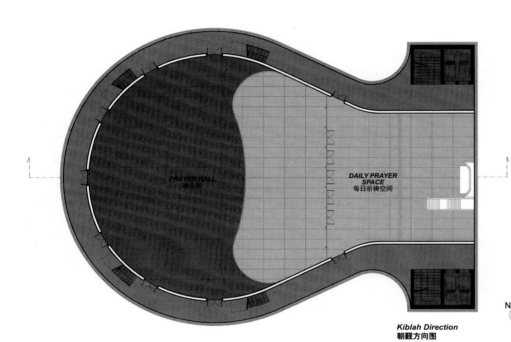

*Kiblah Direction*
朝觐方向图

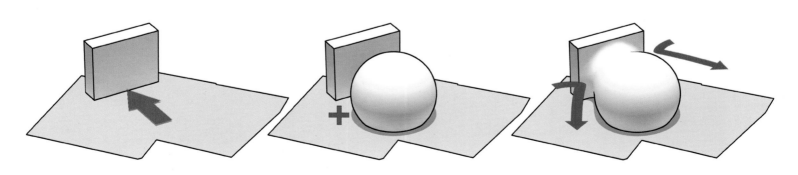

**THE MAIN BLOCK IS PLACED IN THE DIRECTION OF THE MECCA.**
主块被放置在麦加的方向。

**A SPHERE IS ADDED IN FRONT TO CREATE A HUGE DOME FOR MUSLIM FAITHFULS.**
在前面增加一个球体为穆斯林忠心拥护者建立一个巨大的圆顶。

**THE TWO ELEMENTS ARE MERGED CREATING A SINGLE SPACE TO PRAY.**
融合两个元素建立一个单一的祈祷空间。

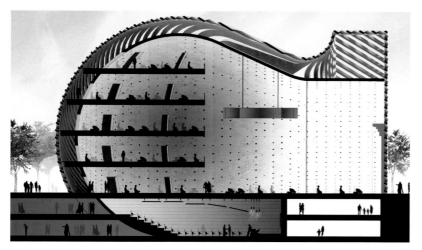
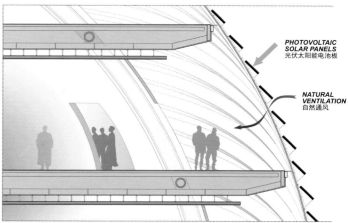

**PHOTOVOLTAIC SOLAR PANELS**
光伏太阳能电池板

**NATURAL VENTILATION**
自然通风

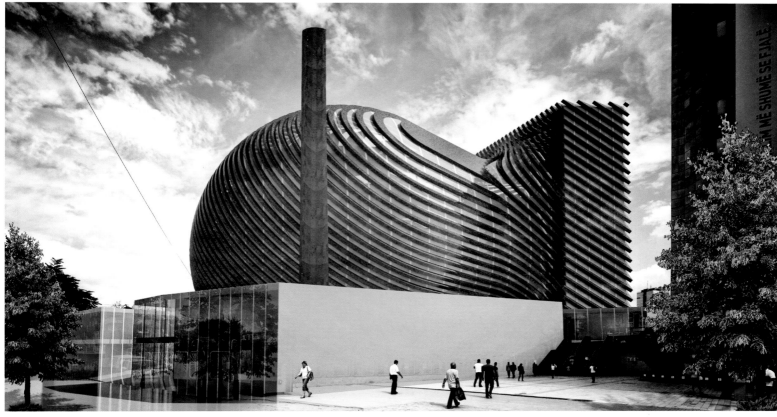

# 金州新区医疗中心
## Jinzhou New Area Medical Center

设计单位：Design Initiatives
开发商：大连医科大学附属第一医院
项目地址：中国大连
建筑面积：184 828m²

Designed by: Design Initiatives
Client: First Affiliated Hospital of Dalian Medical University
Location: Dalian, China
Building Area: 184,828m²

**建筑设计**

公司项目在进行设计时，将重点放在患者和医护人员的体验上。医院主要包括四个主建筑，分别是门诊部、住院部、医药部和教研行政部，中央是一个散步广场。广场分为两层，连通四栋楼的同时，为医生和患者及家属提供宁静开放的户外空间。医院临街面建有临时泊车区，车辆按规格分开进入。大楼还安装有相关的节能系统，例如废水回收、雨水收集等。建造过程中还采用了材料回收利用、预制构建等节能环保手段。设计师还在周围建造了地下停车场、公园、屋顶花园等。整个建筑群是钢筋支架，外覆金属板和幕墙结构建成的。

User Experience Energy Saving System
用户体验 节能系统

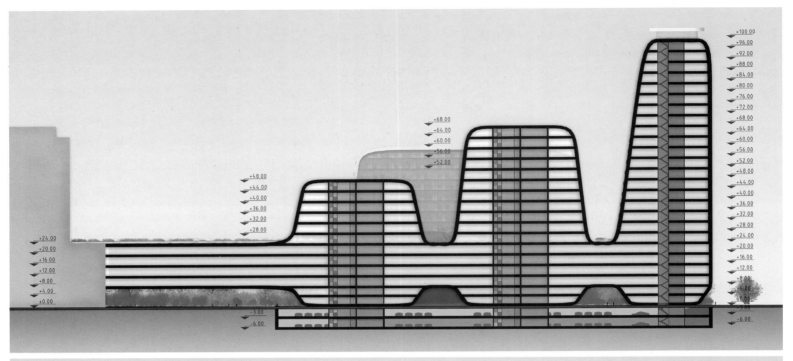

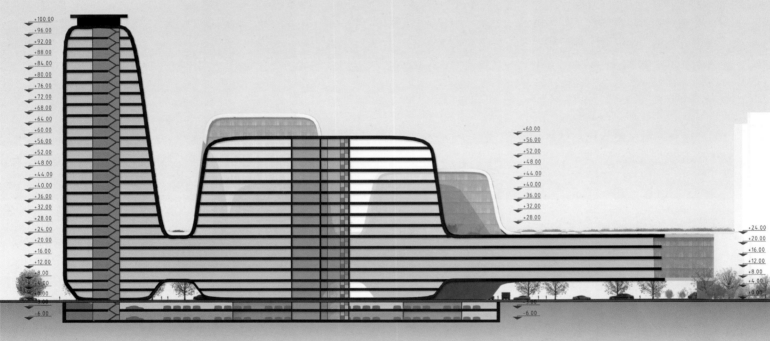

## Building Design

Company during the design project will focus on the experience of patients and medical staff. Hospital consists of four main buildings which respectively are outpatient and inpatient departments, medical department, teaching and research department and the administrative department, and the center is a promenade square.

Square is divided into two stories, connecting four buildings, while providing a peaceful and open outdoor space for doctors and patients and their families. Hospital frontage built temporary parking areas, according to specifications into separate vehicles. The building is also equipped with related energy-saving systems, such as wastewater recycling, rainwater collection and so on. The construction process is made use of recycled materials and prefabricated building structure. Designers also built around the surrounding the underground parking, park, and roof garden and so on. The entire building is a steel frame, built by covering with metal plates and wall structure.

# 中央洛杉矶公立高中视觉和表演艺术的洛杉矶联合校区

## Central Los Angeles Public High School for the Visual and Performing Arts of the Los Angeles Unified School District

| | |
|---|---|
| 设计单位：COOP HIMMELB(L)AU Wolf D. Prix & Partner ZT GmbH | Designed by: COOP HIMMELB(L)AU Wolf D. Prix & Partner ZT GmbH |
| 设计师：Wolf D. Prix | Designer: Wolf D. Prix |
| 开发商：美国加利福尼亚州洛杉矶联合学区 | Client: LAUSD, Los Angeles Unified School District, CA, USA |
| 项目地址：美国洛杉矶 | Location: Los Angeles |
| 占地面积：39 578m² | Site Area: 39,578m² |
| 总建筑面积：31 138m² | Gloss Floor Area: 31,138m² |
| 摄影：Roland Halbe　Duccio Malagamba | Photography: Roland Halbe, Duccio Malagamba |

**项目概况**

该项目是洛杉矶联合学区严格国债资金计划的第二阶段的一部分，它位于洛杉矶市中心格兰大道上，占地面积达39 578m²。该学校将是一个综合性高中，有图书馆、自助餐厅和剧院，有咖啡天窗的上层广场，此外将提供视觉艺术、表演艺术、音乐和舞蹈等课程。

由于该项目位于格兰大道的中心位置，所以在建筑设计过程中，设计师们有意将这所中央高中融入到格兰大道文化走廊，使其成为文化设施的一部分，它将连接迪斯尼音乐厅、音乐中心、柯尔伯恩音乐学校、当代艺术博物馆和天使圣母大教堂。为了履行其公共设施职责，融合进符合格兰大道的其他设施的精神，校园将搭建专业表演艺术剧院，该剧院将用于教育目的，将开放给公众和其他机构使用，并配备了一个完整的舞台、乐池、后台和飞阁楼。

**建筑设计**

建筑标志——国际象棋概念。

奥地利 COOP HIMMELB(L) AU 建筑设计公司的设计理念是,利用建筑的标志符号作为沟通的洛杉矶社会对艺术的承诺。建筑主体被设计成国际象棋数字 3 的模式,一个塔图位于影院飞阁楼的顶部,整个建筑作为一种广泛的、可视性的标志艺术服务于城市和学生,并与大教堂的塔楼的双塔一起将成为城市的新地标。

图书馆通过倾斜、截锥的形式呈现,并被摆在学校院子里的中心位置。内饰方面,锥体提供了一个大的开放空间,由一个圆形的天窗从上面照射,从而为沉思和专注学习提供了一个开放的、动态的,且性格内向和集中的空间。通过其与其他建筑物的对角位置及其倾斜的形式,动态的圆形建筑可指向风景,并且人流可通过学校庭院,从而改变了庭院空间的感知,并为校园内的学生提供方向点。

四个教学楼形成了学校内部庭院的正交周长。功能箱梁形建筑可容纳每一个学院以及其他共享的教育和管理空间。从外部,圆窗是吸引注意力、增强学校与城市之间沟通的一种手段。从内部,圆窗是通过教室内不同的光线条件创造一个活泼的气氛的元素。学校主要入口是通过一个约 25m 宽的的大型开放式楼梯正式展现的,这直接导致在其中心及在后台的剧院和塔楼进入有锥形图书馆的主校院。正门象征性地为学生设置了舞台,作为一个决定性阶段让他们在生活和教育中体验这所学校。

# 国际象棋锥形图书馆

# Chess Conical Library

## Project Overview

The Central Los Angeles Public High School for the Visual and Performing Arts of the Los Angeles Unified School District (LAUSD) is part of phase II of LAUSD's rigorous state bond funded plan. It is located on a 39,578m² site area on Grand Avenue in downtown Los Angeles. The school will be a comprehensive High School and in addition will offer courses in the Visual Arts, Performing Arts, Music and Dance.

Due to its central location on Grand Avenue, the High School will be a part of the cultural facilities along the Grand Avenue cultural corridor, joining the Disney Concert Hall, Music Center, Colburn School of Music, Museum of Contemporary Art and the Cathedral of our Lady of the Angels. To fulfill its mandate to be a public facility in keeping with the spirit of the other facilities on Grand Avenue, the school will build a professional performing arts theater. The theater will be used for educational purposes, will be open to the public and for use by other institutions, and is equipped with a full stage, orchestra pit, back stage and fly-loft.

## Building Design

### Architectural Signs - Chess Concept

COOP HIMMELB(L)AU's design concept is to use architectural signs as symbols to communicate the commitment of the Los Angeles community to Art. Like chess figures three sculptural buildings, a tower figure with spiraling ramp located on top of the theater's fly-loft serves as a widely visible sign for the Arts in the city and a point of identification for the students, and together with the Cathedral's tower the twin towers will become a new landmark for the city.

The Library is formally expressed through a slanted, truncated cone and placed in the center of the school courtyard. Inside, the cone provides a large open space illuminated from above by a circular skylight thus offering an open, dynamic, but introverted and concentrated space for contemplation and focused learning. Through its diagonal position in relationship to the other buildings and its slanted form, the dynamic, circular building directs views and flows of people through the school courtyards, changes the perception of the courtyard space and provides a point of orientation for the students within the campus.

The four classroom buildings form the orthogonal perimeter of the school's interior courtyards. The functional box beam buildings house one academy each as well as other shared educational and administrative spaces. From the outside the round windows are a means to attract attention and enhance communication between the school and the city. From the inside the circular windows are an element for creating a lively atmosphere through distinct light conditions within the classrooms. The main school entrance is formally expressed through an 80 feet wide grand open stair, which leads directly into the main school courtyard with the conical library in its center and theater and tower in the background. The main entrance symbolically sets the stage for the students to experience this school as a decisive stage in their life and education.

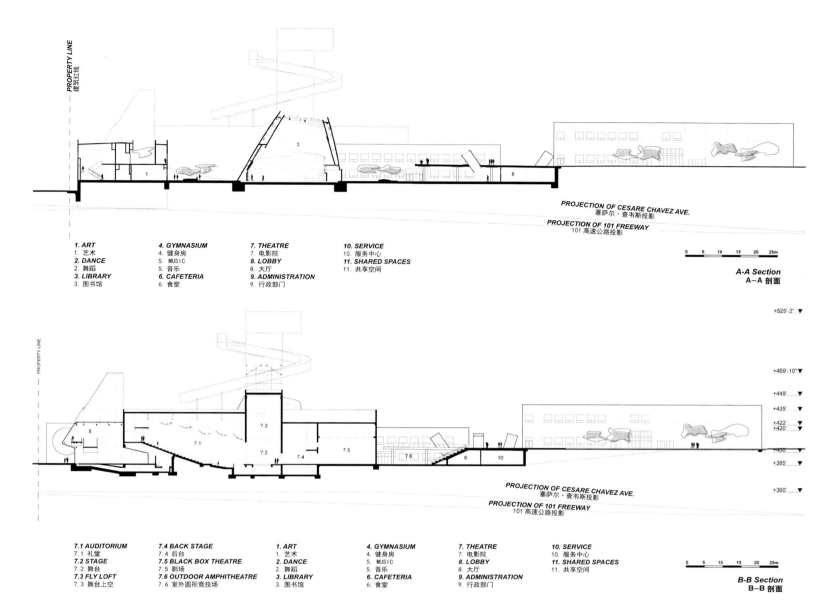

1. ART  艺术
2. DANCE  舞蹈
3. LIBRARY  图书馆
4. GYMNASIUM  健身房
5. MUSIC  音乐
6. CAFETERIA  食堂
7. THEATRE  电影院
8. LOBBY  大厅
9. ADMINISTRATION  行政部门
10. SERVICE  服务中心
11. SHARED SPACES  共享空间

A-A Section
A-A 剖面

7.1 AUDITORIUM  礼堂
7.2 STAGE  舞台
7.3 FLY LOFT  舞台上空
7.4 BACK STAGE  后台
7.5 BLACK BOX THEATRE  剧场
7.6 OUTDOOR AMPHITHEATRE  室外圆形竞技场

B-B Section
B-B 剖面

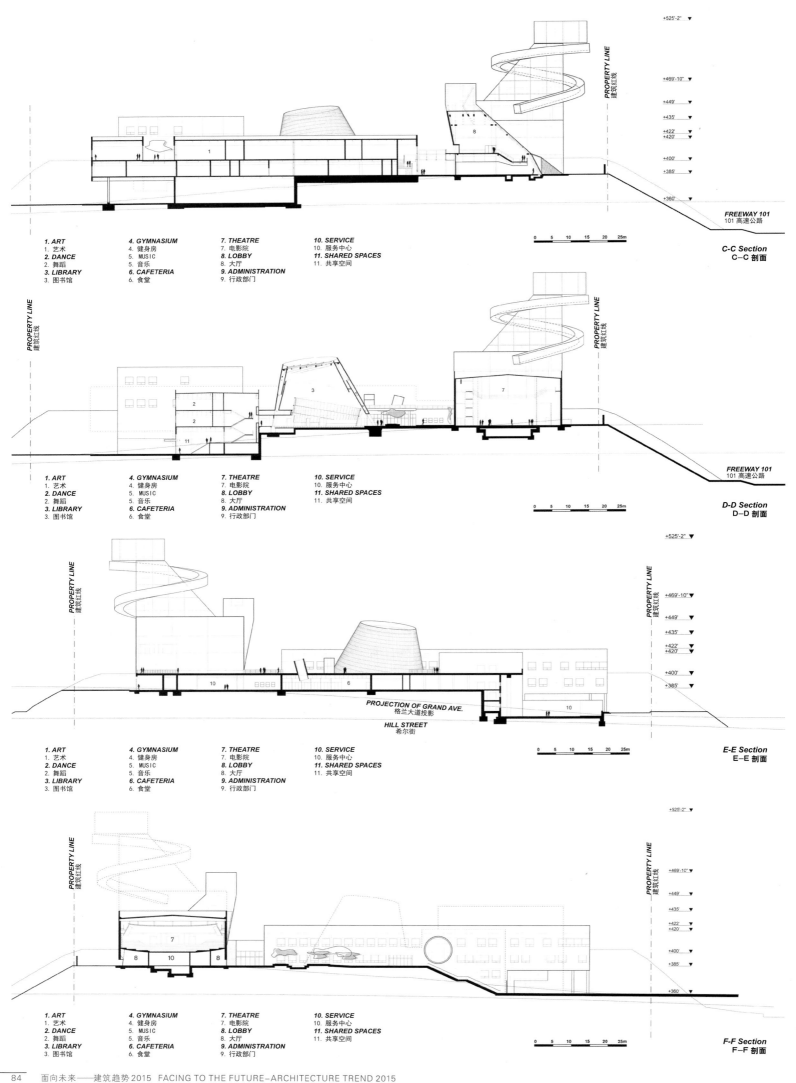

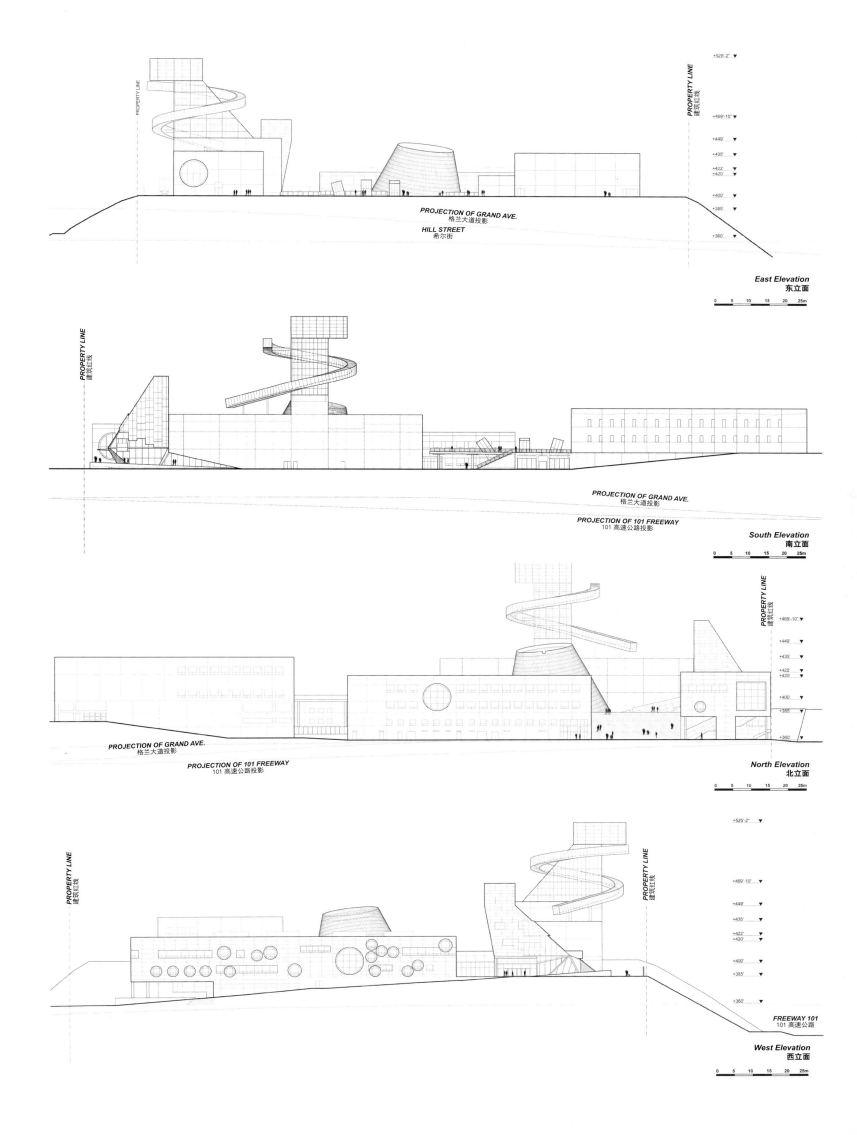

# 哥伦比亚波哥大 EAN 大学 E1 Nogal 校区教学楼

Classroom Building of El Nogal Campus, Universidad EAN

| | |
|---|---|
| 设计单位：Daniel Bonilla Arquitectos | Designed by: Daniel Bonilla Arquitectos |
| 开发商：EAN 大学 | Client: Universidad EAN |
| 项目地址：哥伦比亚波哥大 | Location: Bogota, Colombia |
| 建筑面积：14 016.8m² | Building Area: 14,016.8m² |
| 设计团队：María Alejandra Echeverry　María Paula Gonzálaz　Elizabeth Añaños　Sebastián Chica　Alexander Roa　Guillermo Barahona　Alex Larin | Design Team: María Alejandra Echeverry, María Paula González, Elizabeth Añaños, Sebastián Chica, Alexander Roa, Guillermo Barahona, Alex Larin |
| 摄影：Rodrigo Dávila | Photography: Rodrigo Dávila |

Prism
Random Boxes
Green Shades

棱柱形
错乱盒状体
绿色百叶

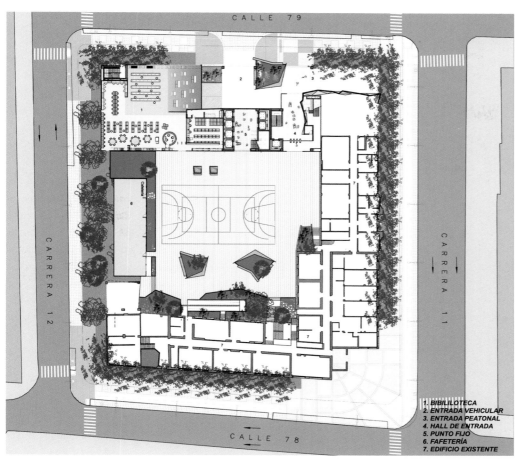

1. BIBILILOTECA
2. ENTRADA VEHICULAR
3. ENTRADA PEATONAL
4. HALL DE ENTRADA
5. PUNTO FIJO
6. FAFETERÍA
7. EDIFICIO EXISTENTE

## 项目概况

该教学楼项目是一项邀请赛的结果，旨在建立一所建筑与公共开放区域同样重要的大学校园建筑。在这里，学生可以聚会、学习和休息，是供教育与娱乐的第二住宅。

## 建筑设计

教学楼设计为七层棱柱形建筑，第一层为通道区和中央图书馆。第二层为标准教室，顶楼则为公共使用的平台。棱柱体建筑设计为许多错乱布局的混凝土盒装构造物，既打破了建筑体量的单调性，也可用作教室和会议室。

## Profile

This classroom building project is the result of a competition by invitation, aiming to set up a university campus where the buildings are as important as the common open spaces. It is a place where students can meet, study, rest, a second home for education and fun.

## Architectural Design

The classroom building is designed as a seven level's prism. The first floor is located the access area and the central library. The second floor are standard classrooms, ending with a rooftop terrace for collective use. The prism is affected by a series of random outcoming concrete boxes that break the volume's monotony while serving as study and meeting rooms.

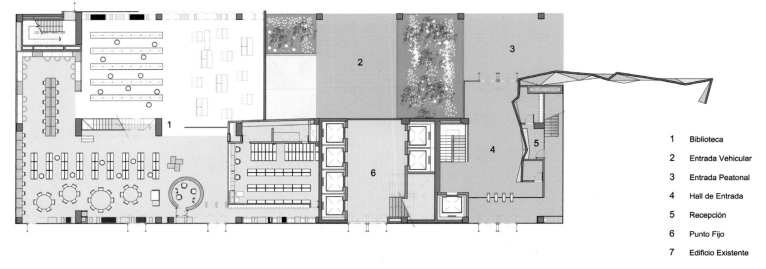

| | |
|---|---|
| 1 | Biblioteca |
| 2 | Entrada Vehicular |
| 3 | Entrada Peatonal |
| 4 | Hall de Entrada |
| 5 | Recepción |
| 6 | Punto Fijo |
| 7 | Edificio Existente |

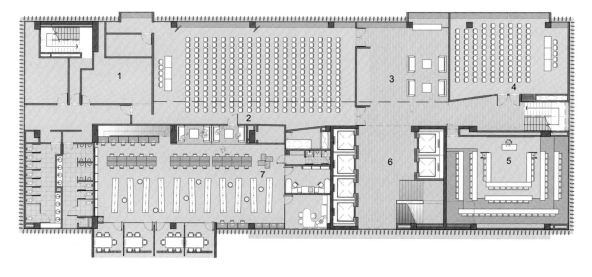

| | |
|---|---|
| 1 | Cuartos Técnicos |
| 2 | Auditorio |
| 3 | Foyer Auditorio |
| 4 | Aulas Combinables |
| 5 | Aula Magistral |
| 6 | Punto Fijo |
| 7 | Biblioteca Piso 2 |

◀ 立面——立面由单一元素构成：微型穿孔绿色百叶窗，使室内免受太阳直射的同时更突出"错乱盒状体"的形态，使得建筑形成独特的形象，宛如一座葱郁的绿光森林，彰显出大学校园的无限活力与生机。

Façade—it is composed of a singular element: stressed micro perforated green shades, protecting the interiors from direct sunlight. This element gives a particular image to the building while highlighting the "random boxes". The whole building is in a unique form, like a verdant forest of green ray, which reflects infinite energy and vitality of university campus.

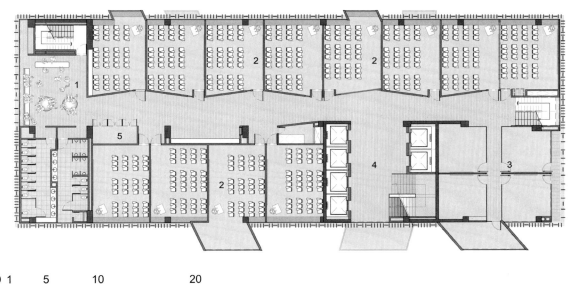

1　Area Refrigerios
2　Aulas
3　Tutorías
4　Punto Fijo
5　Depósitos

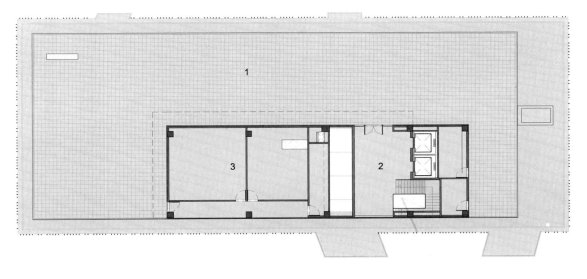

1　Terraza
2　Punto Fijo
3　Cuartos Técnicos

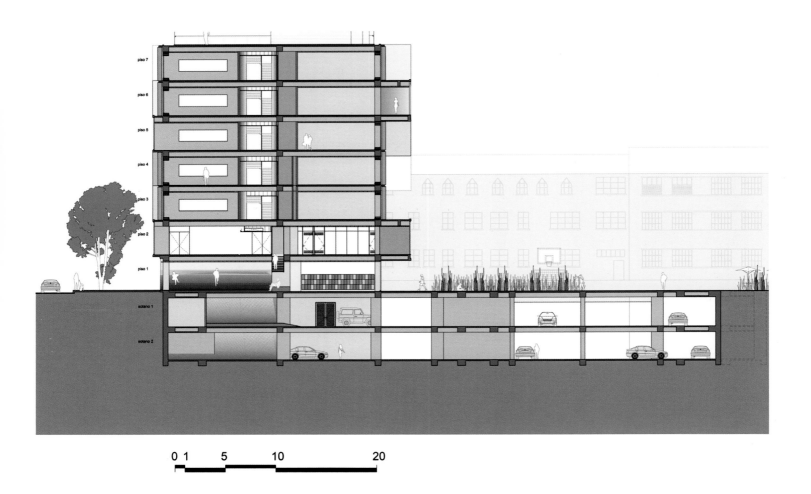

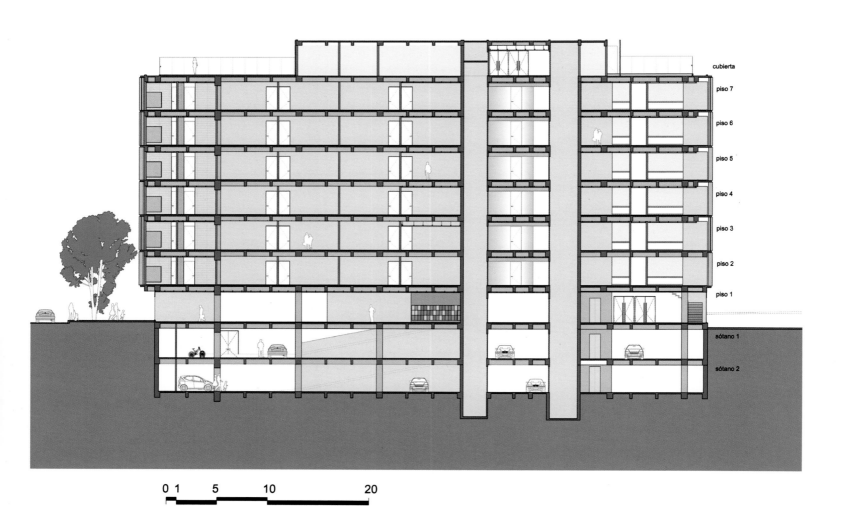

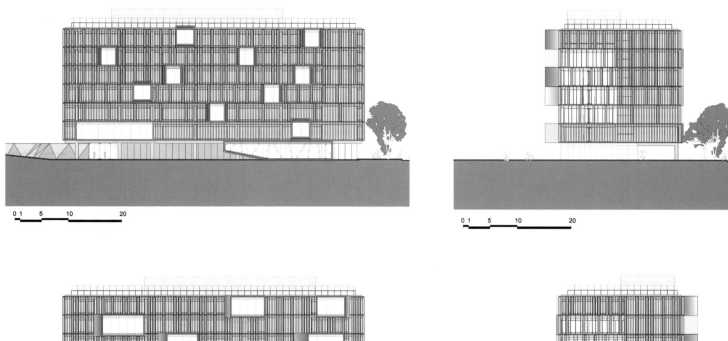

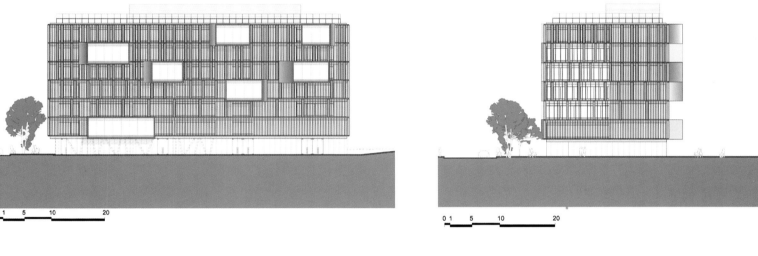

# 比利时科特赖克 Sint-Amand 学院住宅塔楼
## Housing Tower of Sint-Amand College, Kortrijk

设计单位：Philippe SAMYN and PARTNERS, architects & engineers
开发商：VAN ROEY– KORAMIC
项目地址：比利时科特赖克
建筑面积：13 227m²
摄影：POLYGON Graphics cvba

Designed by: Philippe SAMYN and PARTNERS, architects & engineers
Client: VAN ROEY– KORAMIC
Location: Kortrijk, Belgium
Building Area: 13,227m²
Photography: POLYGON Graphics cvba

# 纤细优雅 白色山毛榉木 艺术展示立面

# Slender and Elegant White Beech Façades of Arts

**项目概况**

Sint-Amand 学院场址位于科特赖克内环路 Lys 河畔，这座寄宿学校的塔楼建于 1960 年，其指挥台覆盖额外的教育设施。由于学校中的塔楼已不再满足市场需求，科特赖克城市发展委员会对其重建发动起一场比赛。

**规划设计**

在进行深入分析之后，最终决定规划一座新的塔楼来占据现有塔楼的场地，并且塔楼由于高度增加而变得更宽。设计在建筑功能的基础上规划其最终造型，通过两条略微倾斜的坡道可进入大楼宽敞的入口大厅。

裙楼被划分为半独立式住宅，停车场则设在底层。为避免遮挡后面学校建筑的光线，塔楼规划为向后稍微倾斜。另外，在现有塔楼的裙楼东侧规划建一栋新的塔楼，从而解决教学区域和住宅区域的间距问题，并且新的塔楼会加固现有的裙楼楼体，使得整个建筑区的布局更为合理。

**建筑设计**

建筑塔楼呈东西走向，体量显得纤细而优雅，保证能最大限度地观看到 Lys 河的风景，同时还能看到南面的历史中心。

塔楼内部布局相当灵活，每层可布局 2 套、3 套或者 4 套公寓，公寓将设有露台和双层高的天花板，从露台和窗户可观看壮观的风景。为了提高气温调节能力，露台还设计有玻璃百叶窗。建筑立面覆盖有白色山毛榉木，并以艺术作品加以装饰，这些艺术品将会和邻近的 Buda 岛上的展览品连接起来，使其成为科特赖克城中艺术与设计的展示地，呈现出与众不同的立面效果。

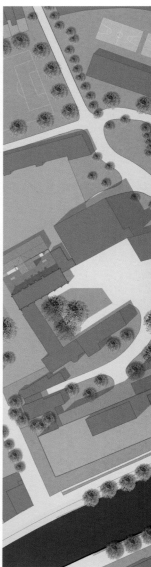

### Profile

The College Saint-Amand site is located inside Courtrai's inner ring road, on the banks of the river Lys. The tower was built in 1960 to house a boarding school; its podium houses additional educational facilities. As the tower at the College no longer fulfills market requirements, the Courtrai Urban Development Board has launched a competition for its redevelopment.

### Planning Design

After an in-depth analysis, an initial decision was taken to build a new tower occupying the current tower's site. The tower would get wider as it increases in height. Form would then really follow function. The building would be entered by two long, slightly sloping ramps, leading to a spacious, transparent entrance hall.

The podium would be divided into semi-detached houses and the car park would be on the ground floor. To avoid blocking sunshine of the school buildings behind it, the tower is inclined backwards slightly. A new tower to the east of the current tower podium is built to resolve the separation between the educational and residential areas. Furthermore, the new residential tower will reinforce the imposing podium, and the tower's position will round off the whole built area logically.

### Architectural Design

The tower will face east-west. In this way, the view over the river Lys will be exploited to the maximum, whilst the historic center will be visible from the south side. From the town, the tower will have a slender and elegant appearance.

The suggested layout for the tower interior is flexible. It is possible to plan for two, three or four apartments per floor. The apartments will have terraces, alternated with double height ceilings. There are magnificent views from the terraces and windows. To improve the temperature regulation on the terraces, the tower will be covered with glass louvers. The façades will be covered with white beech, and can be decorated with works of art. These temporary exhibitions will link in with the exhibitions visible on the neighboring island of Buda. The tower will therefore become a showcase for art and design in Courtrai. A unique façade will be presented.

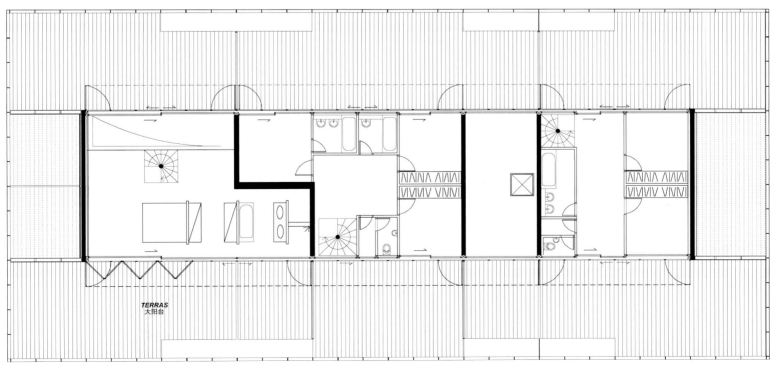

**TERRAS**
大阳台

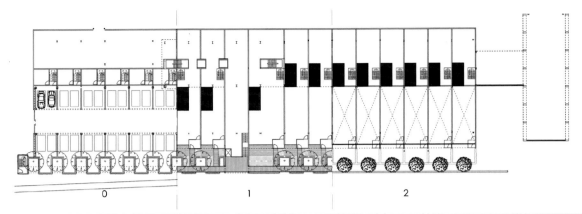
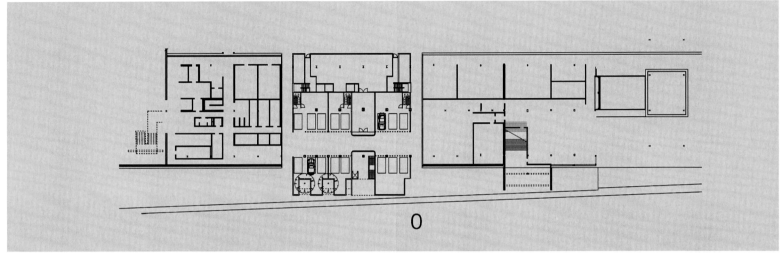
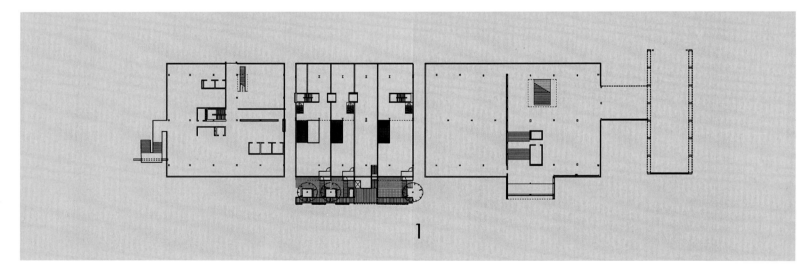
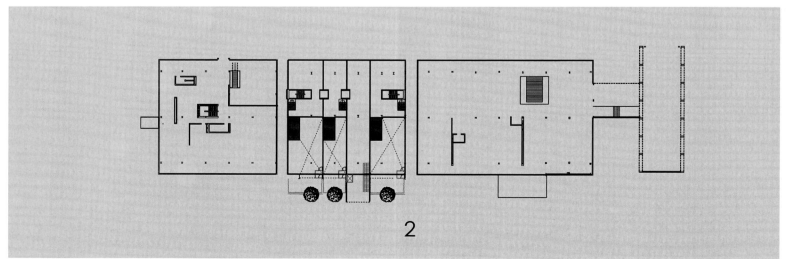

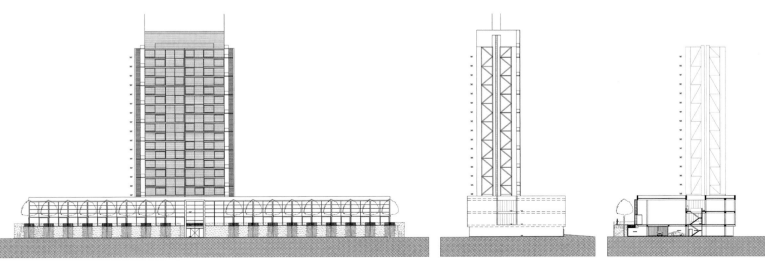

# CUBE BIOMETRIC CENTER　"魔方"
## Cube Biometric Center

设计单位：香港 TheeAe 有限公司
委托方：香港科技园管理
项目位置：香港新界沙田香港科学园科技大道西2号科学园8楼生物资讯中心
总建筑面积：3 140m²

*Designed by: Hong Kong TheeAe Ltd.*
*Client: Hong Kong Science Park Management*
*Location: 8/F, Bio-Informatics Center, No. 2 Science Park West Avenue, Hong Kong Science Park, Shatin, New Territories, Hong Kong*
*Gross Floor Area: 3,140m²*

3D Design
Low Carbon Design

3D 设计
低碳设计

**项目概况**

该项目在融合了 3D 设计理念的前提下，构建了一个创新性的标志性建筑，并形成巧妙的、可持续的低碳设计。该魔方的简单形式将代表科技园成为新的图标，特别是作为综合办公大楼，它转变成成熟的形状和功能，从而成为香港科技园标志性的想法和可持续的建筑。

**建筑设计**

"魔方"在设计过程已经演变。它从它自己的形式转化为另一种形式。为了遵循所需的设计元素之一，该形式是朝向太阳的角度取向，该形式是通过调节与太阳的角度，并利用倾斜的屋顶面积安装太阳能光伏板，最大限度地吸收太阳能。所以，整个建筑的耗电量将通过这个系统进行覆盖，除了第一层的零售和咖啡区。该能量将聚集到安装在四层和五层的电池，并且将密切连接到安装在同一水平的太阳能光伏板。此外，斜立面墙体代表物质使用的可持续发展理念。它采用双层的复合板。墙单元在外墙结合轻质预制复合板，在室内结合中纤板，以及在两者之间结合低辐射双层玻璃。白天它会限制太阳光线，并且将散热率最小化到内部空间。此外，外墙是由对使用客房提供阳台、玻璃开口和建筑核心的空气通风口的功能要求进行转变的。因此，"魔方"连接到另一台立方体（魔方附件）。这是一种过渡形式连接到桥梁平台（现有的建筑设计形式）至主"魔方"。这种联系不仅放大了"魔方"的动态形式，还保持现有的建筑设计与新图标的和谐。

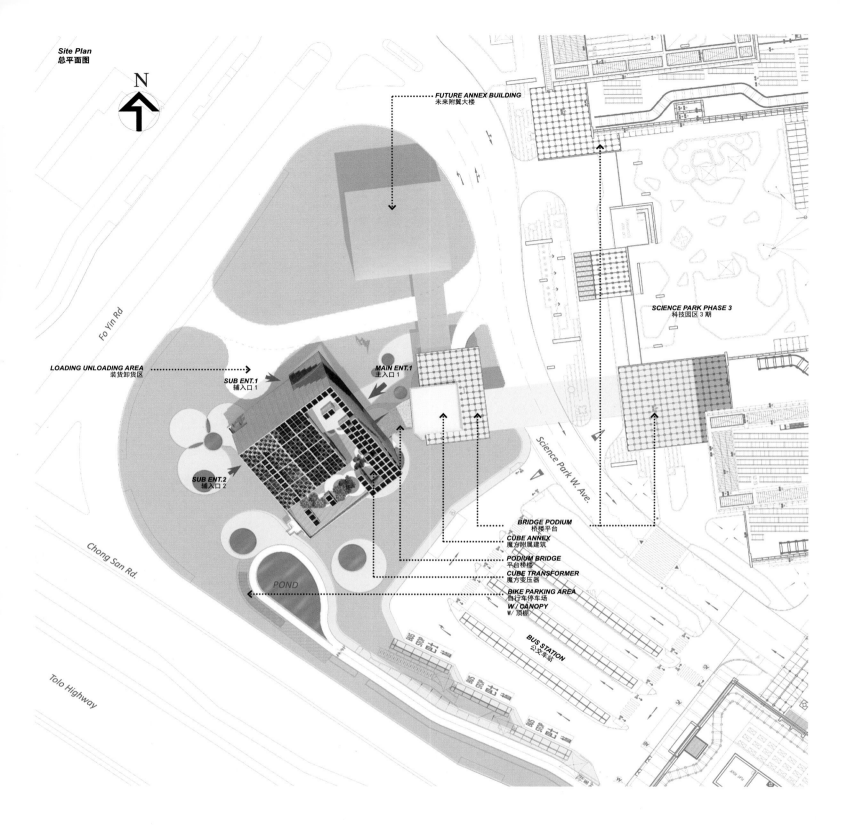

*Site Plan*
总平面图

## Project Overview

As per design requirements of the Phase 3D design ideas competition which are to create an innovative iconic architecture, generate ingenious and sustainable low carbon design, and design the building which unifies the Park's development in harmony with other buildings in the Park, a big challenge is to reveal the gateway of Hong Kong Science Park. The Park is the place where the science and its technology has been studied and developed in Hong Kong.

## Building Design

The "Cube" has been evolved in design process. It is transformed from its own form to another. In order to follow one of the required design elements, the form is oriented toward sun angles to maximize the absorption of the solar energy. The Sun will be utilized to generate electric power through the installed photovoltaic panels on tilted roof area. So, the entire electric consumption of the building will be covered through this system except retail and coffee area at level 1. The energy will be gathered to the battery installed at level 4 and level 5, and it will be closely connected to the Photovoltaic Panels installed on the same levels. In addition, inclined façade walls represent sustainable ideas of the material uses. It uses dual layered composite panels. The wall unit is combined with light weight precast composite panel at exterior, MDF panels at interior, and low-E dual glass in-between. It will limit the sun lights during day time and minimize heat transmittance to the interior space. Furthermore, exterior walls are transformed by its functional requirement for the use of rooms providing balconies and glass openings, and air ventilation openings at building core. Thus, the "Cube" is connected to another cube (Cube Annex). This is a transitional form to link the bridge podium (existing building design form) to the main 'Cube'. This linkage not only amplifies the dynamic form of the Cube, but maintains the existing building design in harmony with the new icon as well.

**Ground Floor Plan Scale 1/300(A1)**
底层平面图比例 1/300（A1）

**Level 1 Foor Plan Scale 1/300(A1)**
一层平面图比例 1/300（A1）

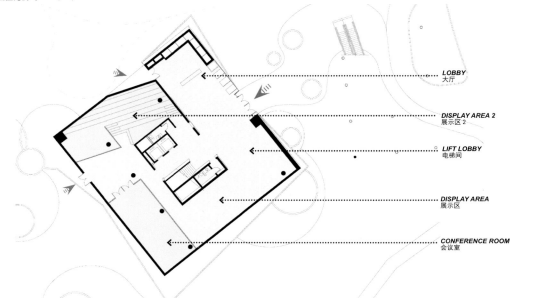

- LOBBY 大厅
- DISPLAY AREA 2 展示区 2
- LIFT LOBBY 电梯间
- DISPLAY AREA 展示区
- CONFERENCE ROOM 会议室

**Level 3 Floor Plan Scale 1/300(A1)**
三层平面图比例（A1）

**Level 4 Floor Plan Scale 1/300(A1)**
四层平面图比例 1/300（A1）

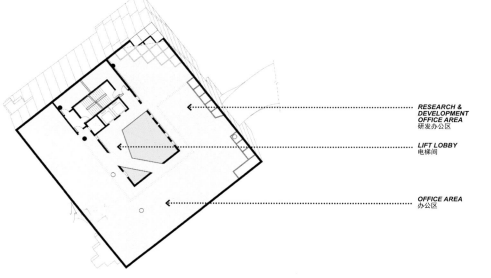
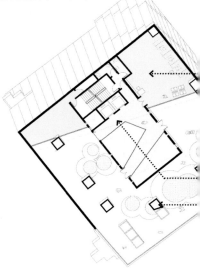

- RESEARCH & DEVELOPMENT OFFICE AREA 研发办公区
- LIFT LOBBY 电梯间
- OFFICE AREA 办公区

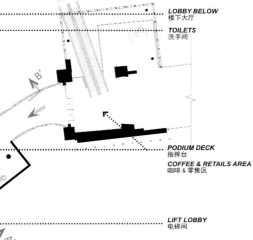
- LOBBY BELOW 楼下大厅
- TOILETS 洗手间
- PODIUM DECK 指挥台
- COFFEE & RETAILS AREA 咖啡 & 零售区
- LIFT LOBBY 电梯间

### Level 2 Floor Plan Scale 1/300(A1)
二层平面图比例 1/300(A1)

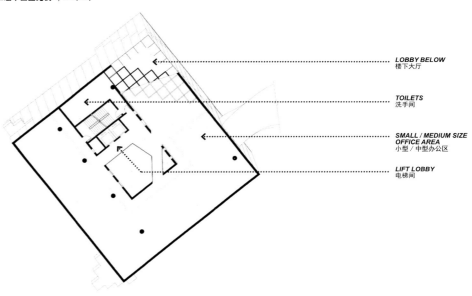
- LOBBY BELOW 楼下大厅
- TOILETS 洗手间
- SMALL / MEDIUM SIZE OFFICE AREA 小型 / 中型办公区
- LIFT LOBBY 电梯间

- RECREATIONAL AREA 娱乐区
- ROOF GARDEN 屋顶花园
- LIFT LOBBY 电梯间
- PLANTING BOX 种植箱

### Level 5 Floor Plan Scale 1/300(A1)
五层平面图比例 1/300(A1)

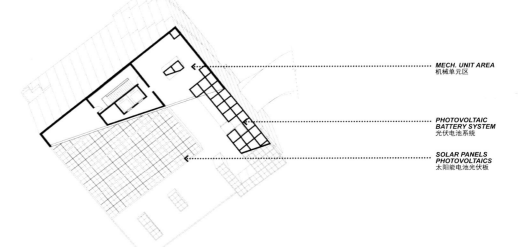
- MECH. UNIT AREA 机械单元区
- PHOTOVOLTAIC BATTERY SYSTEM 光伏电池系统
- SOLAR PANELS PHOTOVOLTAICS 太阳能电池光伏板

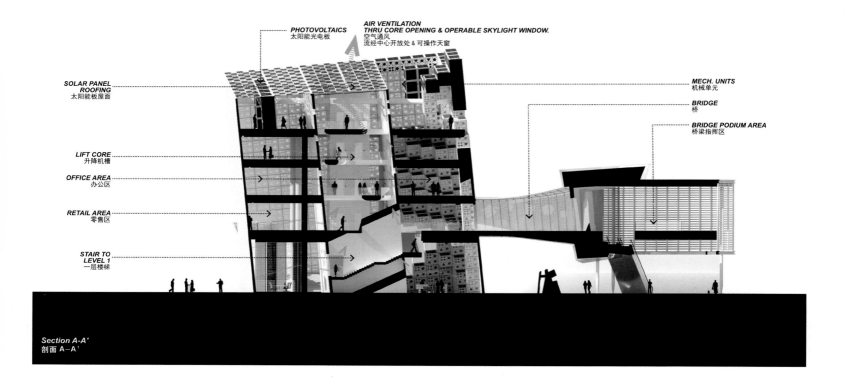

*Section A-A'*
剖面 A–A'

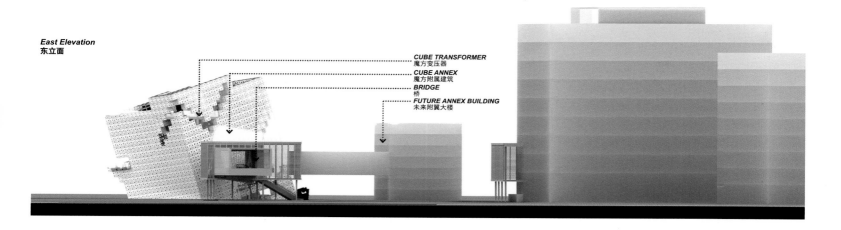

*East Elevation*
东立面

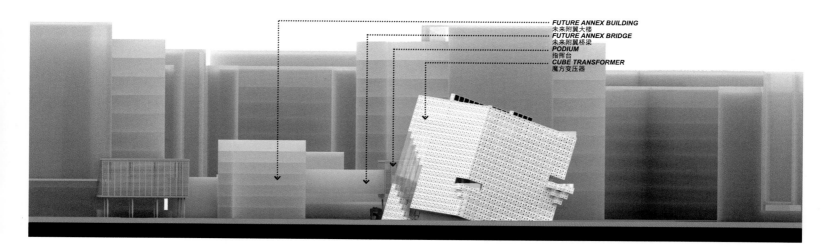

*West Elevation*
西立面

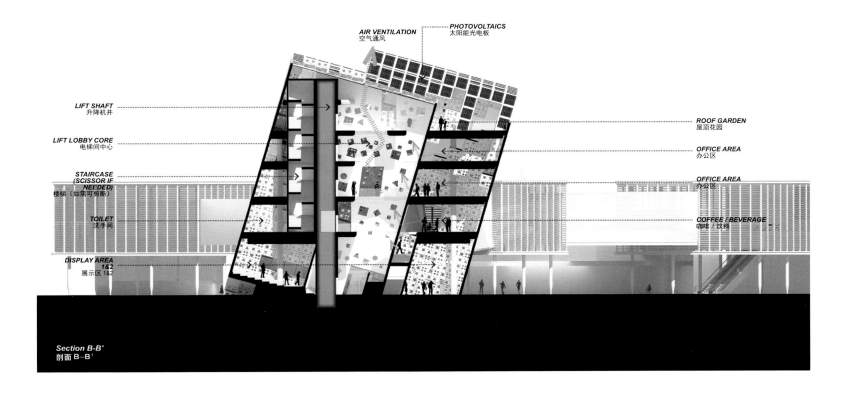

*Section B-B'*
剖面 B–B'

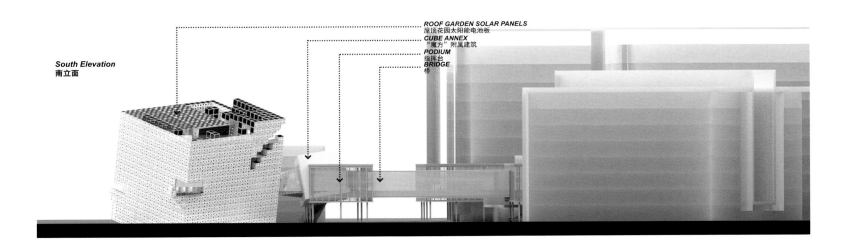

*South Elevation*
南立面

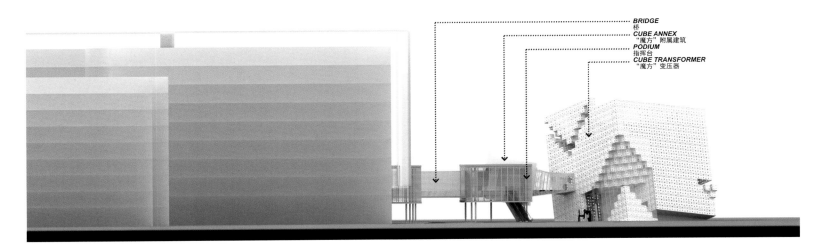

*North Elevation*
北立面

# 丹麦艾尔西诺文化中心
## The Culture Yard

| | |
|---|---|
| 设计单位：AART Architects | Designed by: AART Architects |
| 开发商：艾尔西诺市政府 | Client: Elsinore Municipality |
| 项目地址：丹麦艾尔西诺 | Location: Elsinore, Denmark |
| 建筑面积：13 000m² | Building Area: 13, 000m² |
| 摄影：Adam Mørk | Photography: Adam Mørk |
| 奖项：国际建筑公开赛一等奖 | Award: 1st Prize in Open International Architecture Competition |

**项目概况**

艾尔西诺文化中心由当地的一个老船厂改造而成,其中包括音乐厅、展览区、会议室和公共图书馆,意味着该地已从工业重镇转变为文化中心,希望能通过这个项目联系过去与未来,加强当地社区身份,同时展现着国际面貌并与世界接轨。

**结构设计**

该地历史背景成为设计过程中的主要结构概念,原先的混凝土框架得到了加固,暴露在外的设计让人们回想起其过去的工业特征。铁制楼梯和混凝土元素与现代化的玻璃结构、室内设计相结合,将历史与现在的对比烘托得格外明显。

**Profile**

The Elsinore Culture Yard is transformed from an old shipbuilding yard, including concert halls, show rooms, conference rooms, a dockyard museum and a public library. The Culture Yard symbolizes Elsinore's transformation is an old industrial town to a modern cultural hub. In this way, the yard is designed as a hinge between the past and present, reinforcing the identity of the local community, but at the same time expressing an international attitude, reinforcing the relation between the local and global community.

**Structural Design**

The historic context has thus been the main structural idea in the design process. The original concrete skeleton with armored steel has been reinforced, but left exposed as a reference to the area's industrial past. Thanks to architectural features such as wrought iron stairs and concrete elements, interacting with modern glass structures and interior designs, the contrast between the days of yore and the present becomes evident.

Multifaceted Façade
Wrought Iron Stairs
Concrete Elements

铁制楼梯
混凝土元素
多棱立面

◀ 多棱立面——建筑采用多棱面的立面设计,虽然如同被切成碎片一样,但却仍有很强的一致性。透明巨大的玻璃和钢制的立面面朝着分隔丹麦和瑞典的海峡,也加强了室内外的联系,闪烁出来的光芒和线条塑造了一种宽大的空间感。

尽管立面用上千个线条和三角形组成,但它看上去还是一个巨大的整体,体现出场所和时间的感觉,不仅具有美学效果,还是一道遮阳屏障,减少了建筑对于采暖和制冷的需求。

Multifaceted Façade—particularly striking, when viewed from the seafront and Kronborg Castle, is the multifaceted façade. Like a fragmented, yet strongly coherent structure, the enormous glass and steel façade challenges the historic site and stares unflinchingly across the strait that separates Denmark and Sweden. The transparent façade also reinforces the relation between inside and outside, with the dazzling and dramatic play of lines generating a sense of spaciousness.

Although the façade is made of hundreds of lines and triangles, it appears as one big volume, generating a sense of place and time. The volume also takes the environment into account, since the façade not only functions as an aesthetic and spatial architectural feature, but also as a climate shield, reducing the energy demand for cooling and heating of the building.

**Level -1 1:400**
负一层 1:400

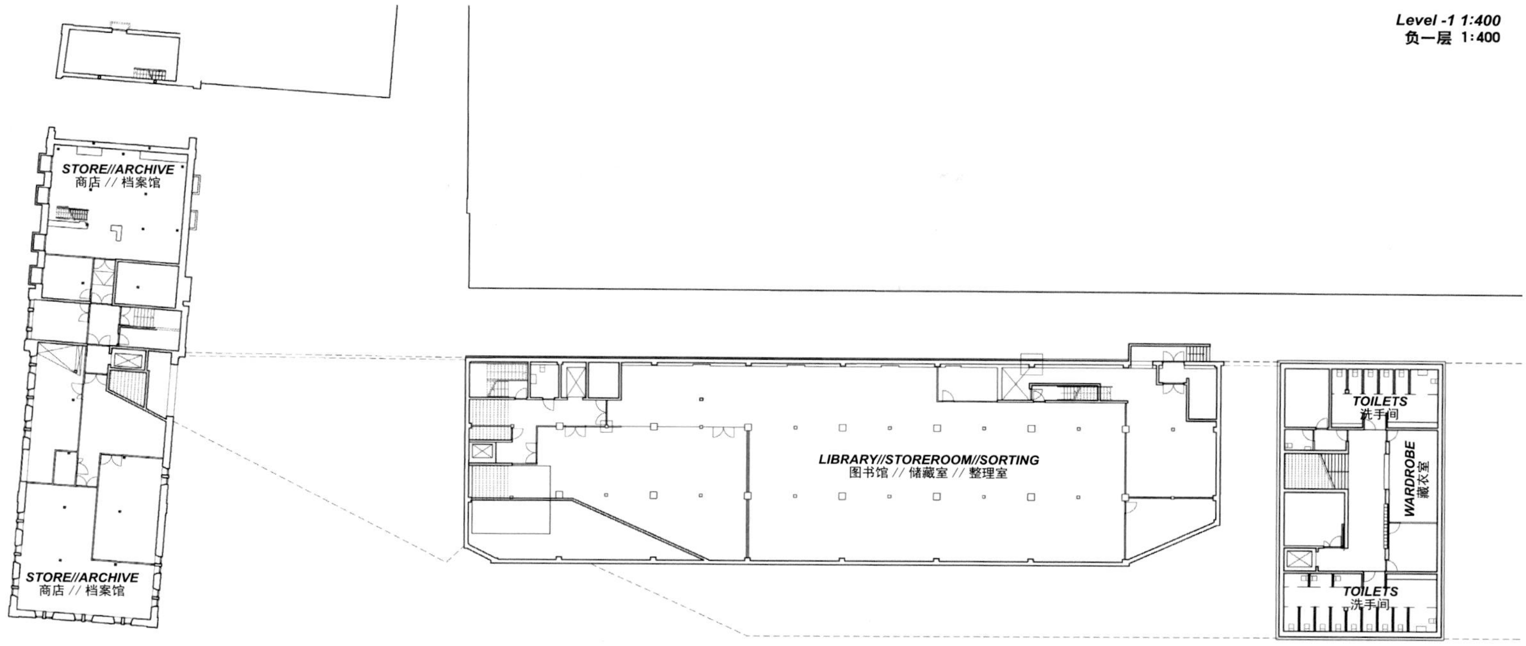

STORE//ARCHIVE
商店 // 档案馆

STORE//ARCHIVE
商店 // 档案馆

LIBRARY//STOREROOM//SORTING
图书馆 // 储藏室 // 整理室

TOILETS
洗手间

WARDROBE
藏衣室

TOILETS
洗手间

**Level 0 1:400**
底层 1:400

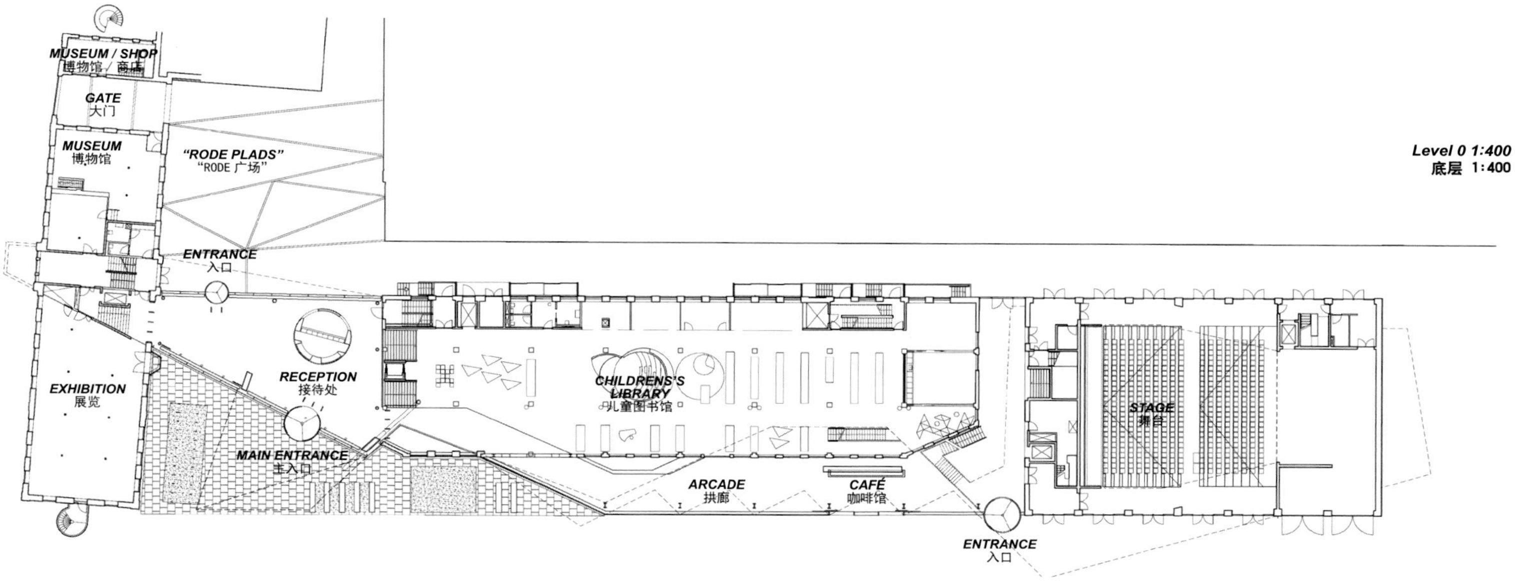

MUSEUM / SHOP
博物馆 / 商店

GATE
大门

MUSEUM
博物馆

"RODE PLADS"
"RODE 广场"

ENTRANCE
入口

EXHIBITION
展览

RECEPTION
接待处

MAIN ENTRANCE
主入口

CHILDREN'S LIBRARY
儿童图书馆

ARCADE
拱廊

CAFÉ
咖啡馆

STAGE
舞台

ENTRANCE
入口

**Level 1 1:400**
一层 1:400

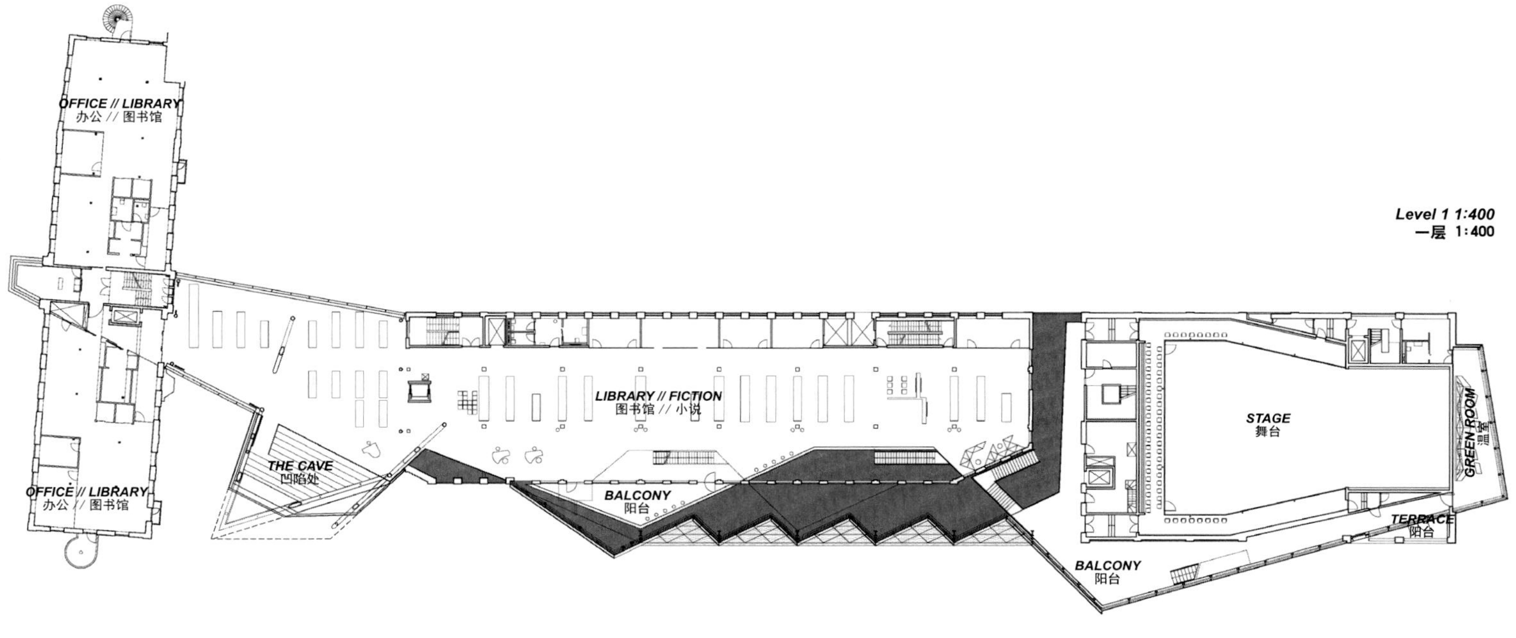

OFFICE // LIBRARY
办公 // 图书馆

OFFICE // LIBRARY
办公 // 图书馆

LIBRARY // FICTION
图书馆 // 小说

THE CAVE
凹陷处

BALCONY
阳台

BALCONY
阳台

STAGE
舞台

GREEN ROOM
绿屋

TERRACE
阳台

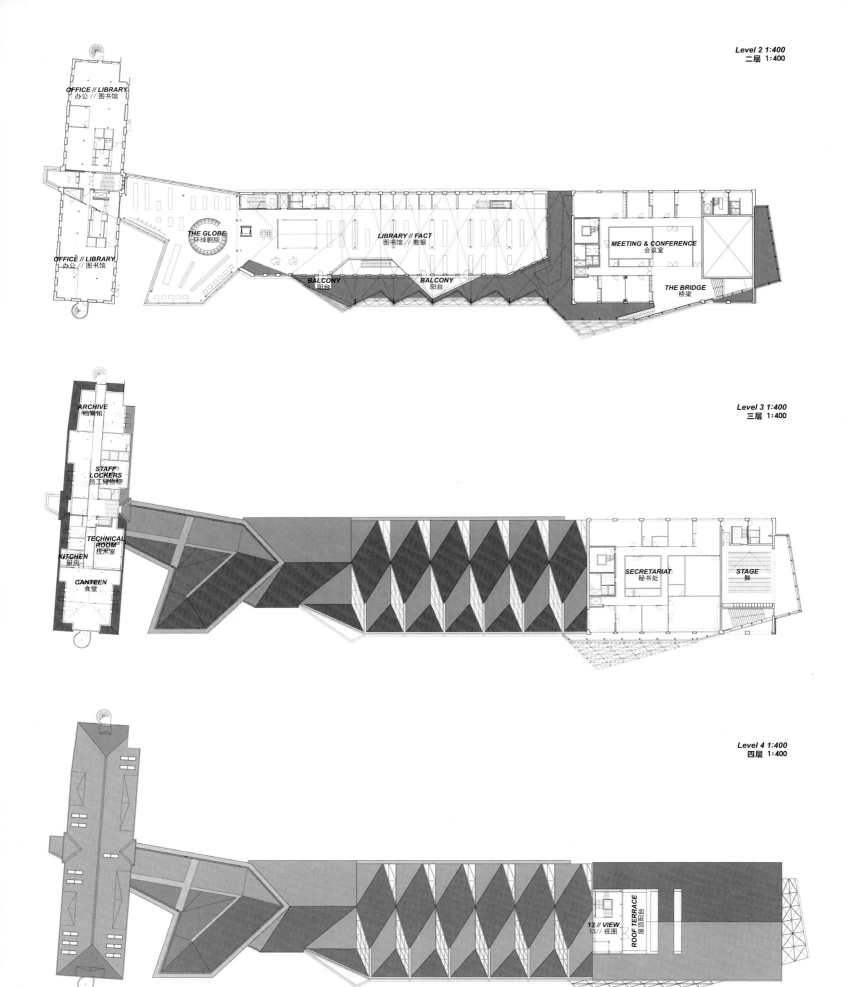

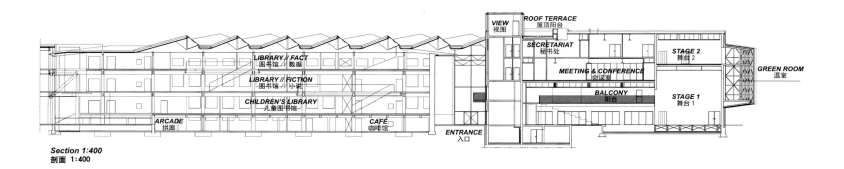

**Section 1:400**
剖面 1:400

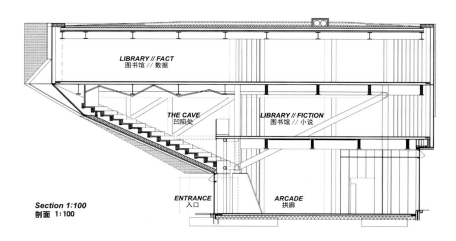

**Section 1:100**
剖面 1:100

118  面向未来——建筑趋势 2015  FACING TO THE FUTURE—ARCHITECTURE TREND 2015

# 瑞典斯德哥尔摩朋友竞技场
## Friends Arena

设计单位：C. F. Møller Architects
开发商：瑞典足球协会
　　　　索尔纳市政府
　　　　Peab　Fabege　Jernhusen
项目地址：瑞典斯德哥尔摩
建筑面积：100 000m²

Designed by: C. F. Møller Architects
Client: Swedish Football Association,
Municipality of Solna,
Peab, Fabege, Jernhusen
Location: Stockholm, Sweden
Building Area: 100,000 m²

碗状
伸缩式屋顶
灵活立面

Bowl
Retractable Roof
Flexible Façade

## 项目概况

"朋友竞技场"是瑞典新的国家体育馆,位于索尔纳火车站附近,地理位置尤为显著,通过铁路上方的人行天桥可到达 Arenastaden 区域。

## 建筑设计

整个竞技场呈碗状,拥有 50 000 席位观看足球比赛,还可供 65 000 人观看音乐会和其他重大活动,而通过"关闭"看台的上层区域,其容量可减少至 30 000 人,可进行更为小型、私密的活动。其合理完善的人流疏散通道可使人群在 6 分钟内疏散而空。

场馆设计有可伸缩式的屋顶,使其能用作各类运动场,无论是草坪、红土、沙土、冰地还是水中运动等,一年四季皆可进行。另外,竞技场外观是依据最佳音响效果而设计的,使其在举办音乐类活动时能让观众享受到极尽完美的视听,而活泼的外立面也能根据场内进行的活动不同而变换颜色和外观。

## Profile

Friends Arena is Sweden's new national stadium, strategically located next to Solna Station, which will be extended, reaching as far as Arenastaden via a new pedestrian and bicycle bridge over the railway tracks.

## Architectural Design

The bowl-like arena can seat 50,000 for football games, while concerts and other events can accommodate up till 65,000 spectators. By shutting off the upper level of the stands, capacity can be reduced to around 30,000 spectators. The logical evacuation routes make the whole arena bowl can be emptied in six minutes.

The arena has a retractable roof making it suitable for all types of playing fields such as grass, clay, gravel, ice or water, all year round. The arena is shaped for optimum acoustics. Even the outside of the arena is flexible, with the ability to change colour and appearance depending on what's happening inside the arena.

# 比利时布鲁塞尔托儿所和福利办公室
## A Childcare Center and Welfare Office, C.P.A.S.

设计单位：Philippe SAMYN and PARTNERS, architects & engineers
开发商：C.P.A.S. – O.C.M.W, BRUSSELS.
项目地址：比利时布鲁塞尔
建筑面积：2 437m²
摄影：Marie-Françoise PLISSART

Designed by: Philippe SAMYN and PARTNERS, architects & engineers
Client: C.P.A.S. - O.C.M.W, BRUSSELS.
Location: Brussels, Belgium
Building Area: 2,437m²
Photography: Marie-Françoise PLISSART.

# Cylindrical Tower
# Modest Materials
# Feeling of Security

# 圆柱形塔
# 低调建材
# 安全感

**项目概况**

项目厂址位于布鲁塞尔 rue de l'Epée 1 000 号，其中大部分属于布鲁塞尔的 C.P.A.S，其余部分则属于布鲁塞尔市。新项目的修建无疑将为周边地区带来福利。

**规划设计**

项目发展区包括公益住房、一家托儿所和一家福利办公室，其布局就像一个接待社会人群的小村庄。托儿所与福利办公室规划建于同一场址，加强了项目的象征价值。并且，由于其功能和地理位置的关系，项目显得质朴而不奢华。

**建筑设计**

福利办公室大楼外形就像一座圆柱形的塔，外直径尽可能地减小，使其与周边建筑大小极为相称。另外，圆柱形的外观和采光也增加了项目的视觉效果。

塔楼台柱下的建筑是一个露台，是通往塔楼和托儿所入口的户外接待区。塔楼主要有各种医药室和社工办公室，经户外楼体和两座电梯可以到达。顶层有一间大型会议室，可观望到室外优美的风景。

大楼使用的建筑材料也非常低调，独立的外墙使用简单的金属板铺设，内墙是用粗糙或贴面的石工砌成，门和窗的框架则都是用淡色的木材制成。

托儿所是一栋位于后方的很低的大楼，外形弯曲，充满着惊喜和乐趣。建筑利用街道上的圆形部署区域，设计有可通往附近建筑的紧急车辆通道。

托儿所的宿舍位于底楼，采用鸟巢和洞穴的建筑形态。从底层通往楼顶设有一条斜坡，斜坡有一个覆盖的露台和一个空中花园，可用来进行运动锻炼。屋顶花园和操场用金属织物来预防大风，整个设计给人一种安全感。

## Profile

The site of this project is located rue de l'Epée in 1000 Brussels. It belongs to the C.P.A.S. of Brussels for the most part and the rest is the property of the City of Brussels. The renovation will certainly benefit the surroundings.

## Planning Design

This new development, including social housing, a childcare center and a welfare office is laid out like a hamlet to favor social and human relations. A childcare center is available in the same premises reinforces the symbolic value of the project. At the same time, given its function and location, the project should steer away from any lavishness.

## Architectural Design

The building of the welfare office will be shaped like a small cylindrical tower. The outside diameter will be as small as possible to blend with the size of the lots in the neighborhood. The cylindrical shape and the resulting lighting add the visual effect of the project.

The zone under the pillars of the tower is a patio acting as an outdoor reception area leading to the entrance of the tower as well as of the childcare center. The stories in the tower house the various medical premises and social workers' offices. They are accessible via an outdoor staircase and two lifts. A large conference room with a panoramic view is located on the top floor.

The use of materials in the building should express certain modesty. The exterior bearing walls are isolated and covered with simple bolted metal plate. The inner walls are made of rough or faced masonry. Door and window frames are in light-colored wood.

The childcare center is a low building erected at the back of the lot. The building has a curved shape, full of surprises and fun, taking advantage of the circular maneuvering area in the street that gives emergency vehicles access to neighboring buildings.

The dormitories in the childcare center are located on the ground floor, using the nest and grotto concepts. The technical premises are in the basement. A ramp runs from the ground floor to the roof, fitted with a covered terrace and a hanging garden, and used for psychomotor exercises. The rooftop garden and playground, protected from the wind by a veil of metal fabric, offer a great feeling of security.

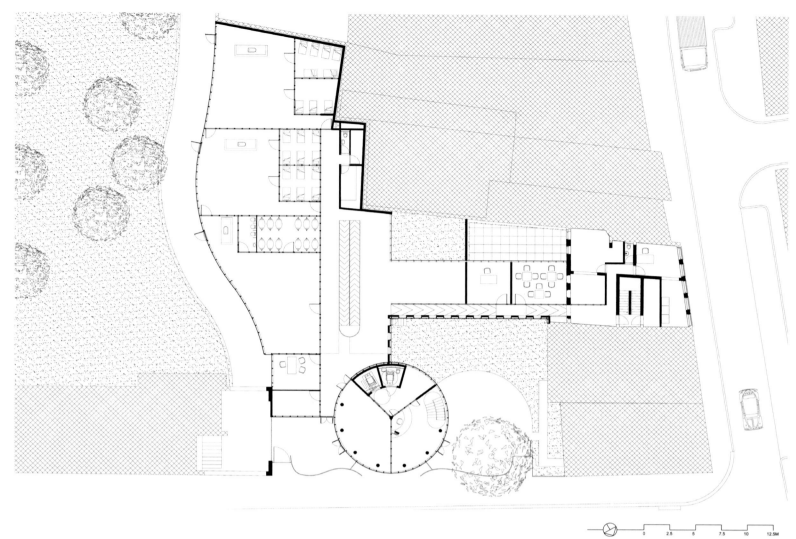

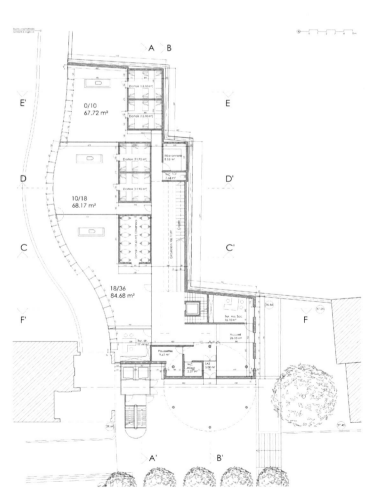

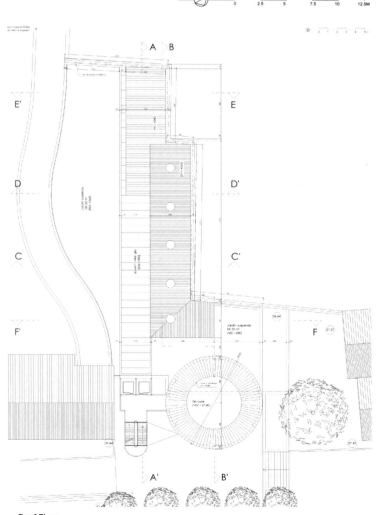

**Ground Floor**
底层

**Roof Floor**
屋顶层

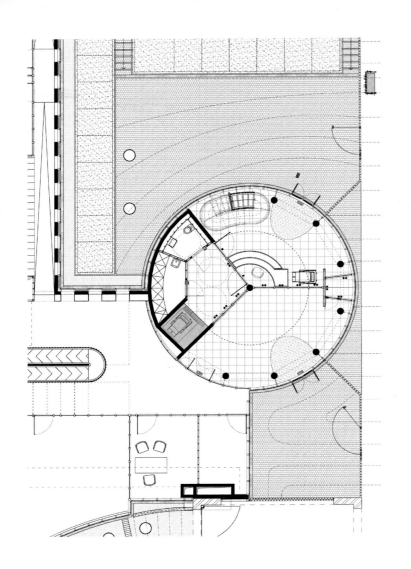
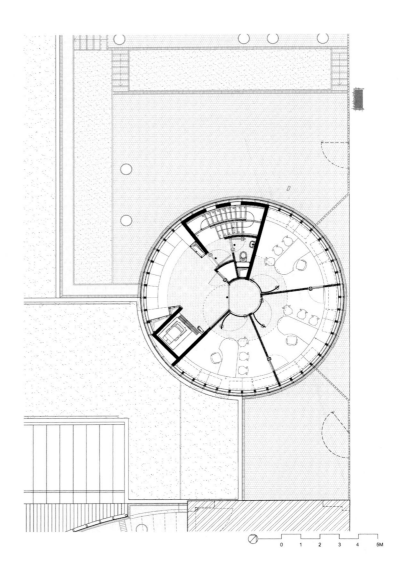

128　面向未来——建筑趋势 2015　FACING TO THE FUTURE-ARCHITECTURE TREND 2015

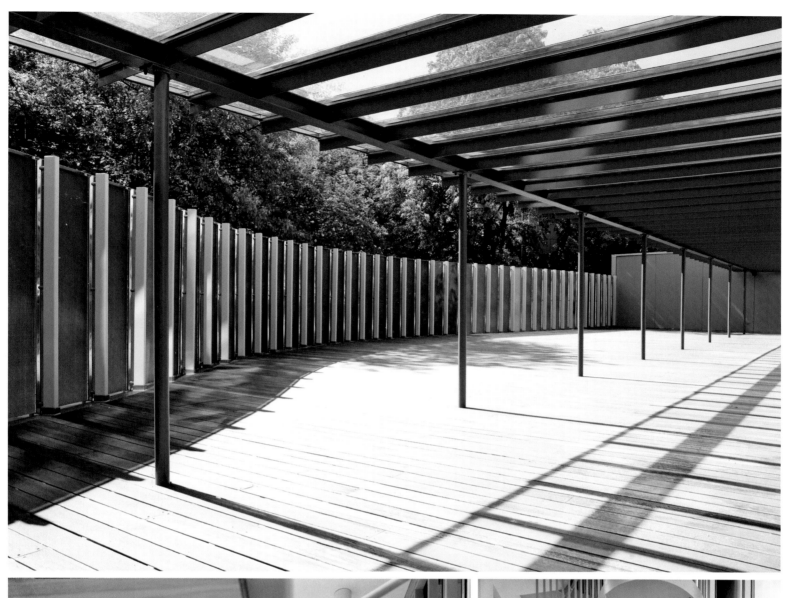

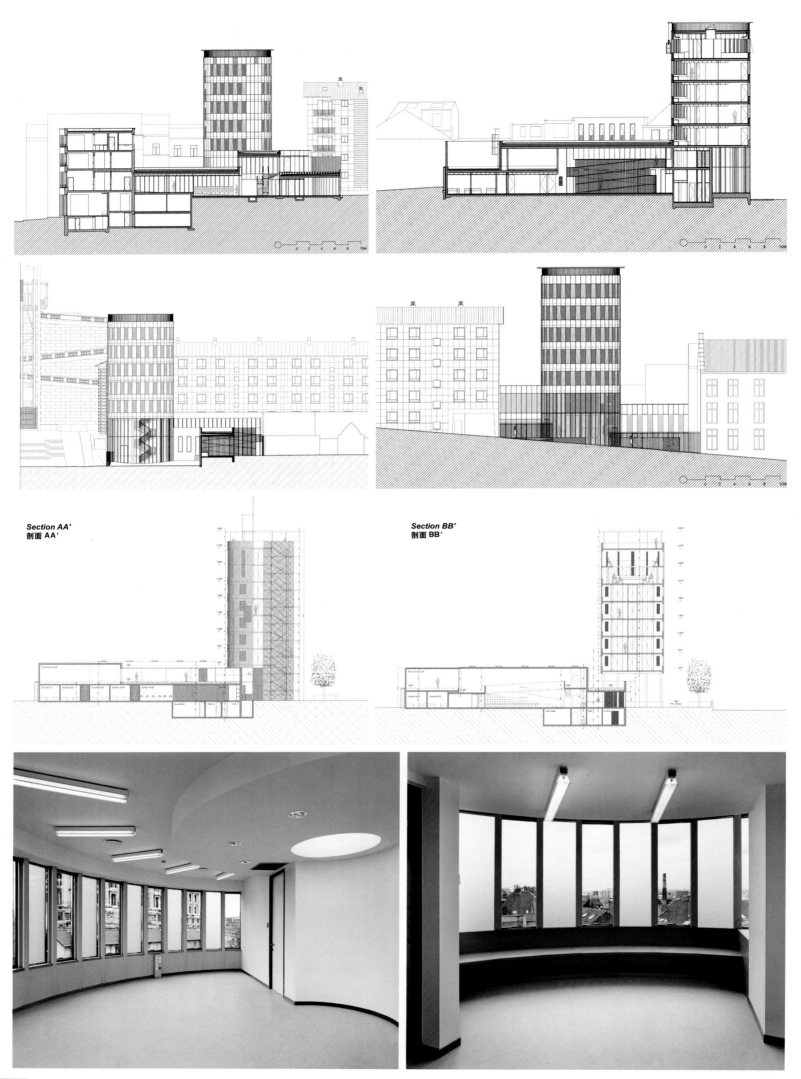

# 德国文登 Drehmo 公司新大楼
## New Building for Drehmo, Wenden

设计单位：wurm + wurm architekten ingenieure
开发商：Drehmo GmbH
项目地址：德国文登
建筑面积：7 500m²
摄影：Ester Havlova　Walter Fogel

*Designed by: wurm + wurm architekten ingenieure*
*Client: Drehmo GmbH*
*Location: Wenden, Germany*
*Building Area: 7,500m²*
*Photography: Ester Havlova, Walter Fogel*

# "阁楼"屋顶 铝金属栅格板 动感立面

# "Attic" Roof Aluminum Expanded Metal Panels Dynamic Façade

**项目概况**

Drehmo 公司新大楼位于德国南端，水管理行业、电厂行业和石油天然气工业中的 sauerland 机电制动器都是该公司的工厂制造的，而这一地区的特殊地形情况也决定了该新大楼形态的特殊消减形式。

**功能布局**

整栋大楼基本的安排包括会馆和装货间附属区域，还有一个全自动的货架储存间，两层的行政大楼作为建筑的侧翼，坐落在会馆的前面，共同组成了一个入口。地下室是工人们的起居室和餐厅，自然光线通过院子和地板，天花板光滑的平面可照射到这个自由的区域。

**建筑设计**

建筑屋顶是一个超大的带状阁楼结构，横跨整座建筑，随着建筑体量的变化，条带的长短也各不相同。阁楼、屋顶拱背和部分墙面材料都是由较轻的铝金属栅格板构成的，更加突出了建筑的造型。

建筑形态遵循场地环境而设计，梯田景观创造的移动自然景观产生了许多不规则的地块，山坡上的轮廓则衍生出不同寻常的平面图。在这里，可以欣赏到山谷的迷人景色。

室内轻型的木质地板和露石混凝土表面形成鲜明对比，宽阔的空间和两层楼的中庭提供了令人惊奇的视觉，倾斜的水泥柱层层镶嵌，在建筑立面上形成动感十足的造型。

## Profile

The new building for Drehmo GmbH is located in the southernmost town of the region of Sauerland in Germany. Electromechanical actuators for water management, the power plant sector and the oil and gas industry are manufactured by the company.

## Function Layout

The basic arrangement consists of the assembly hall with the affiliated areas of the loading bay and automatic shelf storage. The two-story administrative wing situated in front of the assembly hall forms with this an entrance situation. The basement accommodates staff rooms and the canteen, which is naturally lit via a courtyard and expands into this free area with glazing from floor to ceiling.

## Architectural Design

The roof with an oversized, band-like attic spans the entire complex. This band is different in height, as it flows to the building around. The homogeneous materiality of the attic, roof soffits, columns and parts of the wall covering, using lighter aluminum expanded metal panels; the corporeal of construction is emphasized.

The building form follows the site. The moving landscape caused the terracing of the landscape and produces irregularly shaped lot. The unusual floor plan generated by this contour, and the hillside situation with a view over the valley determine the character and exceptional styling.

Inside light wooden floors contrast with the large exposed concrete surfaces. Spacious air spaces and the two-story atrium always enable new, surprising perspectives. Slanted concrete columns, already nested in itself, create dynamic images in the overlay.

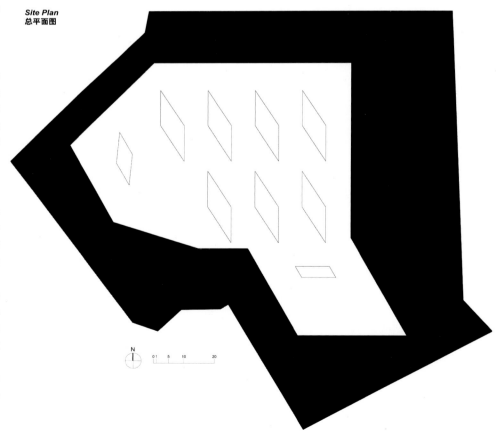

*Site Plan*
总平面图

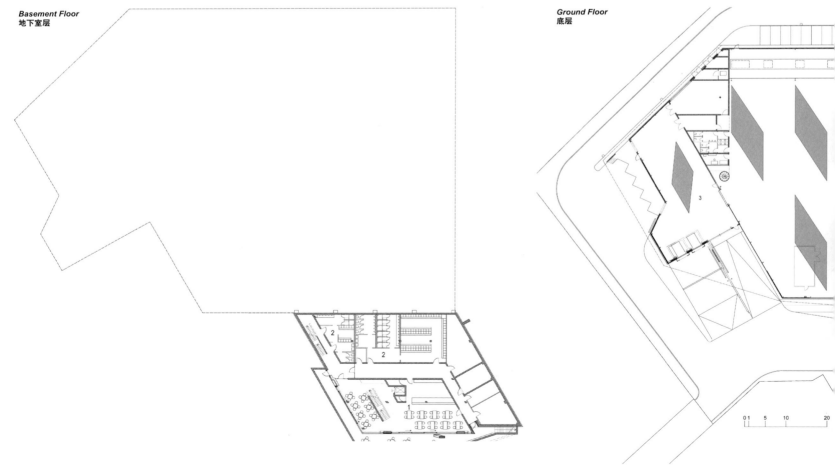

**Basement Floor**
地下室层

**Ground Floor**
底层

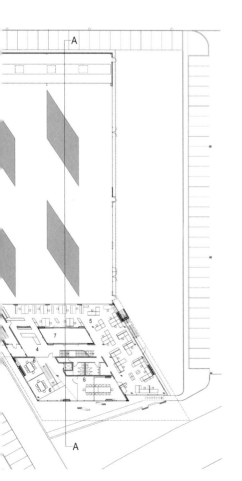

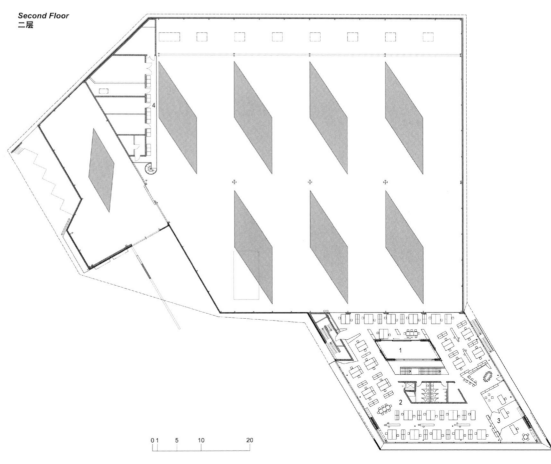

*Second Floor*
二层

木盒子
灯塔
百叶窗

Wooden Box
Beacon
Shutter

# 荷兰阿尔梅勒宠物农场
Petting Farm

设计单位：70F architecture
开发商：阿尔梅勒直辖市
项目地址：荷兰阿尔梅勒
建筑面积：126m²

Designed by: 70F architecture
Client: The municipality of Almere
Location: Almere, Netherlands
Building Area: 126m²

### 项目概况

在荷兰阿尔梅勒,城市中大部分地区都有一座宠物农场,而该宠物农场则位于"den Uyl"公园,是在一个旧农场被烧毁的混凝土基地上重建起来的。整个建筑过程,几乎都只使用了赞助资金。

### 建筑设计

整座建筑呈一个立方体,木质材料的立面使得建筑看起来就像是一个巨大的编织灯笼。建筑师还为建筑的上半部分设计了一个带开放式立面的木盒子,使得整个农场保持不断通风的状态,为室内营造一个良好的空气环境。

建筑的一半设计成马厩,另一半则由厕所、储藏室以及位于二层的办公室组成。当人们纵穿过建筑时,会经过位于篱笆后左、右两边的动物。这栋建筑没有设置门,而是由六个百叶窗代替。到了夜晚,室内的灯光通过通透的立面弥散开来,使得建筑宛如一座公园里的灯塔,在"日出而作,日落而息"的作息规律中,散发着温暖的力量。

### Profile

Most city parts of Almere have a petting farm. The new petting farm is located in the "den Uyl" Park. It's built on the concrete foundation of a burnt old petting farm. The building was finally built using almost only sponsored money.

### Architectural Design

The whole building takes the shape of a cube. The wooden façade makes the building look like a giant lantern. A wooden box with an open façade system is designed for the upper half of the building, allowing for continuous ventilation and a healthy environment.

Half of the building is stable; the other half consists of toilets, storage and on the second floor an office and storage. Walking lengthways through the building, visitors will pass the animals that are contained to the left and to the right behind fences. There are no doors in the building, but there are six shutters. At night, the interior light disperses outwards through the transparent façade; the building becomes a light beacon in the park, which wakes up and goes to sleep every day, emitting warm light.

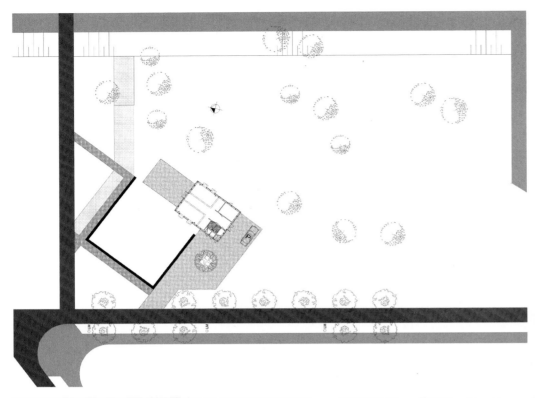

▲ 百叶窗 ——建筑六个大小不一的百叶窗极具特色，两个公用，分别位于建筑较短侧边，四个供动物使用，在建筑较长侧边上各有两个，是为建筑提供活力和生机的源泉。

在早晨，这些百叶窗将手动或自动打开，而当太阳落山，这些百叶窗又将关闭。在百叶窗的开关组合中，也形成建筑富于变换的建筑形态。

Shutters—the six shutters of different sizes are individually unique, two for the public on the short ends of the building and four for the animals, two on either long side of the building. These shutters provide infinite energy and vitality for the building. These shutters will open manually or automatically in the morning, reacting on the upcoming sun, as they will close again at the end of the day, when the sun goes down. With opening and closing of the shutters, the building form is also in continuous changing.

**Ground Floor**
底层

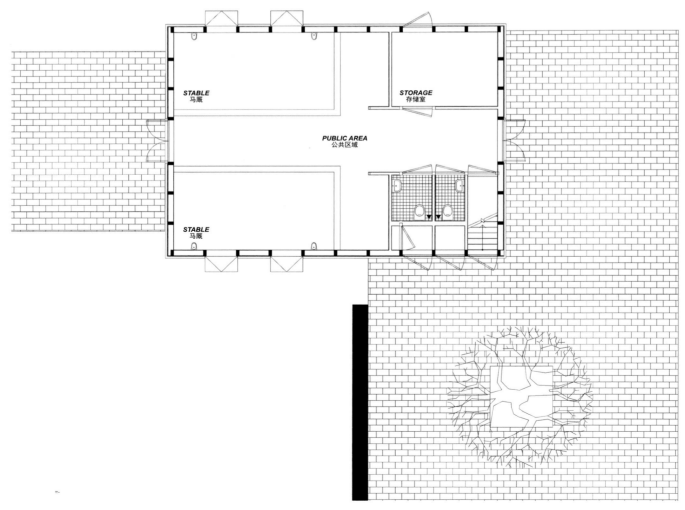

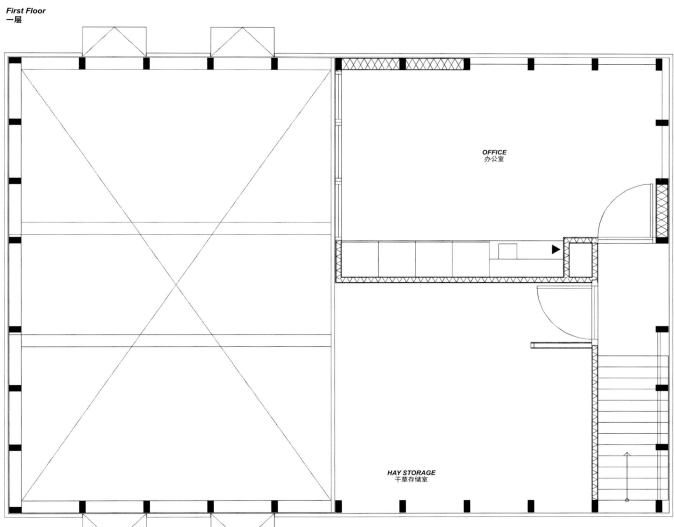

**First Floor**
一层

**OFFICE**
办公室

**HAY STORAGE**
干草存储室

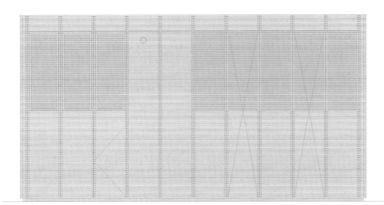
**Back Façade**
背面

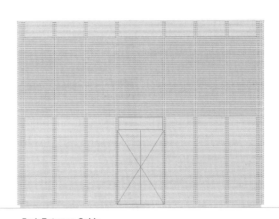
**Park Entrance Gable**
公园入口山墙

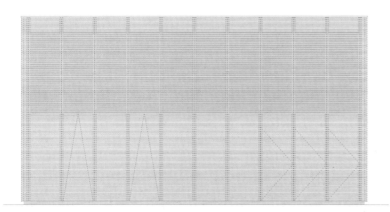
**Front Façade**
正面

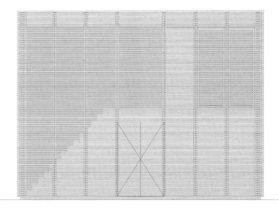
**Main Entrance Gable**
主入口山墙

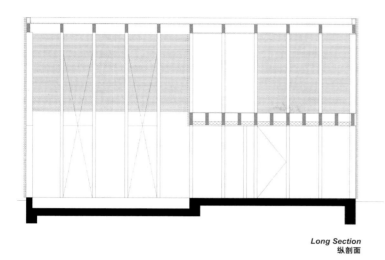

*Long Section*
纵剖面

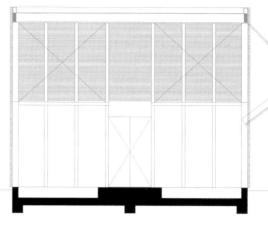

*Cross Section*
横剖面

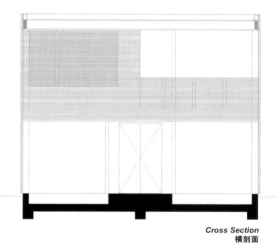

*Cross Section*
横剖面

Pedestrian-oriented
Dynamic Landmark
Bi-level Residence

以人为本
动感造型
双层结构

# 美国加利福尼亚州 F65 中转村
## F65 Center Transit Village

设计单位：Mark Dziewulski Architect
项目地址：美国加利福尼亚

Designed by: *Mark Dziewulski Architect*
*Location: California, U.S.*

**项目概况**

F65中转村是一个交通中转中心，毗邻新的轻轨站，是一项密集的混合型开发项目。它是区域内新总体规划的主要项目，旨在促进周边区域的复兴，创造可识别的城市环境。

**平面布局**

建筑在街道两面并排，形成一幅壮丽的城市街景画，停车场则隐藏在建筑后面，与轻轨站相连。人行道与邻近的场地相连，从而在原先被车辆占据的环境中形成连续性的人行道。在街道交叉口设有公共户外空间，为餐馆和咖啡厅提供席位区，雕塑似的高楼梯连接着住宅层，加强了公共广场的品质，也成为整个场地的标志物。

这是以行人为导向的城市环境，设有户外聚集空间，为区域创造一种社区感。

**建筑设计**

建筑现代化的动感造型体现出其作为城市交通枢纽的独特位置，也反映出工业旧址的特征。设计旨在依托新区域形成动感、夺人眼球、易识别的地标性建筑，反映出波特兰、纽约等城市的工业厂房的特征。

高高的雕塑塔在高速公路上就能看到，成为建筑重要的识别标志。而住宅阁楼为双层结构，旋转楼梯到达一片开阔的空间，可俯瞰下面的双层居住区域。许多结构组件都是暴露出来的，比如抛光混凝土楼板和木质屋顶桁架，创造出建筑独特的结构形象。

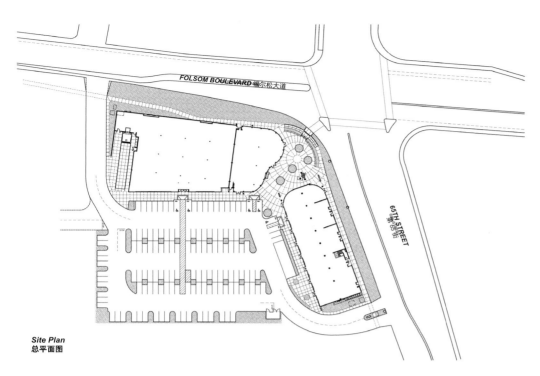

**Site Plan**
总平面图

## Profile

Situated adjacent to a new light station, F65 Center Transit Village is a dense mixed-use development. It is the lead project of a new master plan for the area, aiming to provide impetus for renewal of the surrounding area and create a recognizable urban environment.

## Site Plan

The building lines the two street frontages to create an urban streetscape, with required parking concealed behind. The site plan is driven by strong pedestrian linking to the station and adjacent streets. Sidewalks are connected to the adjacent lots, to create a new sense of pedestrian continuity in what was previously an almost car-only environment. At the street intersection, a public outdoor space provides seating for the restaurants and coffee shop. The public plaza qualities are also reinforced by the high sculpture-like stair linking with the residential levels, which also acts as a landmark for the entire site.

The scheme is a pedestrian-oriented urban environment with outdoor gathering spaces. It is intended to create a sense of community for the area.

## Architectural Design

The modern dynamic forms of the buildings reflect their role as an urban transit hub and also something of their ancestry in the industrial area. The designs are intended to be dynamic and eye-catching, to create a recognizable landmark quality that will anchor the new area. The project also reflects some of the characters of the industrial warehouse conversion going on in other cities, such as Portland and New York.

The sculptural tower can be seen form the highroad, becoming an important identification mark of this building. The residential lofts are bi-level, with a spiral stair leading to an open space that overlooks the double-height living area below. Many of the structural elements are left exposed, such as polished concreted floors and wood roof trusses, creating unique structural image of the building.

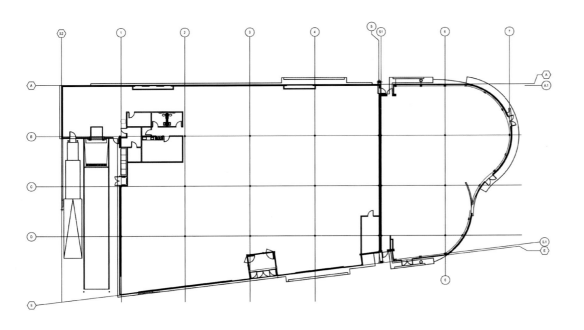

**Building A Plan**
建筑 A 平面图

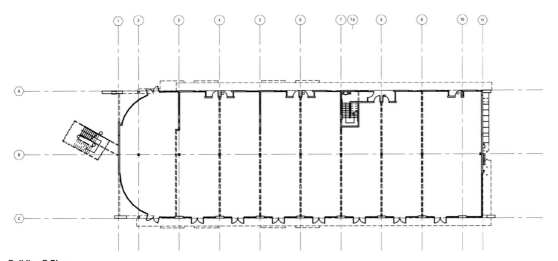

**Building B Plan**
建筑 B 平面图

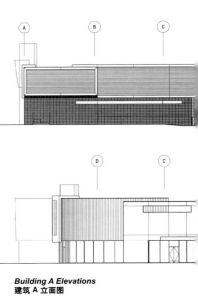

**Building A Elevations**
建筑 A 立面图

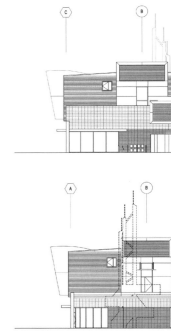

**Building B Elevations**
建筑 B 立面图

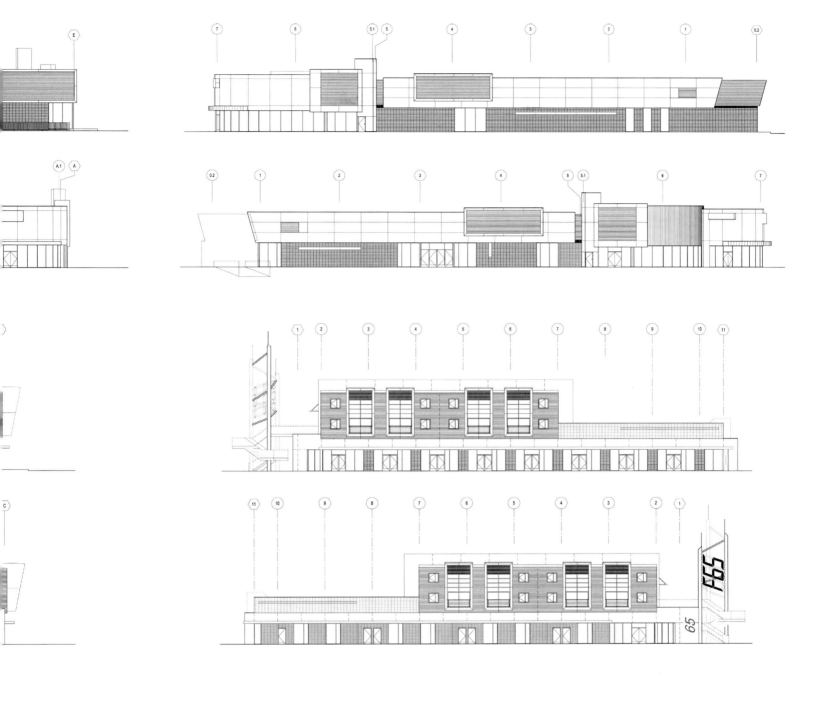
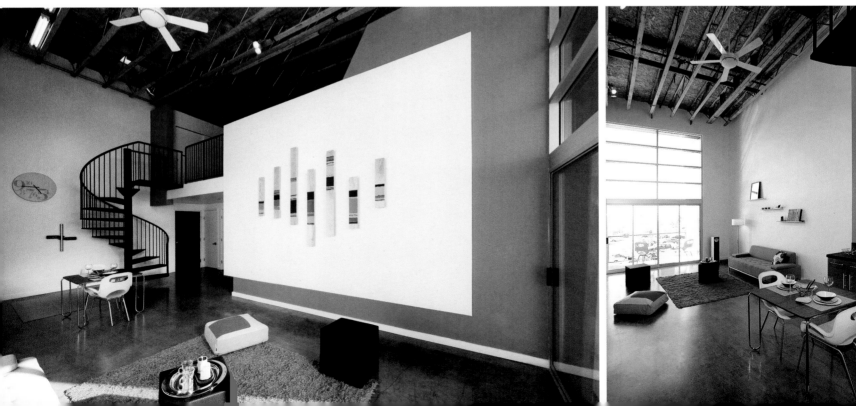

# 荷兰 Zwaluwen Utrecht 1911
Zwaluwen Utrecht 1911

设计单位：NL Architects
开发商：Gemeente Utrecht, DMO
项目地址：荷兰乌特勒支
建筑面积：1 265m²
设计团队：Lukas Haller　　Rachel Herbst
　　　　　Sybren Hoek　　Erik Moederscheim

*Designed by: NL Architects*
*Client: Gemeente Utrecht, DMO*
*Location: Utrecht, The Netherlands*
*Building Area: 1,265m²*
*Design Team: Lukas Haller, Rachel Herbst,*
*Sybren Hoek, Erik Moederscheim*

## 项目概况

两个历史悠久的足球俱乐部——Zwaluwen Vooruit 和 V.V. Utrecht 合并成一个，从而也就诞生了这样一座新的俱乐部建筑。

## 建筑设计

俱乐部方案设计有一个标准截面，由透明、条状部件组成。该部件位于更衣室基座上面，几乎没有窗户，从这一抬升的位置可俯瞰球场全景。

建筑地基是地下墓穴，一个筑堤围绕其四周，挖掘这一大片土地是为了获得必要的日光，使球场的绿草可以一直延伸到露台。由于公共区域比建筑坐落的基座小，建筑拥有一个超大的"游廊"，包围建筑的整个顶楼。筑堤的优势是将经常不引人注目的建筑底层隐藏起来，同时形成一个看台，将露台四周的栅栏面积降到最低。

建筑顶层极具表现力，可伸缩的金属立面通过折页与屋顶连接起来，所有的部件都是可以随意打开或关闭的。另外，建筑侧翼还装有可移动的钢条，在现有位置上起到遮阳的作用。

## Profile

Two soccer clubs with a long history, Zwaluwen Vooruit and V.V. Utrecht merged into one. As such a new clubhouse should be erected.

## Architectural Design

There is a blue print for clubhouses, a typical section. They consist of a transparent, bar-like, part that is often placed on top of a pedestal of changing rooms with hardly any windows. From this elevated position it is possible to overlook the pitches. A panoramic view is guaranteed.

The base, often referred to as the catacombs, is surrounded by a dike. This body of earth is excavated for necessary daylight. It continues the grass of the playing fields all the way up to the terrace. Since the public part is smaller than the plinth it sits on, the building can feature a super sized "veranda" which envelops the entire top floor. The advantage of the dike is that the usually unattractive lower part of the building is hidden. At the same time an instant grandstand comes into being and the fence around the terrace could be reduced to minimum.

The top floor is expressive. The façade of stretch metal is connected to the roof with hinges. All segments can be opened and closed at will. The wings of the building are fixed with removable steel rods. In its current position they serve as sunscreens.

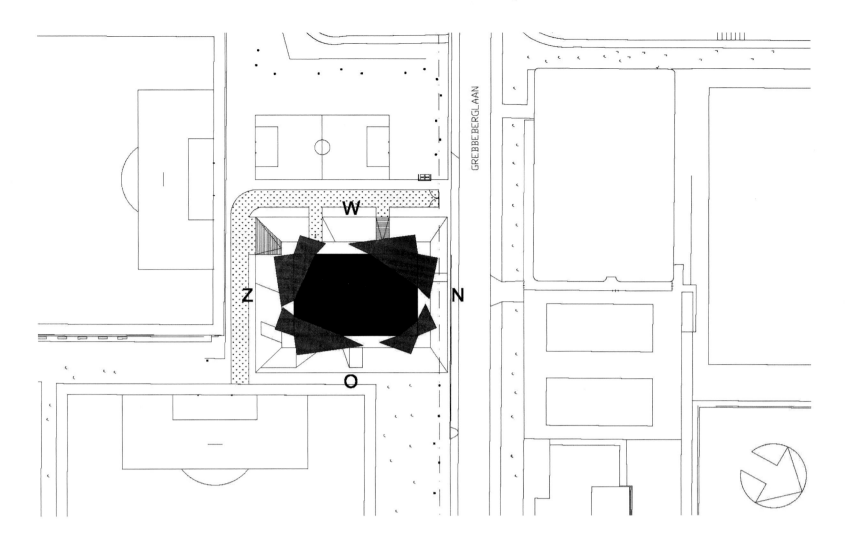

"游廊" 可伸缩立面 移动钢条
"Veranda" Stretch Façade Removable Steel Rods

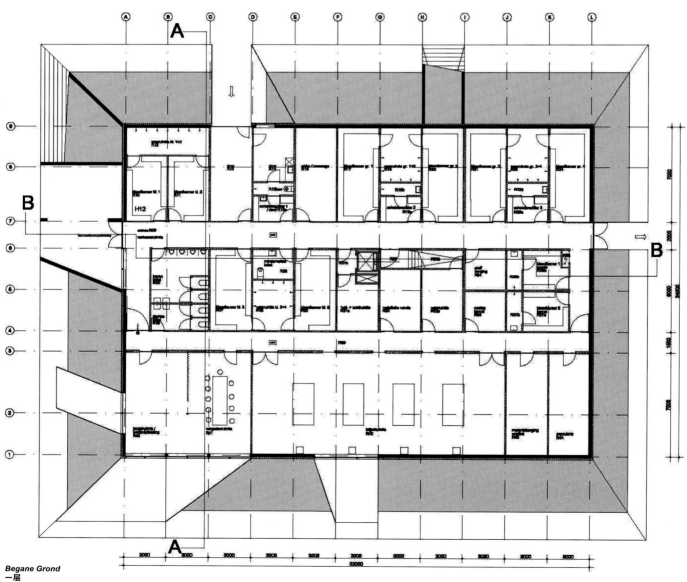

*Begane Grond*
一层

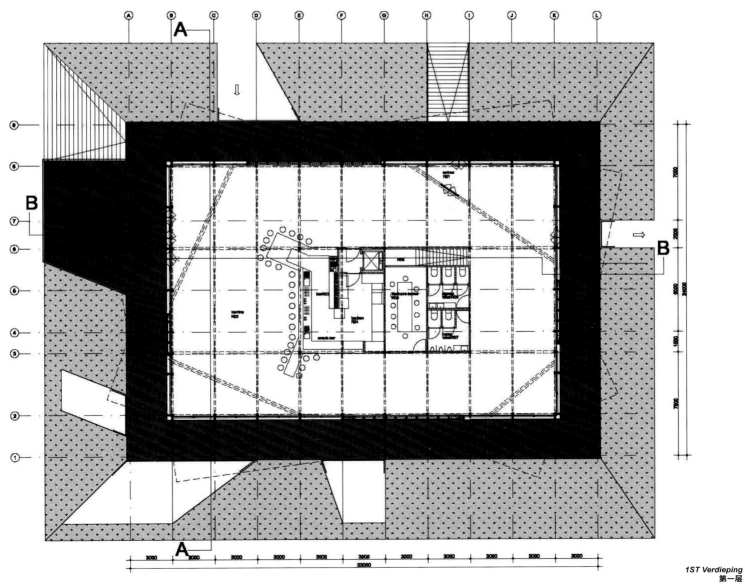

*1ST Verdieping*
第一层

Scheme A
方案 A

0.

1.
THE STACKING OF THE MORE PUBLIC PROGRAM ON TOP OF THE WARDROBES CREATES A TERRACE THAT ALLOWS FOR EXTENSIVE VIEWS OVER THE SURROUNDING SOCCER FIELDS.
衣柜顶端更多公共项目的叠加创建了一个允许看到足球场周围的露台。

2.
THE ADDING OF A DIKE EQUALS THE CHEAPEST POSSIBLE FAÇADE. IT PUTS THE BUILDING ON A TEMPLE-LIKE PEDESTAL.
岩脉的增加可以算是最简单的可能外观。它把建筑放置在一个神殿般的基座上。

3.
BY ELEVATING THE 4 CORNERS, A DIFFERENCE OF SPATIAL OPENING OF THE BUILDINGS UPPER LEVEL IS CREATED FOR EACH INDIVIDUAL CORNER.
通过提升四个角落，在建筑上层为每个角落创建了一个不同的开放空间。

4.
TRACING AND CUTTING ACCORDING TO SCHEME B DIVIDES THE FAÇADE INTO SOLID WALLS AND SEMI-TRANSPARENT, MOVABLE FLAPS.
根据方案B临摹和剪切将立面划分为坚实的墙壁和半透明可移动的襟翼。

THE 'EMPTY' CORNERS ALLOW FOR VIEWS OVER THE SOCCER FIELDS AND THE FLAPS PROVIDE THE TERRACE AND INTERIOR OF THE BUILDING WITH COOLING SHADOWS. OFF-SEASON, THE FLAPS CLOSE THE CORNERS AGAIN AND PROTECT THE BUILDING FROM ANY KIND OF VANDALISM.
"空"角落允许看到足球场地，而且襟翼为建筑阳台和室内提供阴凉处。淡季，襟翼关闭该角落，并保护建筑不受任何破坏。

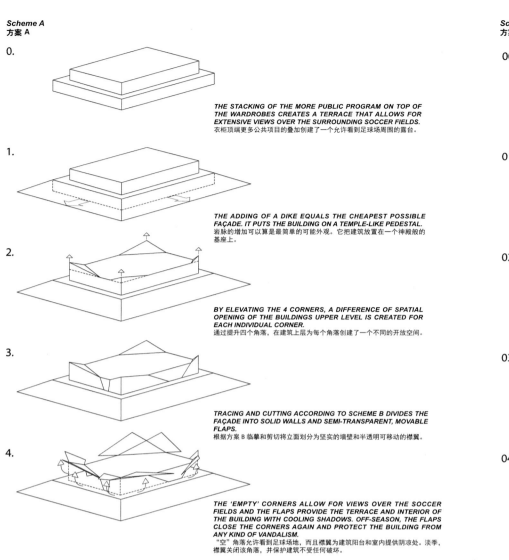

Scheme B
方案 B

Move the Roof
移除屋顶

00.
HERE IT IS
像这样

01.
FOLD DOWN THE 4 WALLS
四面墙向下折

02.
MOVE UP THE 4 CORNERS
四个角落向上

03.
MOVE THE FLAPS
移除襟翼

04.
CLOSE IT
关闭

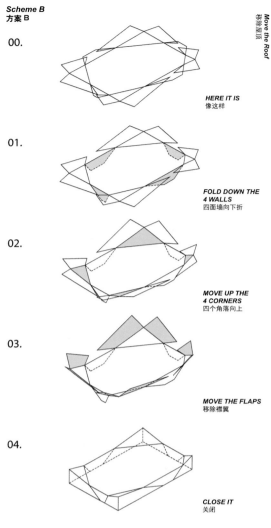

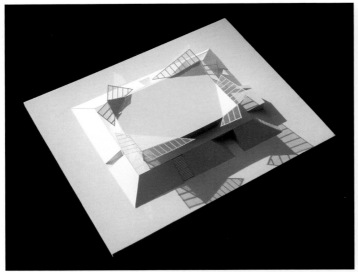
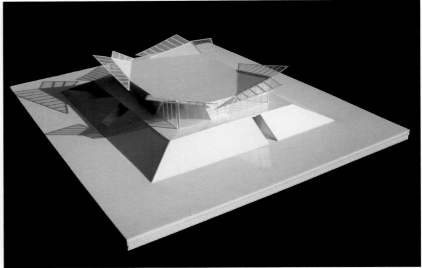

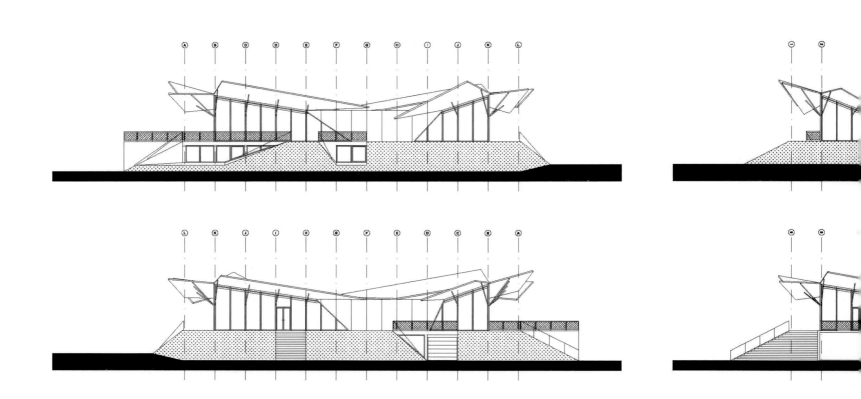

162　面向未来——建筑趋势2015　FACING TO THE FUTURE–ARCHITECTURE TREND 2015

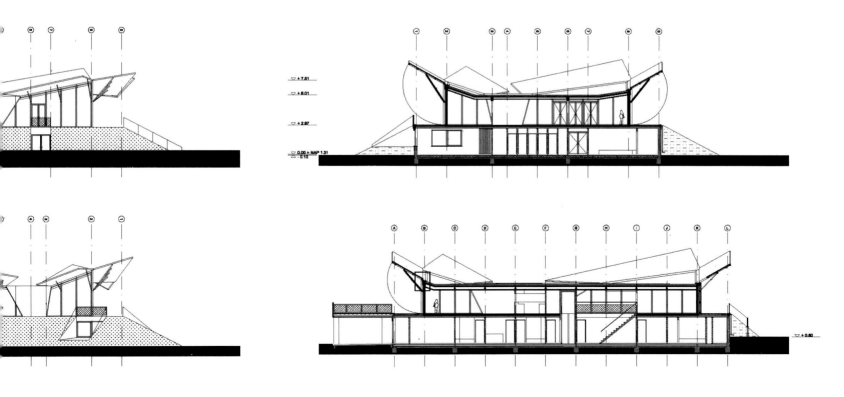

# 韩国京畿道"岛轩"
## Island House

设计单位：IROJE KHM Architects
项目地址：韩国京畿道
占地面积：872.63m²
总建筑面积：628.02m²
设计团队：SuMi Jung　JungMin Oh　Arum Kim　SunHee Kim
摄影：JungSik Moon

*Designed by: IROJE KHM Architects*
*Location: Gyeounggi-do, Korea*
*Site Area: 872.63m²*
*Gross Floor Area: 628.02m²*
*Design Team: SuMi Jung, JungMin Oh, ARum Kim, SunHee Kim*
*Photography: JungSik Moon*

**项目概况**

该项目场地漂浮于河畔，与周边优美的自然景观遥相对望，设计旨在建造一座具有"建筑化自然"气质的休憩场所，使建筑真正成为大自然的一部分。

**建筑设计**

在不规则线条的场地上，这栋造型别致的混凝土建筑最大限度地提高了土地利用效率。整座建筑宛如一座由不规则的多边形混凝土体块和金属网格构成的连绵山脉，和谐地融入"建筑化山脉"的背景之中，使建筑成为山水的一部分。

作为"建筑化自然"的建筑，漂浮的白色多边形盒状结构拥有内嵌式的竹园，形成建筑不同层级的纵向空间，所有的房间都自然地融入如画的景观当中，可欣赏场地周边的优美景色。倾斜的天花板线条、室内跃层式的设计，营造出房屋动感丰富、梦幻绝美的居住空间。

阶梯形屋顶的整体部分层次分明，直通顶层的卧室，不断循环的阶梯形屋顶花园与内庭两侧都连接起来，形成场地内新的连绵不断的绿地景观，与充满水景、花草和水果的内庭完美呼应，浑然天成，宛如一座迷人的梦中花园。

*Plot Plan*
地块平面图

### Profile
The site is floating on river, confronting the graceful natural landscape. The house is a recreation place of "the architectural nature", becoming a part of nature.

### Architectural Design
On the irregular site, the unique concrete building maximizes efficiency of land use. The building is like a continuous mountain composed by irregular polygonal shaped concrete mass and metal mesh, which harmonizes with the context as "the architectural mountain". There is the intention to be a part of the surrounding context of river and mountain.

As "the architectural nature", the floating white polyhedral masses that have the built-in bamboo gardens produce various stories of vertical space. All rooms inside are laid toward the picturesque landscape to enjoy the graceful surrounding scenery. The sloped ceiling line, skip-floor inner space create dynamic and unrealistic living environment.

The whole part of the stepped roof moving upward with various levels is directly linked to the bed rooms in upper floor. The stepped roof gardens are linked to both sides of the inner court filled with water, flower and fruit. The whole roof gardens are circulated as the continuous landscape place and that is the place as "architectural nature" in concept.

**Basement Floor**
地下室层

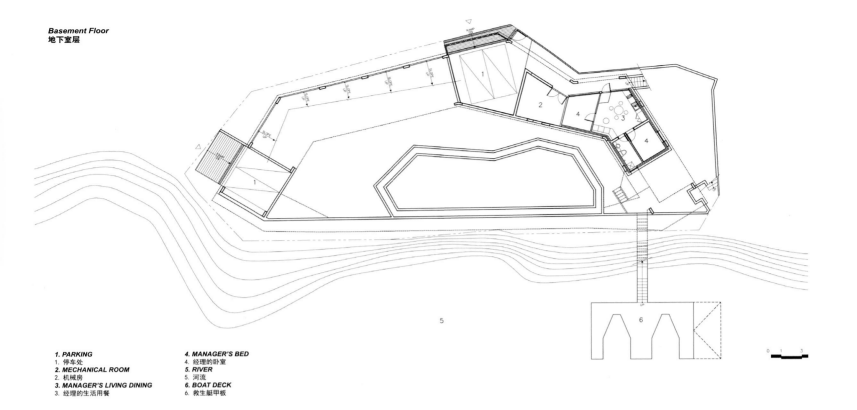

1. **PARKING**
   1. 停车处
2. **MECHANICAL ROOM**
   2. 机械房
3. **MANAGER'S LIVING DINING**
   3. 经理的生活用餐
4. **MANAGER'S BED**
   4. 经理的卧室
5. **RIVER**
   5. 河流
6. **BOAT DECK**
   6. 救生艇甲板

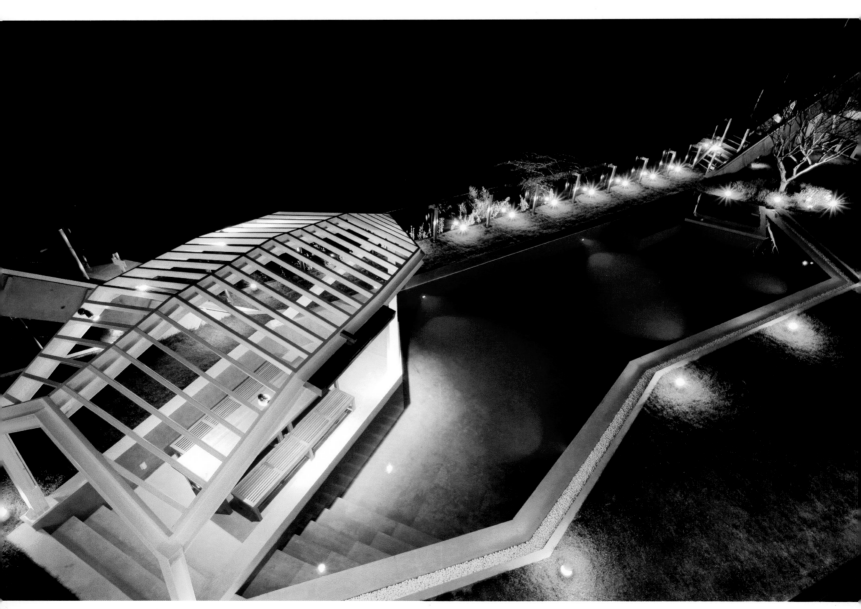

**First Floor**
一层

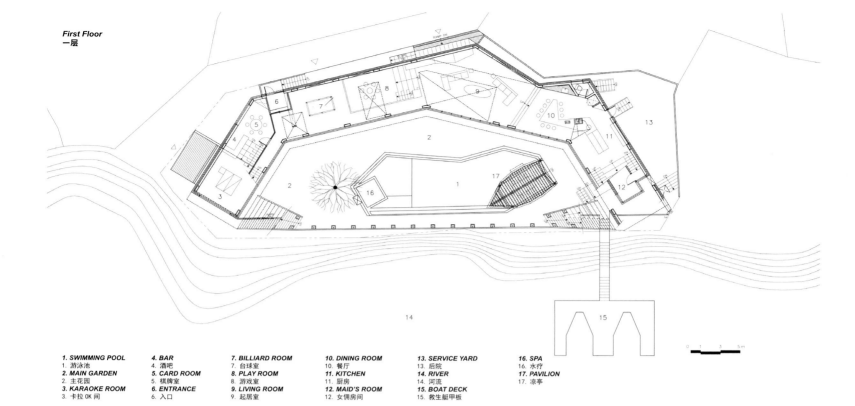

1. **SWIMMING POOL**
   1. 游泳池
2. **MAIN GARDEN**
   2. 主花园
3. **KARAOKE ROOM**
   3. 卡拉OK间
4. **BAR**
   4. 酒吧
5. **CARD ROOM**
   5. 棋牌室
6. **ENTRANCE**
   6. 入口
7. **BILLIARD ROOM**
   7. 台球室
8. **PLAY ROOM**
   8. 游戏室
9. **LIVING ROOM**
   9. 起居室
10. **DINING ROOM**
    10. 餐厅
11. **KITCHEN**
    11. 厨房
12. **MAID'S ROOM**
    12. 女佣房间
13. **SERVICE YARD**
    13. 后院
14. **RIVER**
    14. 河流
15. **BOAT DECK**
    15. 救生艇甲板
16. **SPA**
    16. 水疗
17. **PAVILION**
    17. 凉亭

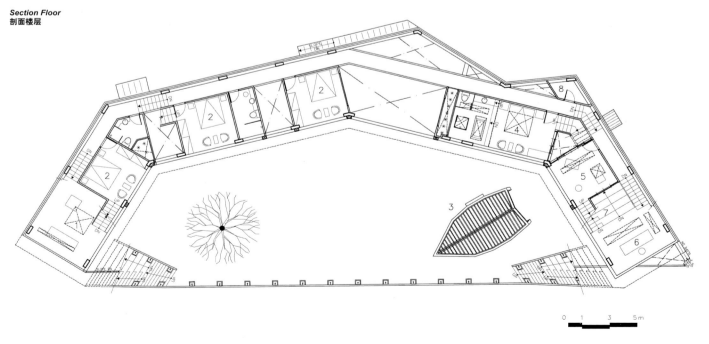

**Section Floor**
剖面楼层

| 1. FITNESS ROOM | 3. ROOF OF PAVILION | 5. CHILD BED ROOM | 7. STAND STUDY AREA |
| 1. 健身房 | 3. 凉亭屋顶 | 5. 儿童卧室 | 7. 自习站立区 |
| 2. GUEST BED ROOM | 4. MASTER BED ROOM | 6. STUDY ROOM | 8. BOILER ROOM |
| 2. 客房 | 4. 主卧 | 6. 自习室 | 8. 开水房 |

**Section 1**
剖面 1

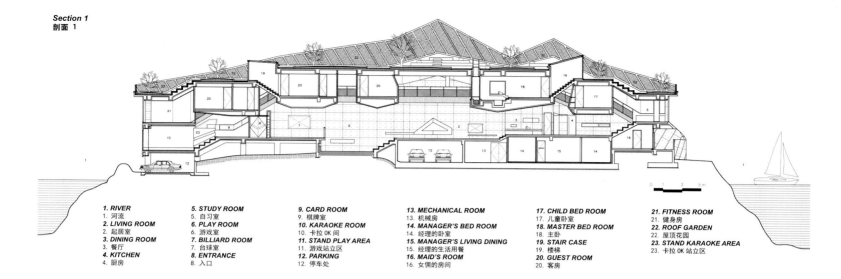

| | | | | | |
|---|---|---|---|---|---|
| **1.** RIVER<br>1. 河流 | **5.** STUDY ROOM<br>5. 自习室 | **9.** CARD ROOM<br>9. 棋牌室 | **13.** MECHANICAL ROOM<br>13. 机械房 | **17.** CHILD BED ROOM<br>17. 儿童卧室 | **21.** FITNESS ROOM<br>21. 健身房 |
| **2.** LIVING ROOM<br>2. 起居室 | **6.** PLAY ROOM<br>6. 游戏室 | **10.** KARAOKE ROOM<br>10. 卡拉OK间 | **14.** MANAGER'S BED ROOM<br>14. 经理的卧室 | **18.** MASTER BED ROOM<br>18. 主卧 | **22.** ROOF GARDEN<br>22. 屋顶花园 |
| **3.** DINING ROOM<br>3. 餐厅 | **7.** BILLIARD ROOM<br>7. 台球室 | **11.** STAND PLAY AREA<br>11. 游戏站立区 | **15.** MANAGER'S LIVING DINING<br>15. 经理的生活用餐 | **19.** STAIR CASE<br>19. 楼梯 | **23.** STAND KARAOKE AREA<br>23. 卡拉OK站立区 |
| **4.** KITCHEN<br>4. 厨房 | **8.** ENTRANCE<br>8. 入口 | **12.** PARKING<br>12. 停车处 | **16.** MAID'S ROOM<br>16. 女佣的房间 | **20.** GUEST ROOM<br>20. 客房 | |

**Section 2**
剖面 2

**Section–3**
剖面 3

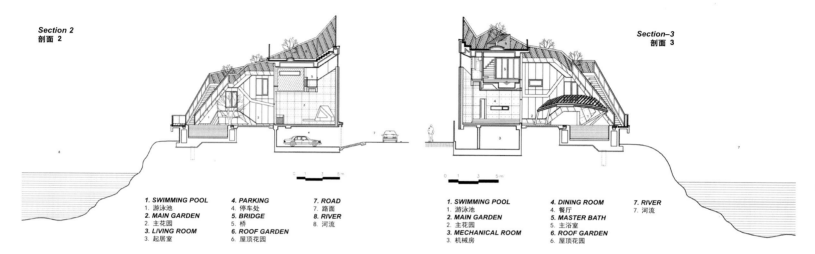

| | | |
|---|---|---|
| **1.** SWIMMING POOL<br>1. 游泳池 | **4.** PARKING<br>4. 停车处 | **7.** ROAD<br>7. 路面 |
| **2.** MAIN GARDEN<br>2. 主花园 | **5.** BRIDGE<br>5. 桥 | **8.** RIVER<br>8. 河流 |
| **3.** LIVING ROOM<br>3. 起居室 | **6.** ROOF GARDEN<br>6. 屋顶花园 | |

| | | |
|---|---|---|
| **1.** SWIMMING POOL<br>1. 游泳池 | **4.** DINING ROOM<br>4. 餐厅 | **7.** RIVER<br>7. 河流 |
| **2.** MAIN GARDEN<br>2. 主花园 | **5.** MASTER BATH<br>5. 主浴室 | |
| **3.** MECHANICAL ROOM<br>3. 机械房 | **6.** ROOF GARDEN<br>6. 屋顶花园 | |

**Roof Floor**
屋顶层

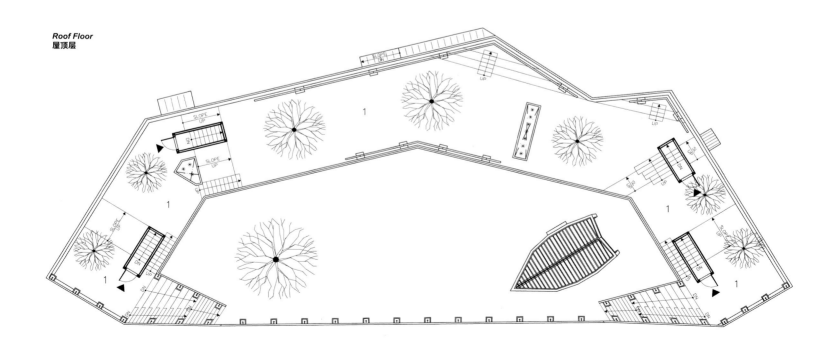

**1.** ROOF GARDEN
屋顶花园

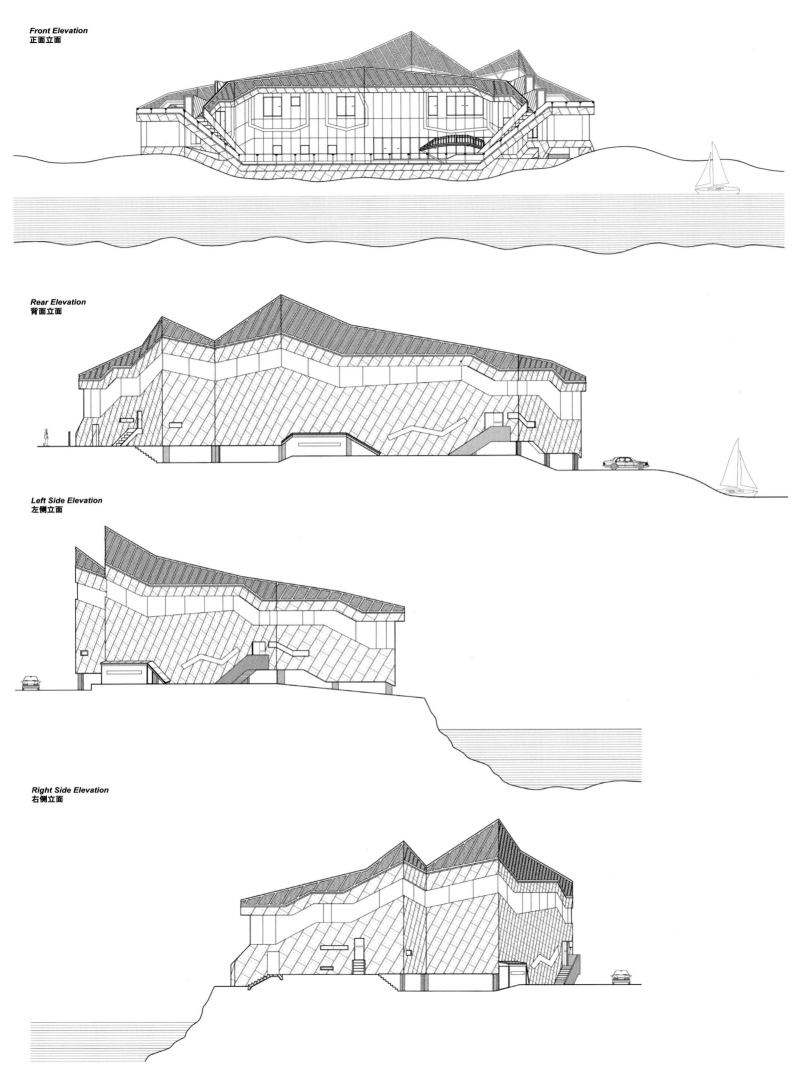

*Front Elevation*
正面立面

*Rear Elevation*
背面立面

*Left Side Elevation*
左侧立面

*Right Side Elevation*
右侧立面

# 湖北十堰大美盛城会所营销中心
## Marketing Center of Dameisheng City Club

设计单位：加拿大 CPC 建筑设计顾问有限公司
开发商：上海华岳投资有限公司
建筑面积：2 741m²

**Designed by: The C.P.C. Group**
**Client: Shanghai Huayue Investment Co., Ltd.**
**Building Area: 2,741m²**

**项目概况**

该营销中心位于十堰市上海路，是坐拥城市黄金区位中心，一揽城市繁华的大美盛城的销售中心，也是整个住宅项目的对外首个展示中心。

**建筑设计**

整个建筑如同从地面自然生长出来，从两端缓缓升起，到顶部高耸成为山体，以起伏的天际轮廓线呼应远近高低变化的山体。这一介于人工和自然之间的形态，其设计灵感来源于山地城市的自然地貌，与自然和谐共生的同时，又与城市紧密相连。

建筑外观造型呈大写字母"M"形，既代表了美盛置业，又呼应了大美盛城"城有大美，于斯为盛"的文化内涵。立面采用铝板和玻璃相结合，在阳光的照耀下，明亮璀璨，体现其现代的时尚感。除此之外，屋面还别具一格地采用绿植装饰，随着建筑独特形态与地面草皮连为一体，使其看起来就像是地面的延伸。

整个营销中心的结构和外观是一个整体，流畅的、完整的、有机的形态使人无法分辨哪里是墙，哪里是屋面，就像城市中一个有趣的景观，而非只是一座建筑。

**Profile**

This bustling marketing center is located on Shanghai Road of Shiyan City, which is the first external exhibition center of the whole residence project.

**Architectural Design**

The whole building gradually rises from two ends, as if growing naturally from the ground. Up to the top, the building responds to the mountains far and near with its rolling contours. This form, between the artificial and the natural, takes inspiration from the natural terrain of the mountainous city. It harmonizes with nature and also connects with city.

The building takes the shape of the capital letter "M", which represents Meisheng Property Company, and responds to the cultural connotation of Dameisheng City "there are lots of beautiful sceneries in the city, especially in this place." Its façade is clad with green plants, continuing the turf on the ground, which looks like the extension of the ground.

The structure and shape of the marketing center is a whole. Its smooth, integral and organic form blurs the distinction of wall and roofing. It is like an interesting landscape in the city, not architecture.

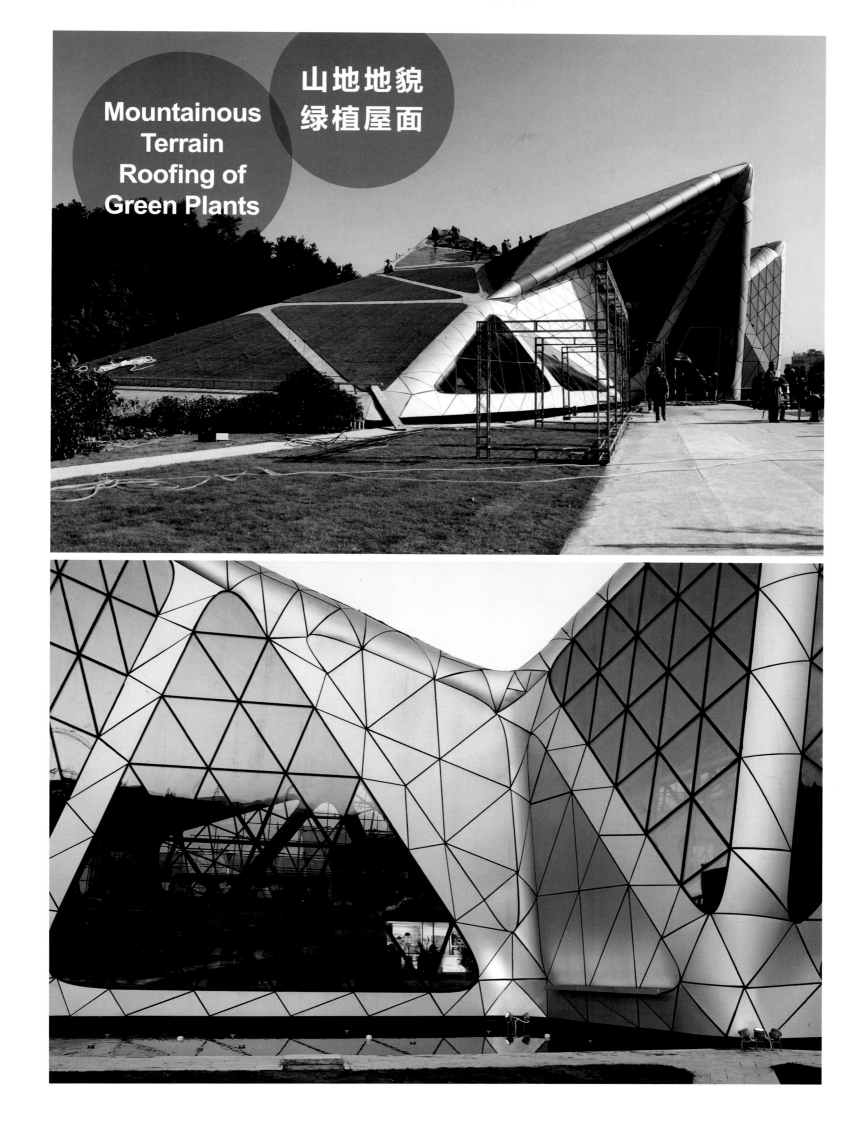

Mountainous Terrain Roofing of Green Plants

山地地貌 绿植屋面

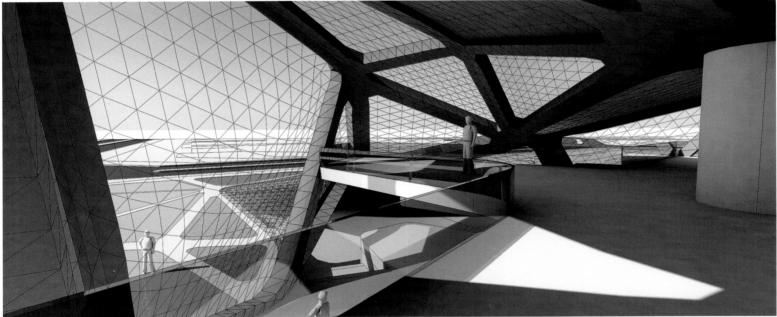

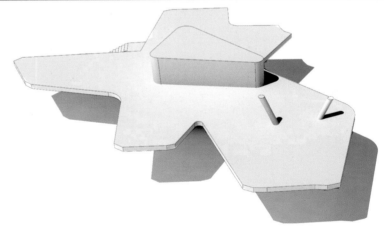

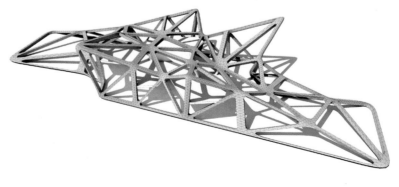

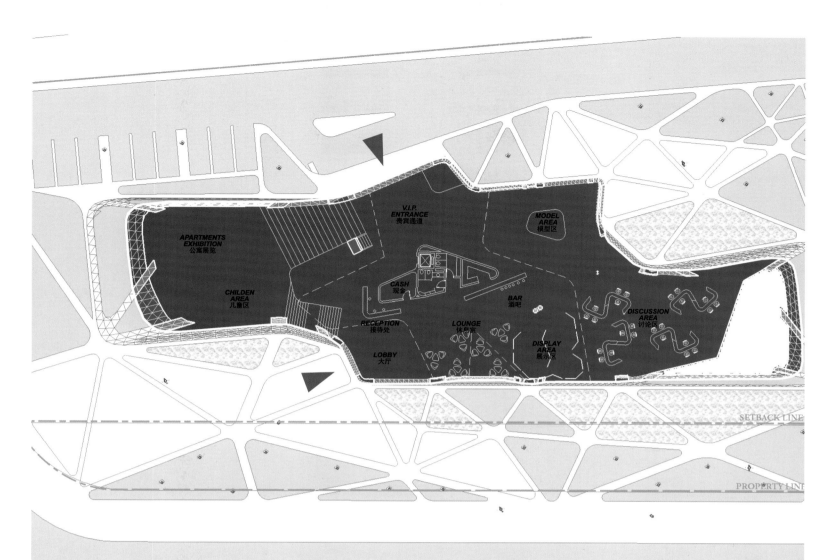

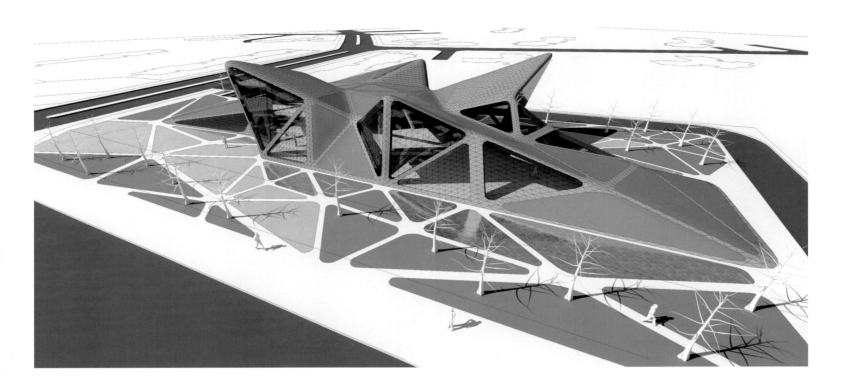
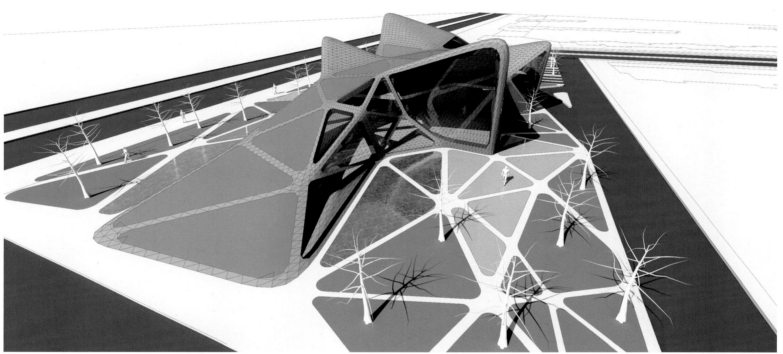
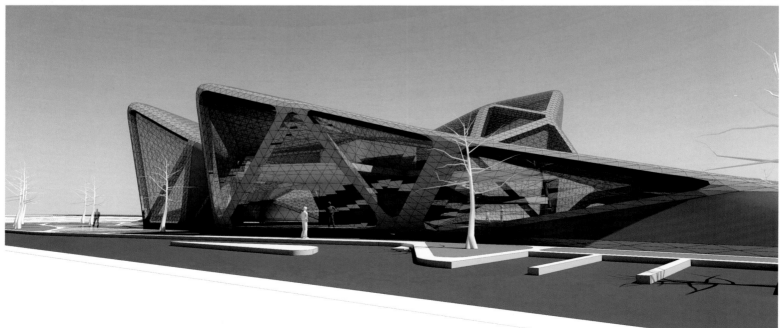

# 辽宁葫芦岛比基尼广场
## Huludao Bikini Plaza

设计单位：北京维拓时代建筑设计有限公司
总建筑面积：3 631.4m²
绿地率：38%
容积率：0.29

Designed by: Beijing Victory Star Architectural & Civil Engineering Design Co., Ltd.
Gloss Floor Area: 3,631.4m²
Greening Rate: 38%
Plot Ratio: 0.29

Casing Construction
Cable Structure

覆土建筑
悬索结构

## 项目概况

兴城比基尼广场位于辽宁省兴城市东北方，是重要的泳衣制造基地，广场建成后每年夏天可举行露天的比基尼大赛及泳装展会。此地地势北高南低，南侧为海滩，南北高差为19m。广场位于海边的宽阔地带，整体建筑造型流畅优雅，犹如一个侧卧着的美女。

## 建筑设计

建筑总体布局结合地形高差，创造大量的覆土建筑。商业上部均有1.2m以上覆土，可种植各种树木，海滨服务则被广场所覆盖，地下车库为开敞景观车库，树木可从地下层透过庭院开口生长到地面层上，中心演艺广场被三处观演看台所围合，利用看台下空间布置演出服务等功能。挑棚为悬索拱结构，外覆的红色膜结构可为看台提供遮阳并丰富海岸天际线。展览中心可提供七千人的平面无柱大空间展览场所。建筑单体分为八个部分：展览中心、1#商业、2#商业、1#海滨服务、2#海滨服务、演出服务、车库和挑棚。

## Project Overview

Bikini Plaza is located in the Northeast of Xingcheng, Liaoning Province, and is an important swimsuit manufacturing base; after the completion of the square, every summer can be held outdoor bikini and swimsuit exhibition race of people. The terrain here is high in north and low in south, and the south side is the beach, with north and south elevation of 19 meters. Plaza is located in the broad areas of the sea, and the overall architectural style is smooth and elegant, like a side of the beauty.

## Building Design

The overall layout of the building combined with terrain elevation, creating a large amount of overburden building. Business has 1.2 meters above the upper casing, planting a variety of trees; ocean services were covered by the plaza; underground garage is open landscape; trees grow from the underground floor through the courtyard opening to the ground floor; Performing Arts Center Plaza is enclosed by the three play stands, using the spatial arrangement under the grandstand to show services and other functions. Canopy is for the suspension arch structure covered with red membrane structures, which can provide shade for the bleachers and rich coastal skyline. Exhibition Center offers seven thousand large column-free space plane exhibition spaces. Single building is divided into eight sections: Exhibition Center, 1# and 2#Business, 1# and 2#Seaside Services, Performance Services, Garage and Canopy Roof.

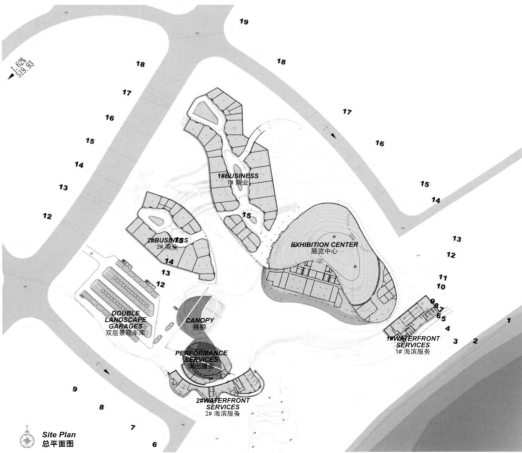

Site Plan 总平面图

# 世界休闲体育大会体育场馆工程
## World Conference Leisure Sports Stadium Engineering

设计单位：广州市景森工程设计顾问有限公司
开发商：莱西市城乡建设局
项目地址：山东省青岛市莱西市扬州路以西、北京路以北（莱西市体育中心院内）
占地面积：133 141m²
总建筑面积：46 356 m²

**Designed by:** Guangzhou Jingsen Engineering Design & Consulting Co., Ltd.
**Client:** Laixi Urban and Rural Construction Bureau
**Location:** West of Yangzhou Road and North of Beijing Road (inside Laixi City Sports Center), Qingdao Laixi City, Shandong Province,
**Site Area:** 133,141m²
**Gross Floor Area:** 46,356m²

### 设计理念

2010年,莱西"葫芦"在中国非物质文化遗产的名单上首次亮相时就引起了极大的关注度,而世界休闲体育大会体育场馆的设计理念也是来源于此。设计师们认为用莱西"葫芦"来表现他们的设计语言和理念是再合适不过的了。建筑借外墙镂空的形式来反映葫芦雕刻的语言,靠近地面的底部开口较多,为增加采光面积,增加了采光面积,并且避免了建筑顶部受到过多的阳光直射。

### 建筑设计

设计师们考虑体育场和体育馆内部座位的数量,同时平衡两个建筑行人的流量,在对于中心广场的处理上也加入了一些有机形态的建筑语言。通过弯曲的流线,将人群缓冲,然后由两个广场轴线将人群带入建筑。同时将外立面设计的理念遵从无背面的理念。包括建筑的屋顶都带有与外墙呼应的设计图案和语言。为了达到观众席上座率高且舒适的目的,体育场两侧的看台都呈直线摆放,同样在体育馆座位设计上也采用同样的理念。两个场地的C值都可以达到90以上,可保证场地有足够的标准承办任何国际级的赛事。该体育馆的外形设计主要以简洁、大方、高效为原则,同时兼顾使用者和观众的安全性和舒适性。一方面,为了平衡两个建筑之间的人行流量,设计师在对中心广场的处理上加入了一些有机形态的建筑语言——通过弯曲的流线,先将人群缓冲,然后借助两个广场轴线将人群带入建筑体内。另一方面,为了达到观众席上座率高且舒适的目的,体育场两侧的看台都是呈直线摆放。 除此之外,考虑到大会结束以后的后续活动,设计师还增加了一个小尺度的活动广场,里面包含了餐厅和咖啡厅等,进一步加强了该建筑的互动性和灵活性。

## Design Concept

In 2010, Laixi gourd in China Intangible Cultural Heritage List on the first time appeared to cause a great deal of attention, while the World Conference Leisure Sports Stadium design is derived from this. Designers believe that with Laixi gourd to express their design language and philosophy is a perfect good. Building façades with hollow form reflect the language of gourd carving, and opening near the bottom of the ground is more to increase the lighting area and keep top from too much direct sunlight exposure.

## Building Design

Designers consider the number of inside seats of the stadium and the gymnasium, while balancing pedestrian flow between two buildings; in the center of the square for the treatment also joined the organic form of architectural language. By the bending flow line, the crowd is buffered and then by two square axes will crowd into the building. While the concept of the façade design follows the concept of no backside. What's more, construction roofs and exterior walls echoed with design patterns and language. In order to achieve the aims of an efficient and comfortable auditorium seats, the two sides of bleachers of the stadium were are put in a linear display, the same seats in the stadium design is also used the same concept. The C value of the two sites can reach more than 90, which can ensure that the sites have enough standard to host any international competitions. The exterior design of the stadium mainly takes the principles of simplicity, elegance and efficiency, while takes into account of the safety and comfort of the users and audiences. On the one hand, in order to balance the flows of pedestrian and traffic between the two buildings, the designers have added some organic forms of architectural language on the processing of the central square—through the curved flow, it makes the crowd flow buffer, then with the two square axis to make the crowd into the building. On the other hand, in order to achieve the purpose of a high and comfortable attendance auditorium, the stands on the sides of the stadium are placed with straight. Besides, the designers also added a small-scale event plaza taking into account the follow-up activities after the end of the general assembly, which includes restaurants and cafés to further strengthen the interactivity and flexibility of the architecture.

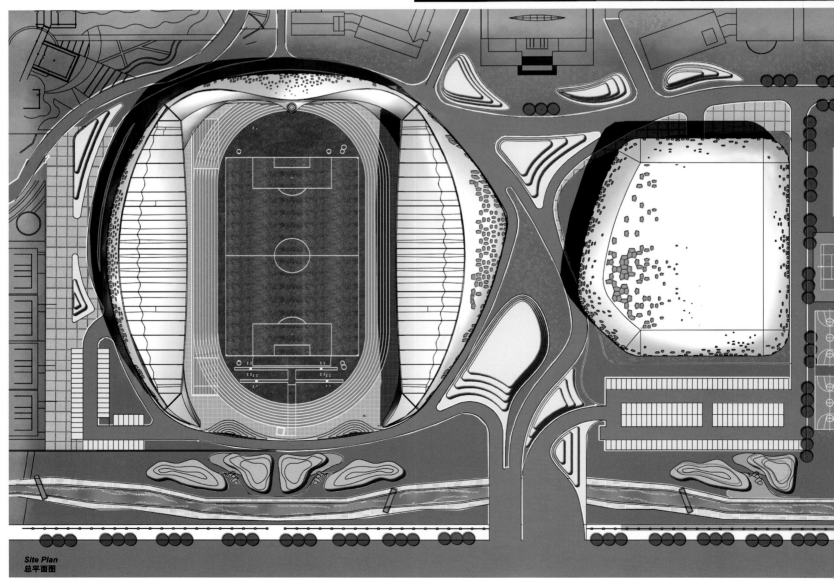

**Site Plan**
总平面图

## Stadium Ground Floor Plan
体育场底层平面图

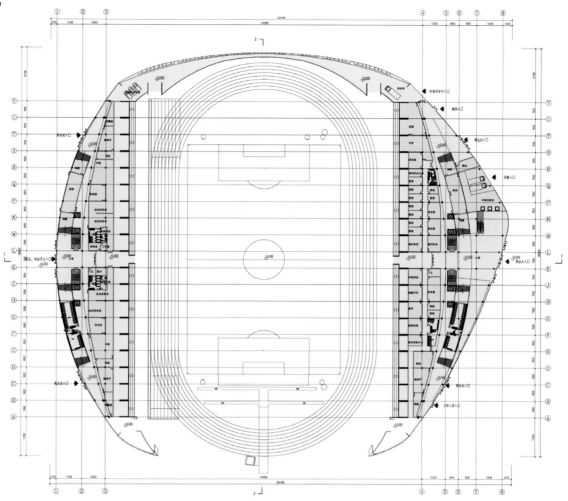

## Stadium Secon Floor Plan
体育场二层平面图

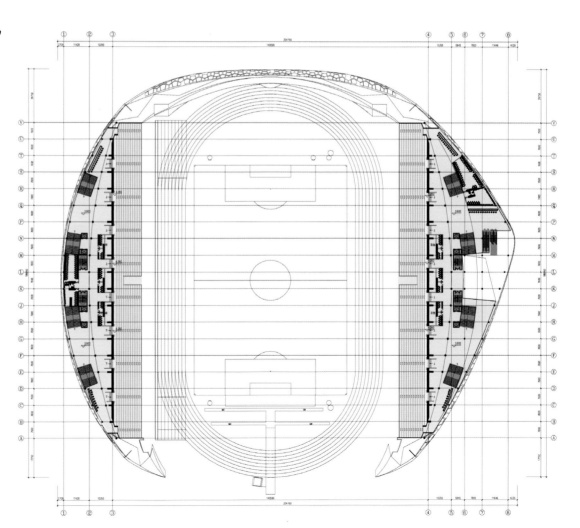

*Arean Ground Floor Plan*
体育场首层平面图

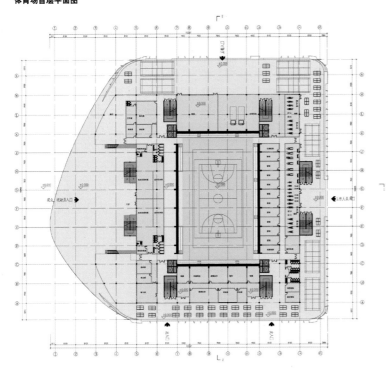

*Arean Secon Floor Plan*
体育场二层平面图

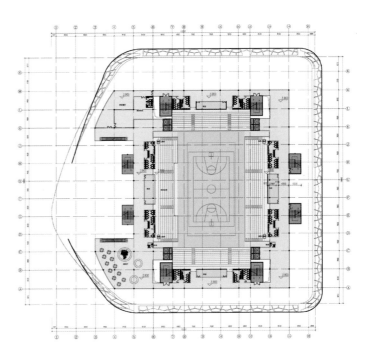

**Build Up Analysis**
结构分析

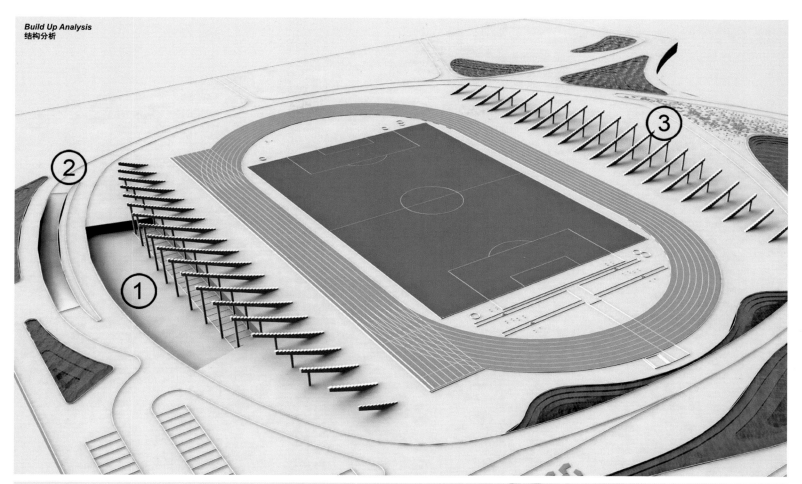

| | | |
|---|---|---|
| **1. VIP CARPARK**<br>1. 贵宾停车场 | **5. STANDS**<br>5. 观众席 | **9. CARVED SKIN**<br>9. 外墙 |
| **2. VIP CARPARK RAMP**<br>2. 贵宾停车场坡道 | **6. MAIN CONCOURSE**<br>6. 主门厅 | **10. LED SCREEN**<br>10. LED 显示屏 |
| **3. RAKERS**<br>3. 倾斜支撑梁 | **7. VIP CONCOURSE**<br>7. 贵宾门厅 | **11. GAMES FLAME**<br>11. 火炬 |
| **4. STAIRS TO MAIN CONCOURSE**<br>4. 主门厅楼梯 | **8. PUBLIC RESTAURANT**<br>8. 餐厅 | **12. ROOF**<br>12. 遮雨棚 |

*Stadium Functionality Analysis*
体育场功能分析

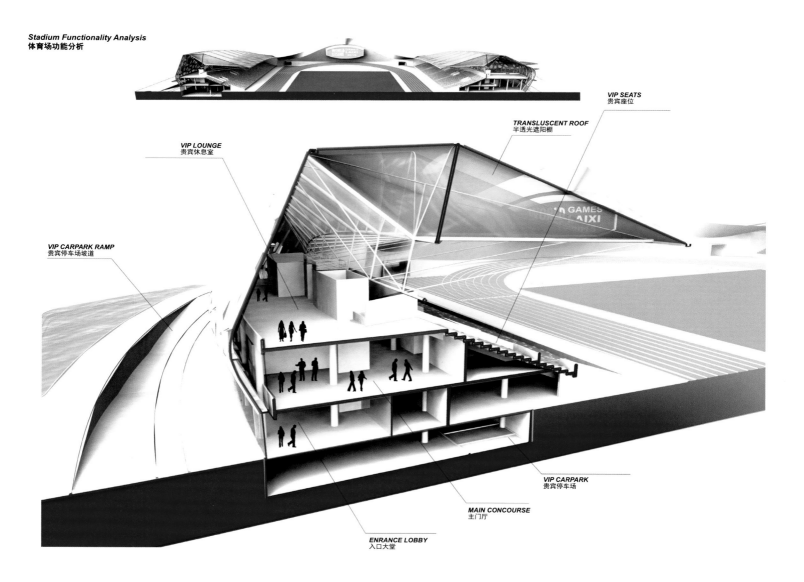

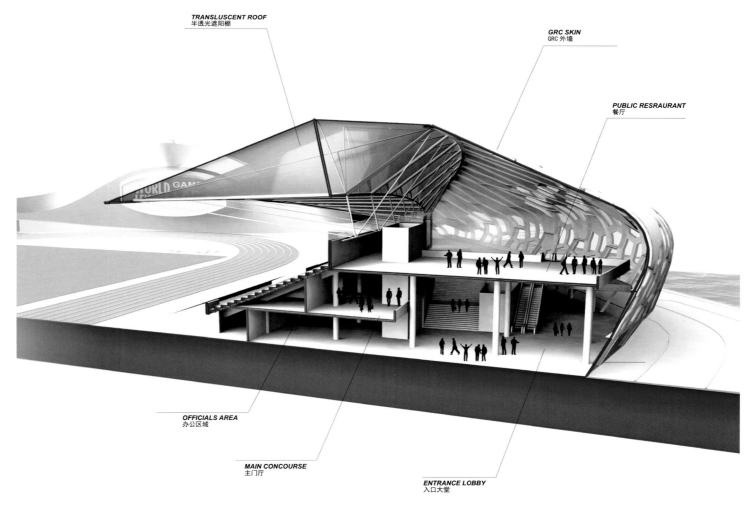

*Section 1-1*
剖面图 1—1

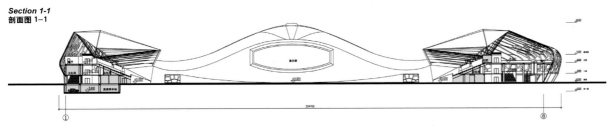
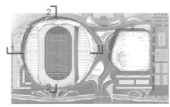

*Section 2-2*
剖面图 2—2

*South Elevation*
南立面

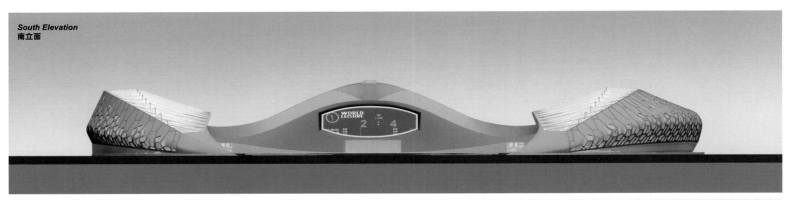

*East Elevation*
东立面

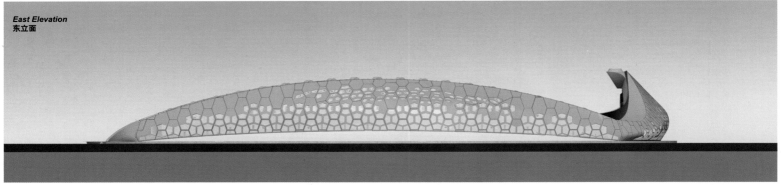

*North Elevation*
北立面

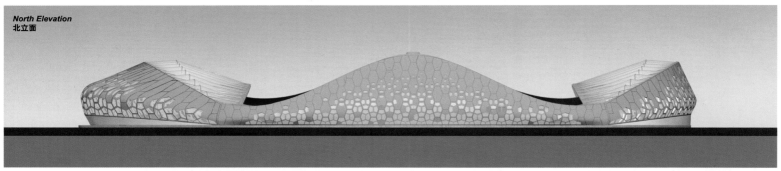

*West Elevation*
西立面

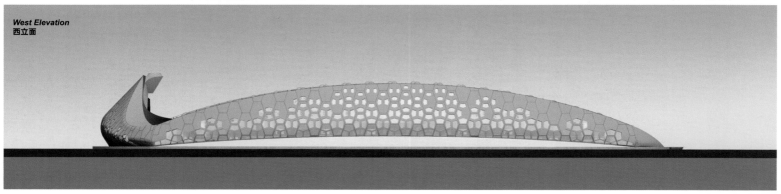

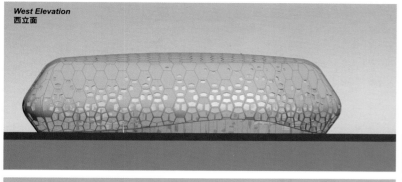
**West Elevation**
西立面

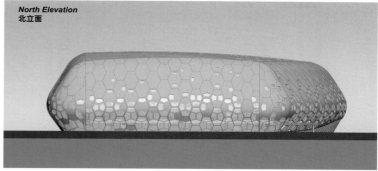
**North Elevation**
北立面

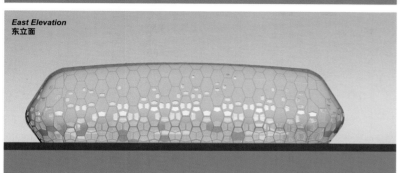
**East Elevation**
东立面

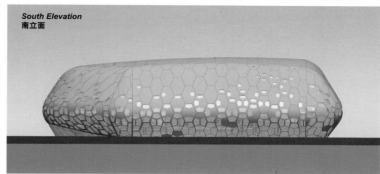
**South Elevation**
南立面

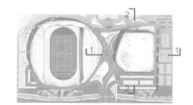

**Section 1-1**
剖面图 1—1

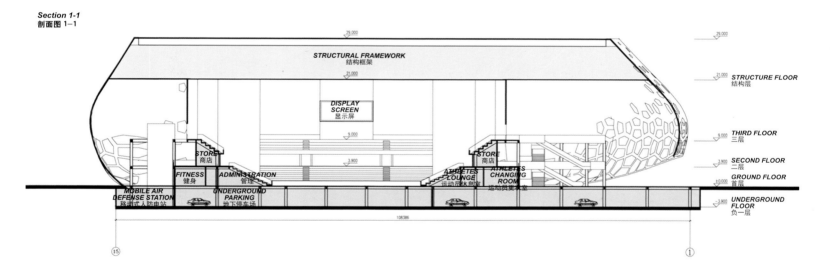

**Section 2-2**
剖面图 2—2

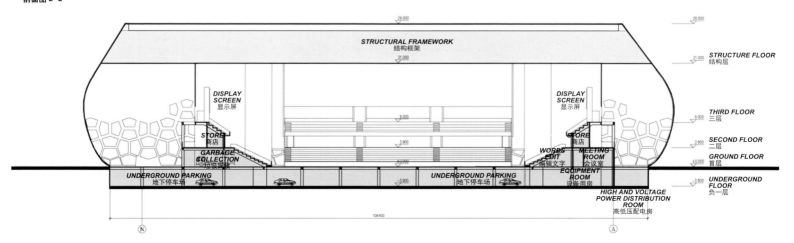

# 荷兰特克塞尔博物馆
## Netherlands, Texel Museum

设计单位：梅卡诺建筑事务所
开发商：斯希尔德 Maritiem & Jutters 博物馆
项目地址：荷兰特克塞尔
建筑面积：1 200m²
摄影：Christian Richters and Mecanoo

Designed by: Mecanoo architecten
Client: Maritiem & Jutters Museum, Oudeschild
Location: Texel, The Netherlands
Building Area: 1,200m²
Photography: Christian Richters and Mecanoo

**项目概况**

特克塞尔岛位于瓦登海，是最大的荷兰瓦登群岛，每年约有一万左右的游客来岛上参观，这里只有通过飞机、轮船或渡轮才可抵达。然而，很少人会熟悉特克塞尔及其与荷兰东印度公司有关的辉煌历史。"我们的生活由大海提供，也由大海带走"——这是特克塞尔人常念于口的话。几百年前，特克塞尔人就知道利用搁浅船舶上的浮木建造自己的房子和谷仓，因此，木质建筑是当地最具特色的传统建筑，如何在传统技艺上创造出新的特色，是该项目需要解决的首要问题。

**建筑设计**

该博物馆最具特色的设计就是那四个非常俏皮的、彼此联系的山墙屋顶和木质外观的建筑体块。透过木板前面的玻璃幕墙，可以看到博物馆户外诱人的风景，还有北荷兰省洁净的天空，以及咖啡馆悠闲的游客们。在建筑内部，透过木板间隙投射过来的日光和阴影的线性模式创造了一种注入光线和住所的氛围，给冰冷的建筑营造了一种神秘的、温馨的空间。

建筑分为上、下两个空间，上层的是博物馆咖啡厅，下层是博物馆展示品，也就是特克塞尔 Reede 面包车，18m 长、4m 深的巨大模型，该模型将在最大限度上展现荷兰的黄金时代，还原当时的盛况。上、下两层通过光线的对比和不同的空间体验，形成两个完全不同的世界。游客在地下室可以通过人造光线、绘制投影和动画效果，如临其境地感受当年数十艘船舶停泊在瓦登岛的海岸的浩大场面。上层坚固的钢框架和可移动的玻璃陈列柜，通过明亮的自然光线的照射，创建了一种透明的效果，给人一种明亮、畅快的舒适感。

## Project Overview

The island of Texel is situated in the Waddenzee and is the largest of the Dutch Wadden Islands. Every year a million or so tourists visit the island, which is only accessible by plane, boat or ferry. Few however will be familiar with the glorious history of Texel and its links with the Dutch East India Company. 'The sea takes away and the sea provides'—this is a saying that the people of Texel know so well. For hundreds of years they have made grateful use of driftwood from stranded ships or wrecks to build their houses and barns. Therefore, the wooden architecture is the most distinctive traditional architecture in the local, and how to create a new feature in the traditional skills of the project is the most important issue to be resolved.

山墙屋顶
玻璃幕墙
人造光线
自然光线

Gable Roof
Glass Curtain Wall
Artificial Light
Natural Light

**Building Design**

The museum is designed with four playfully linked gabled roofs which are a play on the rhythm of the surrounding roof tops which seen from the sea, resemble waves rising out above the dyke. From within, the glass façade in front of the wooden boards allows an inviting view of the outdoor museum terrain and of the famous North Holland skies to visitors of the museum café. Inside the building the boards cast a linear pattern of daylight and shadow creating an atmosphere infused with light and shelter.

The building is divided into two spaces, which the upper is museum café and the lower is the museum exhibits, that is an eighteen-metre long, four-metre deep model of the Reede van Texel, displaying in great detail the impressive spectacle of the dozens of ships anchored off the coast of the Wadden Island. By comparing the upper and lower levels of light and different spatial experience, it forms two completely different worlds. Tourists in the basement by artificial light, rendering and animation projection, immersive experience its position like the vast scenes that dozens of vessels moored in the year Wadden coast of the island. Upper sturdy steel frame and removable glass showcase, through irradiation bright natural light, create a transparent effect to give a bright and fast comfort.

*Ground floor*
底层

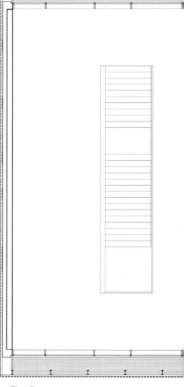

*First floor*
一层

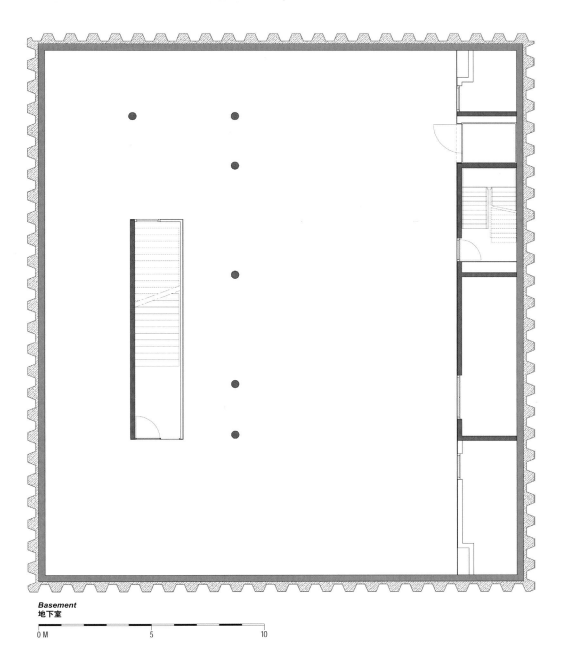

**Basement**
地下室

0 M　　　　5　　　　10

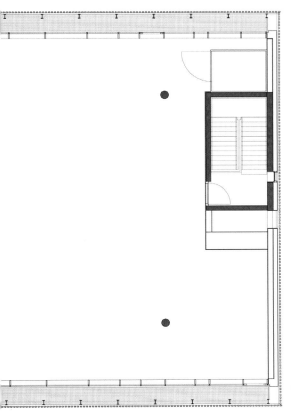

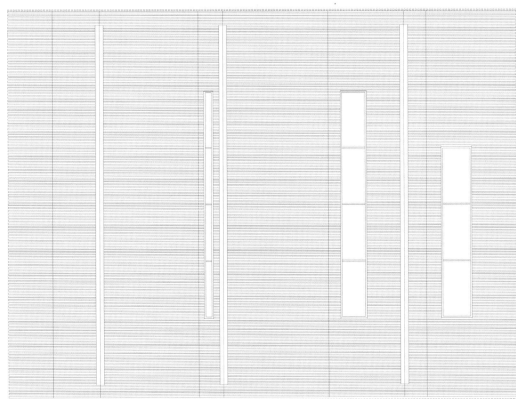

**Roof**
屋顶

**Section**
剖面

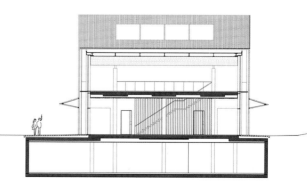

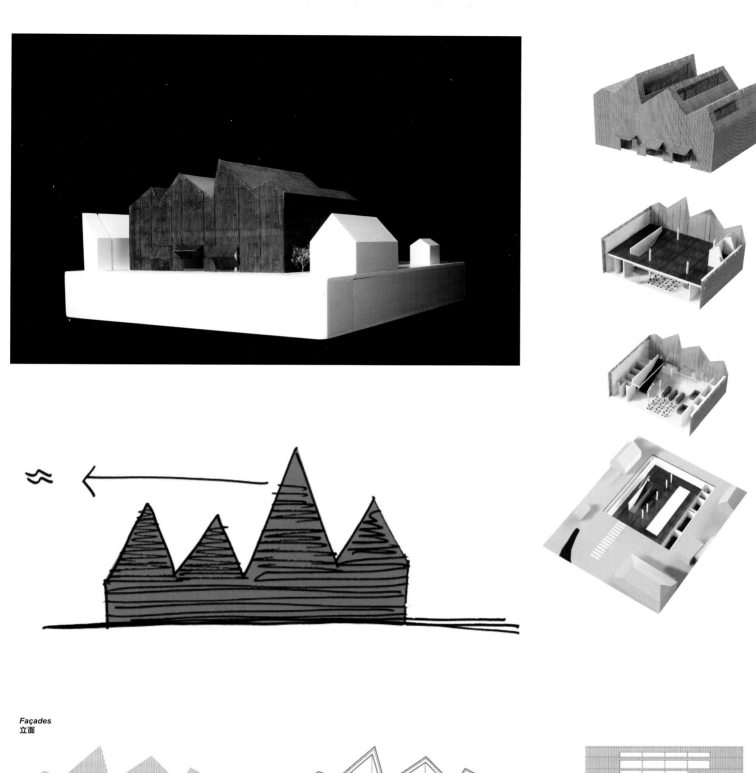

**Façades**
立面

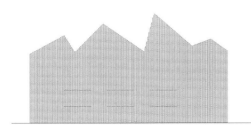

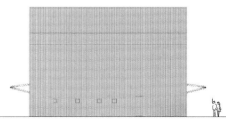

# V Lesu 销售办事处
## Sales Office "V Lesu"

项目地址：俄罗斯莫斯科
建筑面积：1 400m²
摄像：Alexey Naroditsky
设计团队：Anton Mossine　Vera Kazachenkova
　　　　　Olesya Sokolova　Stas Kirichenko

*Location: Russia, Moscow*
*Building Area: 1,400m²*
*Photography: Alexey Naroditsky*
*Design Team: Anton Mossine, Vera Kazachenkova*
*Olesya Sokolova, Stas Kirichenko*

**建筑设计**

该建筑的设计理念是构建一个简单的"快乐与性感的办公室"，建筑整体结构简单而清晰，给人一种明快、爽朗的感觉。建筑"外膜"是聚氯乙烯产品，外形呈圆顶起伏的铆接气球状，特色鲜明，抓人眼球。建筑内部分为三个区域：最大的中部"泡沫"区设有一个1：1公寓模型，较小的两个区域分别是公司员工的办公区和为来访客户配置的会客室（配有咖啡厅）和儿童游戏室。

**Building Design**

The design concept of the building is given a brief to construct a "fun and sexy office", and the overall structure of the building is simple and clear, giving a crisp and bright feeling. The outer membrane of the building is made of PVC fabric, and the shape is presented the domes into an undulating series of riveted balloons, with the distinct feature to attract people's attention. The internal building is organized into three zones: the largest central "bubble" houses an exhibition area with 1:1 apartment models. Branching off this are two smaller 'bubbles' with a sales area for managers in one and a relax area, complete with café and playroom for children, in the other.

铆接气球状
功能分区

Riveted Balloons
Function Division

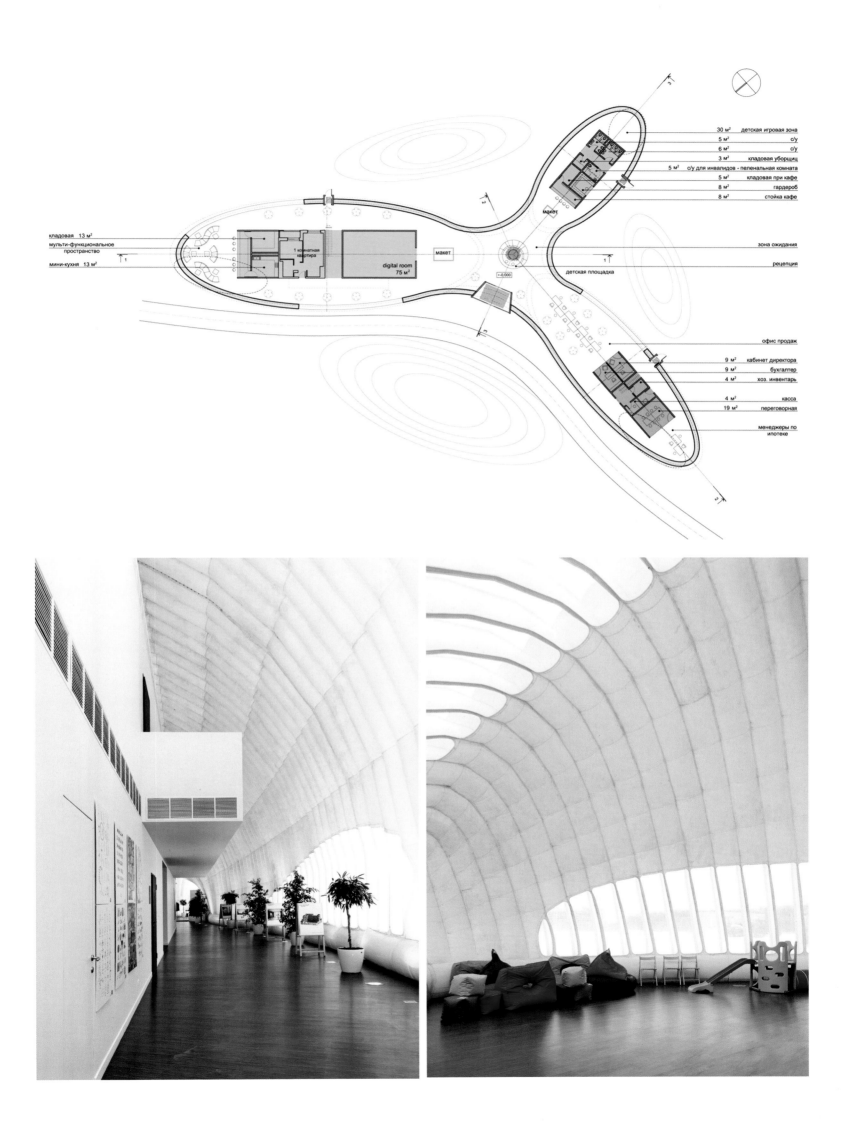

# 瑞典哈拉斯树上酒店
## Tree Hotel in Harads

| | |
|---|---|
| 设计单位：Tham & Videgard Arkitekter | *Designed by: Tham & Videgard Arkitekter* |
| 开发商：Brittas Pensionat  Britta Lindvall  Kent Lindvall | *Client: Brittas Pensionat, Britta Lindvall, Kent Lindvall* |
| 项目地址：瑞典哈拉斯 | *Location: Harads, Sweden* |
| 摄影：Åke E:son Lindman | *Photography: Åke E:son Lindman* |
| 设计团队：Andreas Helgesson  Julia Gudiel Urbano  Mia Nygren | *Project Team: Andreas Helgesson, Julia Gudiel Urbano, Mia Nygren* |

**项目概况**

树上酒店位于瑞典北部小镇哈拉斯,在北极圈以南60km,由芬兰建筑师萨米·林塔拉设计。整个项目由六座各异的树屋组成,树屋分别建在6棵松树上,面积53m²,重20t,离地面6m,是世界上最大的树屋,而"镜立方"是其中之一。

**建筑设计**

这是一间悬挂在树干上的轻型铝结构建筑,建筑整体形状为规格4m×4m×4m的正方盒状体,各面墙壁全是镜面玻璃,被形象地称为"镜立方"。

镜面玻璃围合而成的外墙,反射着周边和天空的景色,使建筑仿佛就像是隐身一般,完全交融在森林环境当中。建筑内部全由胶合板构成,透明的窗户给予酒店360°全景视角,坐在这个隐身的房间内,住客可以自由地观赏森林每个角落的风景。

另外,为了防止飞鸟撞上透明的立面墙,设计师特别采用一种只有鸟类才能看见的紫外线红色胶片,嵌在玻璃板之中,以此提醒鸟类避让。这座世界上独一无二的有趣酒店,势必会带给客人与天地亲近,与自然和乐共处的绝妙体验。

"隐身"
低碳
镜立方

"Invisible"
Low-carbon
Mirror Cube

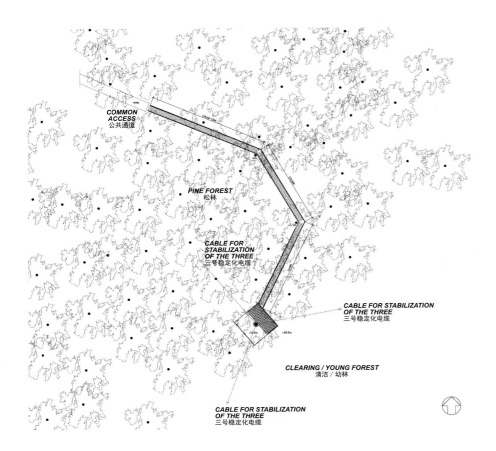

## Profile

The Tree Hotel is located in Harads, north of Sweden, about 60km south of the Arctic Circle. The whole project is composed of six different rooms built on six pine trees, with a total area of 53m$^2$ and a weight of 20t. The hotel is 6m far from the ground with the largest tree rooms in the world, which includes the "mirror cube".

## Architectural Design

This is a shelter up in the trees, a lightweight aluminium structure hung around a tree trunk, and a 4m×4m×4m box clad in mirrored glass, which is called "mirror cube".

The exterior mirrored glass reflects the surroundings and the sky, creating a camouflaged refuge, so the room should be invisible nearly in the forest. The interior is all made of plywood and the windows give a 360 degree view. Sitting in the invisible room, the guest can freely enjoy every corner of the forest.

To prevent birds colliding with the reflective glass, a transparent ultraviolet colour is laminated into the glass panes which are visible for birds only. The unique interesting hotel certainly provides perfect experience of closing to the nature for guests.

*Entrance Plan*
入口平面图

*Roof Plan*
屋顶平面图

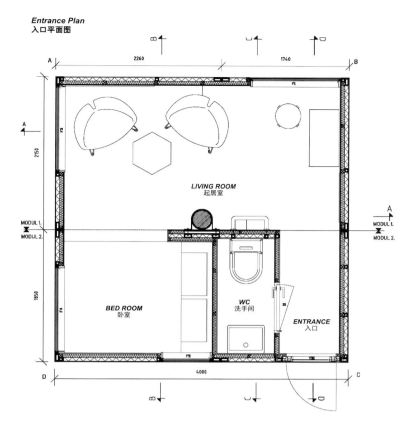
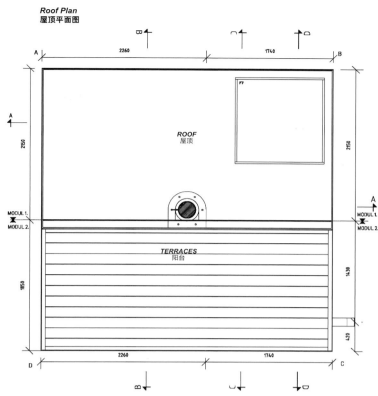

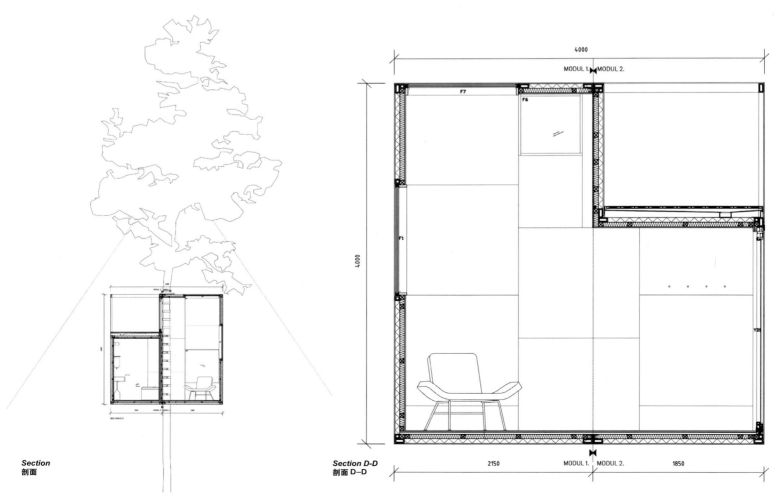

**Section**
剖面

**Section D-D**
剖面 D–D

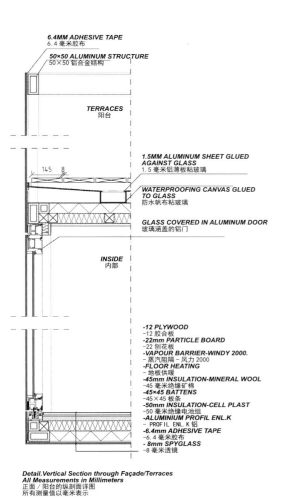
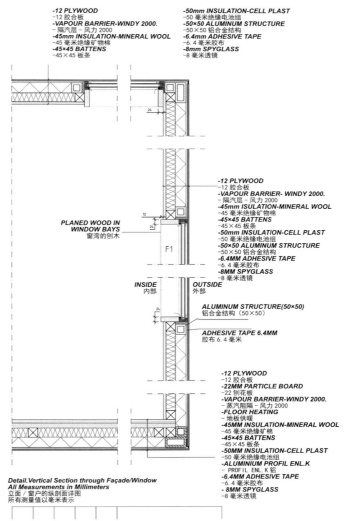

*Detail. Vertical Section through Façade/Terraces*
*All Measurements in Millimeters*
正面／阳台的纵剖面详图
所有测量值以毫米表示

*Detail. Vertical Section through Façade/Window*
*All Measurements in Millimeters*
立面／窗户的纵剖面详图
所有测量值以毫米表示

# 格鲁吉亚第比利斯和平之桥
## The Bridge of Peace

设计单位：Architetto Michele De Lucchi S.r.l.
开发商：格鲁吉亚政府
项目地址：格鲁吉亚第比利斯
摄影：Gia Chkhatarashvili

Designed by: architetto Michele De Lucchi S.r.l.
Client: Georgia Government
Location: Tbilisi, Georgia
Photography: Gia Chkhatarashvili

Suspended Roof
Trapezoidal
Glass Units

悬浮式顶盖
梯形玻璃

**项目概况**

这座和平之桥横跨第比利斯的库拉河,将 Bericoni 老城区与 Rikhe 区连接起来。作为该区的标志性建筑,建筑旨在以不同的方式来表达人们的团结。

**建筑设计**

该桥包括一条人行道和一个悬浮式的带有曲线形剖面的顶盖,仅有的支撑物就是河畔上的四根柱子,穿过人行道和悬浮顶盖的是库拉河路堤,从河的任意一侧都可到达人行道,而经过四架楼梯则可到达路堤。

顶盖由钢管和梯形的玻璃单元构成,看起来像单一的薄膜结构。桥上的拱都是由同样的抛物线构成,纵剖面则是通过有效的平衡力固定起来。桥中心形成一个"广场",是理想的社交和文化聚集地,从这里人们可以观赏到第比利斯壮观的美景。整座桥梁以其独特的外部结构和功能,成为城市的新地标。

## Profile

This bridge crosses the river Mtkvari in Tbilisi, connecting the older district of Bericoni to that of Rikhe. As a symbolic architecture of this area, the bridge aims to represent union of the people in different ways.

## Architectural Design

The bridge comprises a footway and a roof with a sinusoidal profile that seems suspended. Its only supports are four pillars on the river banks. Crossing the two arteries that run along the Mtkvari embankments, the footbridge is accessible from either side of the river, and from its embankments by means of four stairs.

The roof seems a single membrane created by a framework of steel tubes and trapezoidal glass units. The cross-arches are all generated by the same parabola. The longitudinal profile is also the result of the most effectively balanced forces. Created at the center of the bridge is a "piazza", an ideal social and cultural meeting-place, from which the population and tourists can enjoy splendid views of Tbilisi. With its unique exterior structure and functions, the bridge becomes a new landmark of this city.

# 一"技"之长
## ——建筑技术新发展

# Excellent "technique"
## —new development of architecture technology

建筑史上每一次材料、结构以及设备的进步，都为建筑学带来巨大的发展潜力，也为建筑构造的发展提供了基本的物质基础和创新动力。

在建筑发展的历程中，如果罗马人没有发现火山灰的作用，就无法建造43.3m跨度的万神庙；巴黎机械馆由于采用了先进的三铰拱结构和钢材，其跨度达到了史无前例的115m；在亚特兰大奥运会的主场馆，由于张拉弦结构的应用，乔治亚穹顶的用钢量仅为30kg/㎡。当然，如果没有奥蒂斯发明电梯，高层建筑就没有可能，如果电梯速度不能提升到10m/s以上，当代的摩天大楼也只能停留在图纸上；如果没有空调系统的发明，人们就无法安然地坐在玻璃大厦中办公，在舒适的环境中进行各种演出和体育比赛；如果没有玻璃幕墙的出现，也就不会产生轻盈通透的建筑效果；如果没有智能建筑系统的应用，当代超大规模的机场就很难高效、安全地运作。

从可持续发展的要求来说，建筑学要走向绿色生态，应合理利用资源，减少对环境的影响。在资源和能源方面，注重绿色建材、节能技术的运用，重视可再生资源和能源的开发；在环境层面上，减少对自然环境的破坏，提升内部环境的质量；在技术层面上，有以先进设备为代表的主动式节能技术，也有以设计和建造为核心的被动式节能技术。在可再生能源的开发利用领域，太阳能、地冷、地热风能、生物能的使用正在逐渐兴起，在建筑构造中就出现了上述技术与建筑一体化的课题。随着被动式设计策略的引入，太阳房、双层皮幕墙、通风塔、光管、可变遮阳等技术为建筑构造的发展打开了新的领域；生土建筑、竹木建筑、草砖建筑的重现带来的传统建筑更新也重新成为建筑构造面临的话题。

总之，技术进步对于建筑学的影响是巨大的，它不仅对建筑学的发展产生直接的影响，还会通过技术和社会发展的复杂作用的过程，对建筑学不断提出新的要求。简而言之，技术进步对建筑构造的影响应该统一在可持续发展的前提下——就是对绿色生态建筑原则的遵循。

Every improvement upon the buildings in materials, structures and equipment, brings a huge development potential for the architecture, and also provides the basic material foundation and innovative impetus for the development of the building construction.

In the course of building development, if the Romans did not find the effect of volcanic ash, people would not build the Pantheon with 43.3m span; the Paris Machinery Hall with the advanced three hinged arch structure and steel, reached an unprecedented 115m span; the Atlanta Olympic Games' main stadium, due to the application of tension chord structure, the amount of steel for the Georgia dome is only 30kg/m2. Of course , if there was no elevator invention by Otis, the high-rise buildings would be not possible ; if the elevator speed can not be raised to 10m/s or more, the contemporary skyscrapers would only be stay in the drawing; Without the invention of air-conditioning systems, people can not safely sitting in a glass building for the office, and can not conduct a variety of performances and sports in a comfortable environment; If there is no appearance of the glass curtain walls, it will not produce light and airy architectural effect; If there is no application of the intelligent building systems, the oversized contemporary airports are difficult to be efficient and safe operation.

From the requirements of sustainable development, the architecture should, should rationally use resources and reduce the impact on the environment in order to going be green and ecological. In terms of resources and energy, it should focus on green building materials, the use of energy-saving technology, and emphasis on the development of renewable resources and energy; in terms of environmental level, it should reduce the damage to the natural environment, improve the quality of the internal environment; in terms of technical level, there are active energy-saving technologies with advanced equipment as the representative, and passive energy-saving technologies with design and construction as the core. In the field of development and utilization of renewable energy, the uses of solar energy, cold and geothermy, wind energy and bio-energy are gradually emerging. There is an issue in the building construction integrated with these technologies and buildings. With the introduction of passive design strategies, the sun room, double-skin curtain wall, ventilation tower, fluorescent tubes, variable shade and other technologies for the development of building construction opens up new areas: the update of the traditional architecture like the earth architecture, bamboo architecture, straw bale construction also brings to reproduce again which becomes a topic that the architecture construction faced.

All in all, the impact of technological advances is huge on the architecture. It is not only a direct impact on the development of architecture, but also will constantly propose the new demands to the architecture through a complex role processing of the development of technology and social. In short, the influence from the progress of the technology on the building structure should be unified in the context of sustainable development—it is to follow the principles of green ecological architecture.

# 加拿大蒙特利尔 Bota Bota 水疗中心
Bota Bota, Spa-sur-l'eau

设计单位：Sid Lee Architecture
开发商：Bota Bota, Spa-sur-l'eau
项目地址：加拿大蒙特利尔

Designed by: Sid Lee Architecture
Client: Bota Bota, Spa-sur-l'eau
Location: Montreal, Canada

# 渡船
# 光影共生
# 多领域技术

# Ferryboat
# Coexisting Light and Dark
# Multi-discipline Techniques

**项目概况**

Bota Bota 位于蒙特利尔古港 McGill 大街尽头,是一艘渡船翻新改造而成的浮动式斯堪的纳维亚温泉水疗中心。这艘渡船在 20 世纪 50 年代常常来往于 Sorel 和 Berthier 之间,新的建筑设计,为项目带来了全新的视觉享受。

**建筑设计**

建筑设计以"光影共生"为主题,使得其参观者沉浸在光影交错的环境中,既能让人们在室内闭目养神,又为顾客提供了一个视野开阔的室外场所。在建筑 5 个楼层之间穿梭,参观者可从每个楼层观赏到城市风景,完全忘记自己身在船中。

建筑设计有 678 个舷窗,日光通过舷窗射入治疗室,使得天和水、光和影的过渡成为可能。

**技术设计**

Bota Bota 水疗中心项目的独特之处在于多个领域的技术融合:房屋建筑、海洋建筑、室内设计、工业设计、房屋与造船工程等。如此多领域的融合对于解决技术和科技挑战至关重要,建筑设计与严格的概念框架,以保证建筑在海洋中的稳定性和浮力。另外,坚实的综合结构,也加强了建筑的稳固性,从而有利于营造出放松舒适的温泉体验环境。

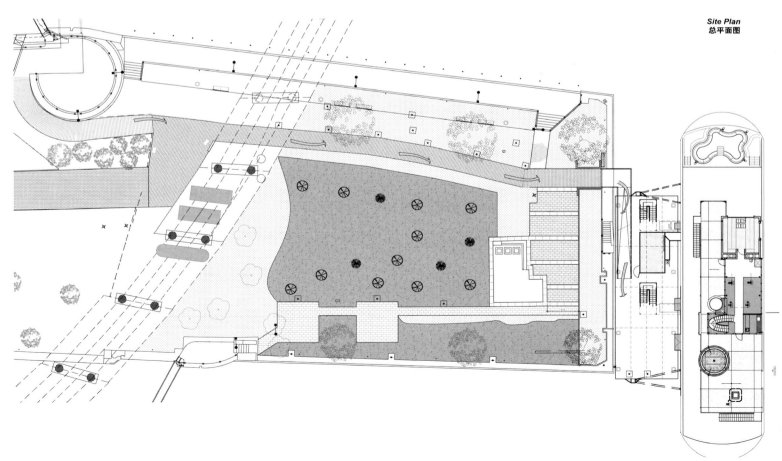

*Site Plan*
总平面图

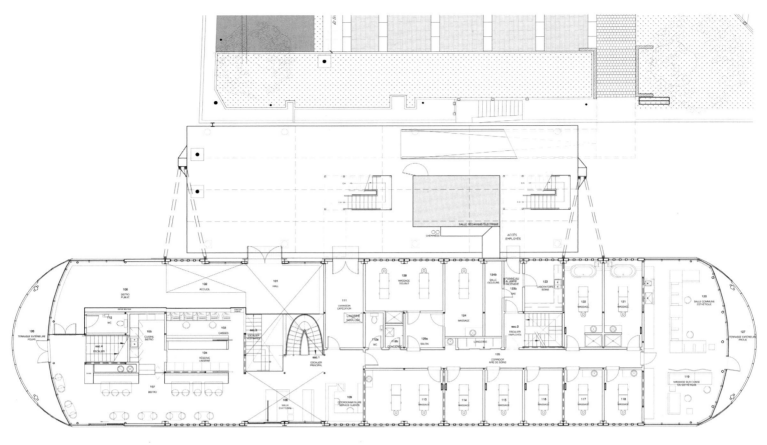

*Main Bridge*
主桥

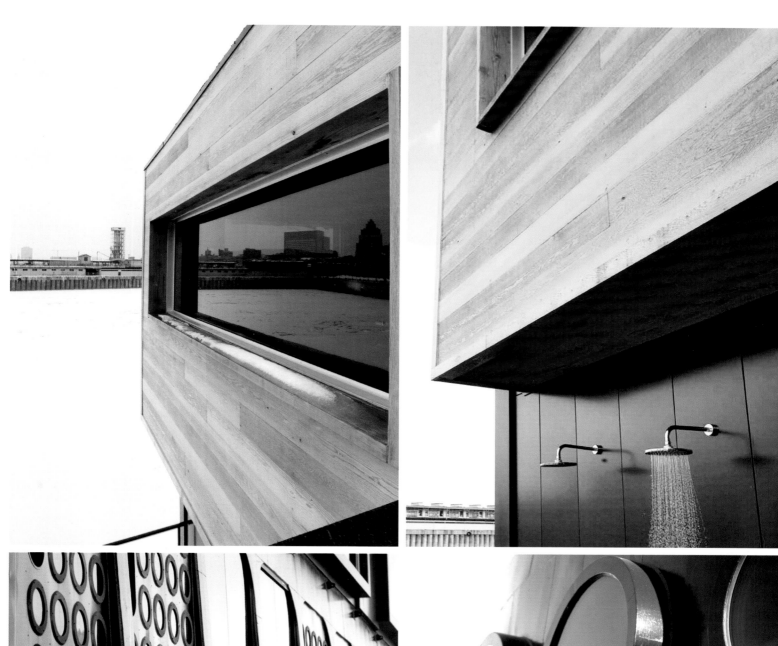

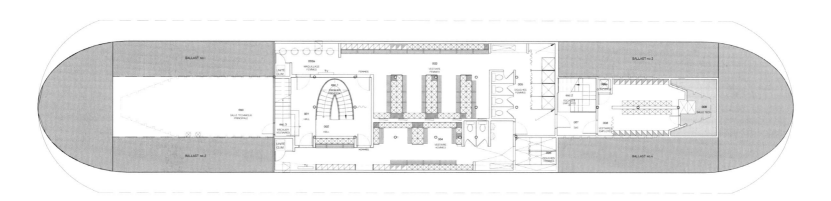

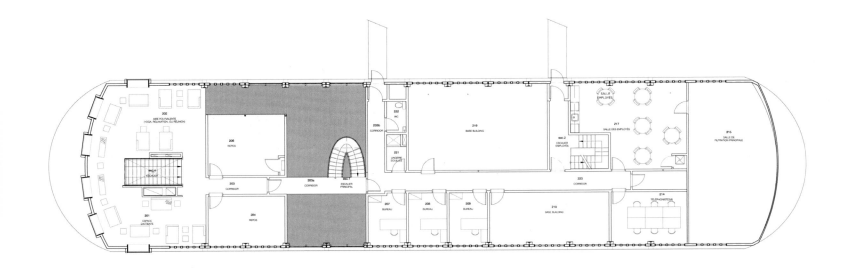

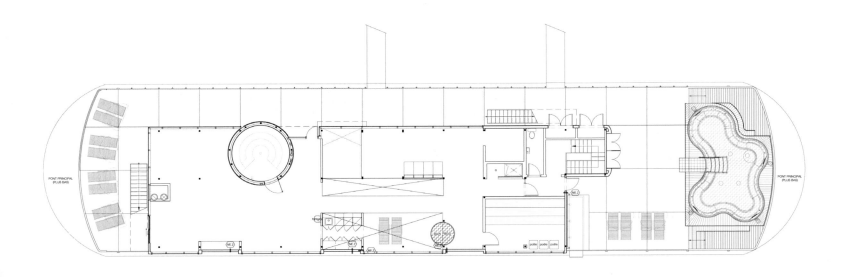

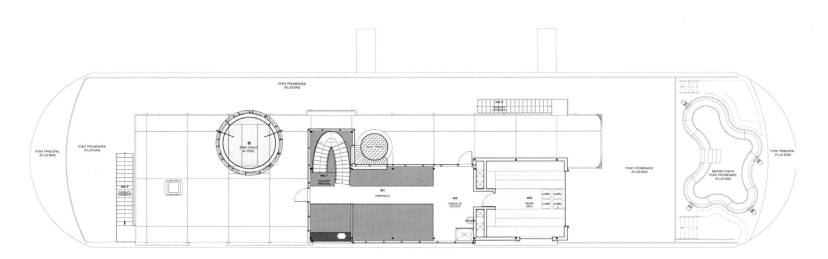

*Définie Par Mmallet*

## Profile

Located at the foot of rue McGill in the old port of Montreal, Bota Bota is the new name given to a ferryboat that used to link Sorel and Berthier in the 1950s and that has been renovated into a floating Scandinavian spa. The redesign brings new visual identity to the project.

## Architectural Design

Taking "coexisting light and dark" as its theme, the Bota Bota immerses visitors in an environment of light and dark. The architects set out to create an indoor space conducive to introspection and an outdoor space affording spectacular views of the city, from upper decks. Visitors forget they're on a boat as they transition through the five different levels, discovering the city from each one.

678 portholes dot the boat, allowing daylight to penetrate the treatment room. As such, the transition from water to sky and dark to light is made possible.

## Technical Design

The uniqueness of the Bota Bota, spa-sur-l'eau project is due to the fact that many fields of expertise were brought together. The project represents a fusion of disciplines: building architecture, naval architecture, interior design, industrial design, as well as building and naval engineering. This mingling of multiple disciplines in this ambitious project was necessary to tackle the technical and technological challenges, such as integrating structures that are usually found on solid ground. Naval architecture required a rigid conceptual framework in order to respect notions of stability and buoyancy. Furthermore, a second important challenge consisted in creating is a spa experience that would be conducive to relaxation.

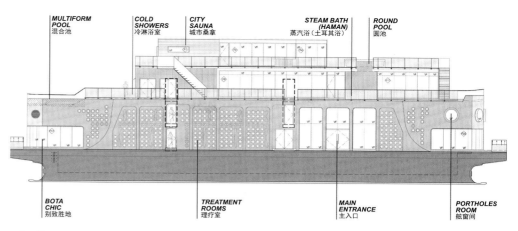

*Port Elevation*
左舷立面

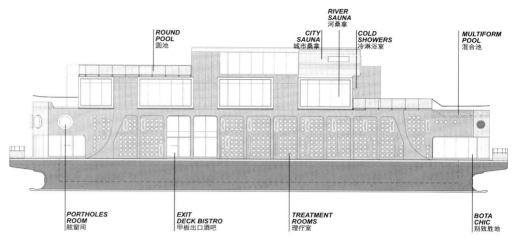

*Starboard Elevation*
右舷立面

# 上海帝斯曼上海总部
## DSM Headquarters, Shanghai

设计单位：de Architekten Cie  
开发商：皇家帝斯曼  
项目地址：中国上海  
总建筑面积：25 000m²

*Designed by: de Architekten Cie*  
*Client: Royal DSM*  
*Location: Shanghai, China*  
*Gross Floor Area: 25,000m²*

**项目概况**

帝斯曼中国园区位于上海浦东新区张江高科技园区，包括所有位于上海的帝斯曼办事处和研发中心，这一融合将为帝斯曼在中国本土培育更强大的创新能力。该园区共600名员工，成为帝斯曼集团在欧美以外规模最大、最重要的研发基地。

**功能布局**

该综合体包含3栋建筑，每栋有7层，其中办公楼设于接待区、员工餐厅，另两栋为实验室和训练中心研发部门。底层一条宽阔、略弯曲的走廊将3栋大楼连接起来。在运河一侧，走廊插入大楼间波浪状的景观花园，给游客带来一种友好、绿色的印象。

后面的走廊向经理餐厅敞开，面朝一个典型的场地，综合体下面是停车场，可直通每栋大楼。

**建筑设计**

3栋大楼规模和材料都相同，中间大楼的末端与另两栋倾斜方向相反，形成动感的组合形象。建筑外墙小块方格的设计可以准确地将每栋楼的立面调整至太阳的位置，还可保证内部空间获取必要的太阳光路径。

外立面水平板条的运用使办公大楼别具一格，立面上较暗的青铜色使建筑更具魅力。建筑外围设有装置核心和楼梯井，使这3栋大楼具有灵活的标准楼层，这些装置可供将来使用，还可供出租。

**生态技术**

帝斯曼园区全面贯彻可持续发展的理念，成为上海首个获得美国LEED金质证书的建筑。设计运用了多种技术，比如建筑绝佳的朝阳方向、景观屋顶、水循环、太阳能板、自然通风和单独的光照调节等。设计还运用了局部可再生材料，比如竹墙、毡、水磨石等。

场地布局是该建筑设计的一个重要部分，设计师在该区域内设计了一个有机废物的堆放区，并进行了分类处理，此外，该建筑区域内还被当地的绿色植被所覆盖，使其成为一座真正的绿色环保建筑。

方格外墙
水平板条
生态环保

Grid of
Curtain Walls
Horizontal Slats
Ecological and
Environmental

**Third Floor Plan**
三层平面图

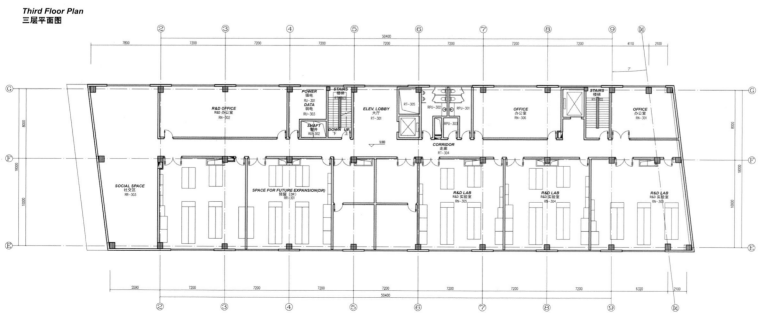

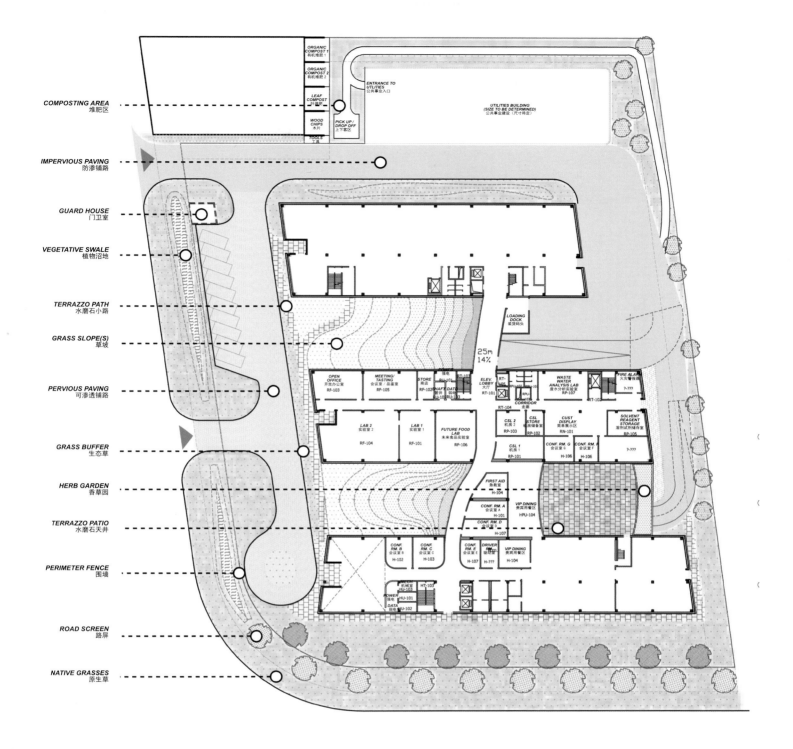

## Profile

The DSM China Campus is located in Zhangjiang High Tech Park of Pudong New District in Shanghai, bringing together all Shanghai offices and research and development center of Dutch multinational DSM at one location. Therefore the merging at one location intends to boost the R&D competences of DSM in the region. The campus houses about 600 employees and is now the largest and most important research facility of DSM outside Europe and the United States.

## Function Layout

The complex comprises three individual buildings each with seven levels: an office building with reception areas and staff restaurant as well as two buildings for the R&D department with laboratories and training center. At ground floor level a wide, slightly curved corridor connects these three buildings. On the side of the canal, the corridor cuts into the undulating landscaped gardens between the blocks, ensuring a friendly, green impression to visitors.

At the backside, the corridor opens onto the executive restaurant which faces a representative court. Under the complex there is a parking garage with direct access to the individual buildings.

## Architectural Design

All three buildings are almost identical in size and materialization. The ends of the middle block tilt in the opposite direction compared to the other ones, which creates a dynamic image. The small-scale grid of the curtain walls makes it possible to accurately adjust the façades of each building to the position of the sun and specific requirements for daylight access of the spaces within.

The office building is distinguished by the application of horizontal slats in the curtain wall structure and due to the darker bronze color of its façade, it has a more representative aura. All three buildings have flexible standard floors with installation cores and stairwells along the northern perimeter. This enables alternative future uses and the possibility to rent (parts of) the individual buildings to several tenants.

## Ecological Technology

The DSM Campus is a very sustainable design and the complex is the first in Shanghai to be awarded a LEED Gold certificate. Different techniques are used for this purpose, such as an optimal orientation of the buildings to the sun, landscaped roofs, water recycling, solar panels, natural ventilation and individual regulation of lighting. Here, possible local and recycled materials are used, such as bamboo walls, felt and terrazzo.

The site layout is an integral part of the design, and the architects has designed a room for composting of organic waste in this region and made a classified process. In addition, the region also is covered by local green vegetation to make the building be a truly green and environmental type.

"自行车"
几何形态
参数化设计

"Bikes"
Geometry
Parametric Design

# 澳大利亚悉尼哈雷·戴维森澳大利亚总部
New Headquarters for Harley Davidson, Australia

设计单位：Tony Owen Partners
项目地址：澳大利亚悉尼

Designed by: Tony Owen Partners
Location: Sydney, Australia

**项目概况**

这栋哈雷·戴维森澳大利亚新总部由悉尼建筑事务所 Tony Owen Partners 设计，位于悉尼风景秀丽的莱恩湾，是这一带新落成的莱恩河商业园的标志性建筑。

**设计灵感**

设计从自行车身上寻找灵感，从建筑线条上能看到引擎框架的几何形态，旨在设计一栋能反映哈雷·戴维森独特性的建筑。设计过程中，并不是复制这种几何形态，而是模仿了这种流动性和风格，同时也不是片面采用其外观，而是通过建筑线条的流动性表现出高雅、动感的建筑形态。

**建筑参数化设计**

建筑在外观和复杂的几何形态的建造上都采用了 3D 建模和参数工具，由于几何体相当复杂，因此了解结构与金属外墙之间的相互作用尤为重要。建筑设计将每个结构、立面元素制作模型，使其能准确地知道空间中每一个元件间的相互联系。这一技术也使得建筑成为一个结构，使它比之前看起来更直观、更容易理解。

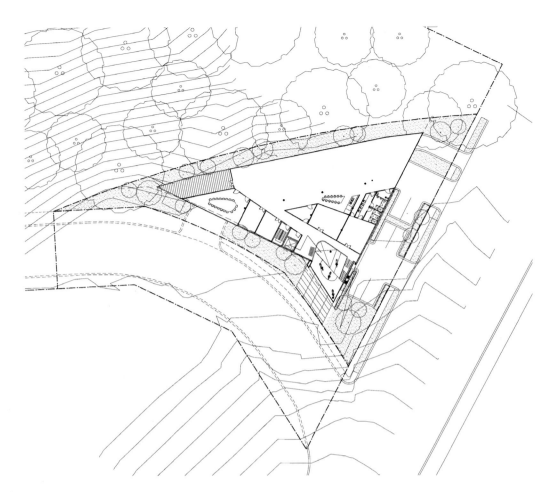

### Profile
The new Australian headquarters of Harley Davidson is designed by Tony Owen Partners. The building is located in Lane Cove and forms an iconic gateway to the new Lane Cove River business park.

### Design Inspiration
The design takes inspiration from bikes. The geometry of the engines forks and frames can be seen in the lines of the building. The building does not copy them; however, it suggests the movement and style. Rather than using the shape literally, the design expresses elegance and aerodynamics of this movement in the lines of the building.

### Parametric Design
The office is increasingly using 3-D modelling and parametric tools both in the generation of formal solutions and in the development and construction of complex geometries. Because the geometry is complex, it is important to know how the structure interacts with the metal cladding. By modelling every element of structure and façade the architects know exactly how each piece related to another in space. This technology makes building a more straight-forward structure, much easier than it might have been in the past.

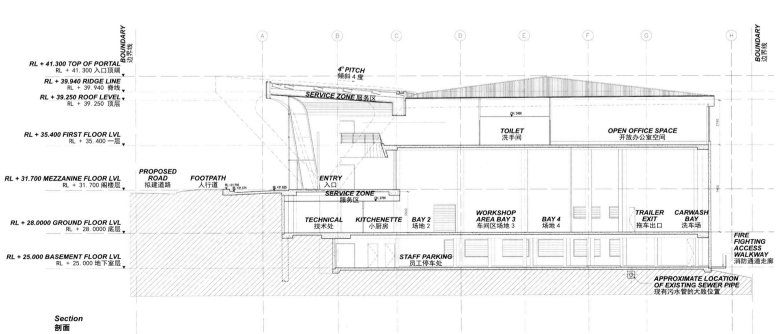

*Section*
剖面

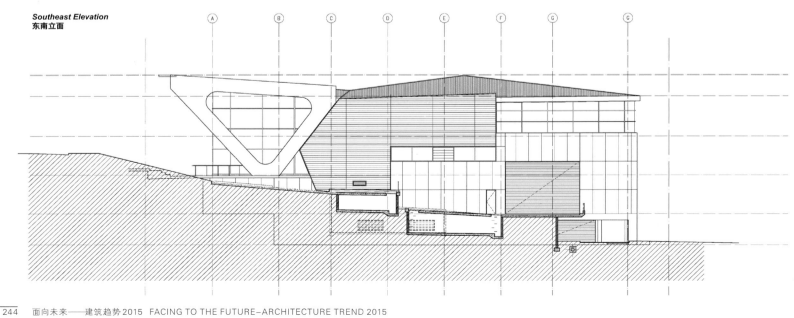

**Southeast Elevation**
东南立面

**Southwest Elevation**
西南立面

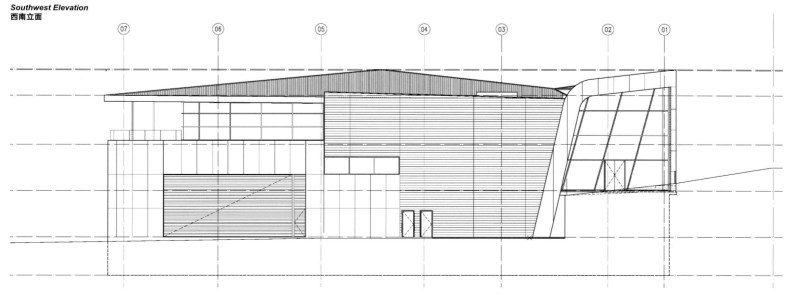

# 西班牙拉斯帕尔马斯 El Lasso 社区中心
## El Lasso Community Center

设计单位：Romera y Ruiz Arquitectos S.L.P
开发商：Excmo. Ayuntamiento de Las Palmas de Gran Canaria
项目地址：西班牙拉斯帕尔马斯
建筑面积：498m²

Designed by: Romeray Ruiz Arquitectos S.L.P
Client: Excmo. Ayuntamiento de Las Palmas de Gran Canaria
Location: Las Palmas, Spain
Building Area: 498m²

## 项目概况

El Lasso 社区中心位于拉斯帕尔马斯一个被忽视的社区，建造在花园和海景之间。有限的场地面积、现有的街道和山谷的地形共同决定了建筑的造型和空间。

## 建筑设计

建筑在山谷地形中设计有凸起的墙用来防止建筑的腐蚀，内部花园由围合式的弧形外墙、建筑形式和树荫共同形成一个开放的公共区域，从而满足社区居民对于渴望获得一处休闲场所的需求。

在海景和花园之间，纵向的露台可观看到大西洋美景，还能充当一个过滤自然光线、北风和受限的景色"过滤器"。随着时间的推移，俯瞰着大海的建筑立面充满着活泼的色彩和影子，使其成为优美景观的偏振光。随着影子不断变换的立面就像一位表情丰富的演员，而不是一个静止的被动式的元素。

从平面上看来，建筑造型是一堵厚重的墙，空间都被切割出来，而稍微呈斜坡状的纵剖面上有两条不同楼层的通道，将这些空间完美地连接起来。这座富有农业气息的建筑，就像迦纳利岛景观的化身，带着神秘的色彩，伫立在大西洋海岸。

## Profile

The El Lasso Community Center, a neglected neighborhood of Las Palmas, is constructed between a garden and a viewpoint towards the sea. The limited site, existing street and valley topography define the form and space of the architecture.

## Architectural Design

The walls raised in valleys are designed to avoid erosion. The inner garden invents an open collective dominion formed by an enclosing curved wall, the built form and the tree shades, to provide a leisure space for the residents.

In between the sea-views and the garden, the longitudinal built terrace watches the Atlantic Ocean, acting as a filter of light, spaces, northern breezes and contained views. As time goes on, the lively colors and casted shadows flood the façade that overlooks the sea, turn it into a landscape of polarized light. The façade turns into an actor rather than a passive element that just stands and shows.

On plan, the built form is a heavy wall on which spaces are carved and oriented to catch snapshots of the exterior. The building section gently negotiates the sloping topography allowing for two accesses at different levels to connect the spaces. Architecture that almost with the ease of farming becomes content of the territory and incarnation of the Canarian landscape, full of color, dry mud walls, torrents and drains.

"过滤器"
偏振光

"Filter"
Polarized Light

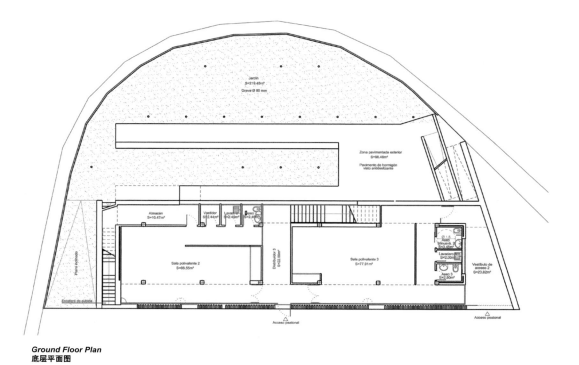

**Ground Floor Plan**
底层平面图

**First Floor Plan**
一层平面图

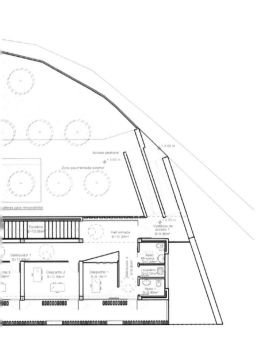
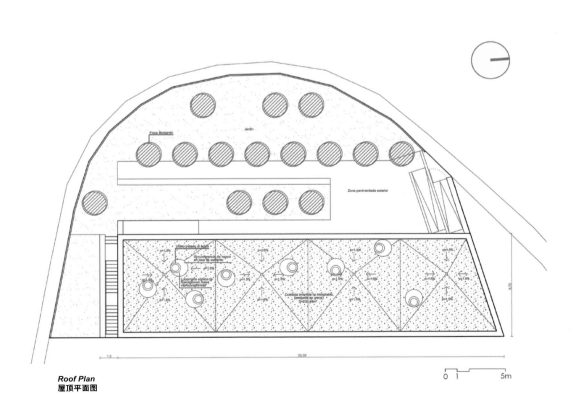

Roof Plan
屋顶平面图

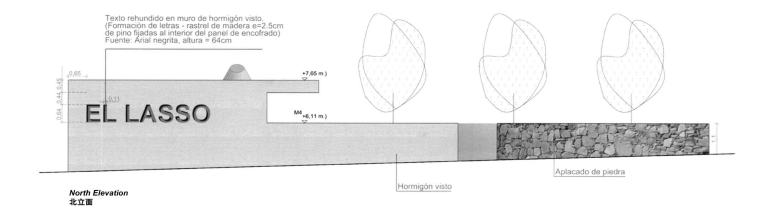

**North Elevation**
北立面

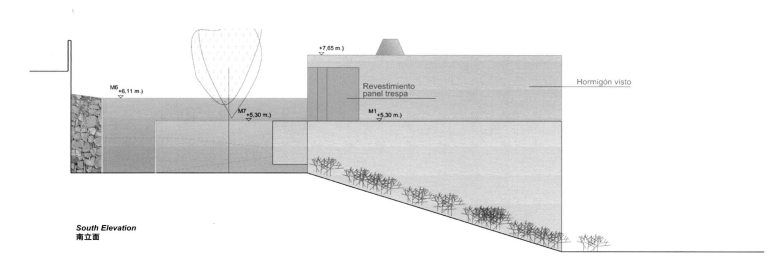

**South Elevation**
南立面

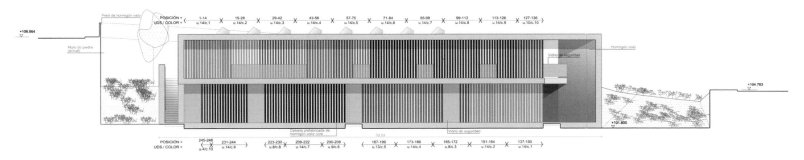

**East Elevation**
东立面

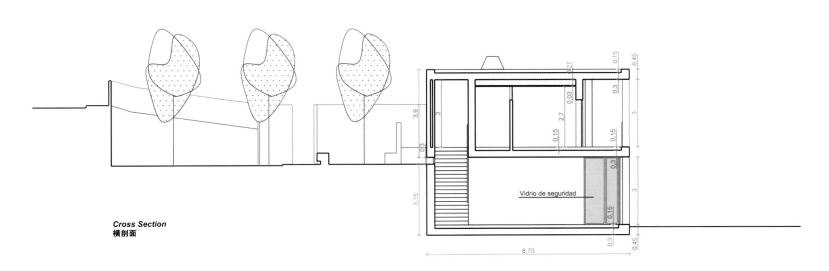

**Cross Section**
横剖面

# 英国伦敦设计研究实验室十周年纪念亭
## [C]space DRL10 Pavilion

设计单位：Synthesis Design + Architecture
开发商：英国建筑联盟设计研究实验室
项目地址：英国伦敦
建筑面积：100m²

*Designed by: Synthesis Design + Architecture*
*Client: Architectural Association, Design Research Lab*
*Location: London, Britain*
*Building Area: 100m²*

**项目概况**

这一精致的"物体建筑"作品在建筑协会举办的"AADRL十点零"展览项目比赛中荣获第一名，而此次比赛正是为了庆祝英国建筑联盟设计研究实验室成立10周年所举办的。

**建筑设计**

这座外观醒目的亭子远远地就能吸引人们的注意，设计主要采用纤维加固混凝土面板作为覆面材料。当你更近距离接触时，建筑通过将复杂的曲线、结构特性、合理的功能布局融入单一的连续性形态中，展示出其不确定性。而当你环绕亭子行走时，建筑表面则从不透明变成透明，形成绝妙的三维叠纹。

建筑表面是闭合的，但也能为行人提供通道，室内外空间的区分也因此变得模糊。

**技术设计**

亭子内的连接系统采用简单的十字连锁，由一组氯丁橡胶垫片将其加固，并与奥地利纤维加固混凝土技术部门密切合作，进行了大量材料试验，该设计才得以进行。

在整个构造方案确定前，经过6周时间反复进行了16次设计模型分析，同时对单个元素进行数字建模、大量的快速成型、比例模型、全尺寸物理实物模型，并对整个装配的公差等进行试验，最终建筑由850个独特的剖面安装在标准的13mm厚的平板上。

曲线
数字建模
纤维加固混凝土

Curve
Digital Modeling
Fiber-C

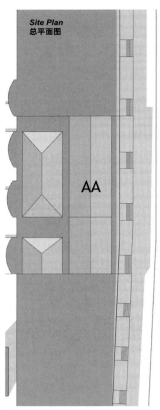
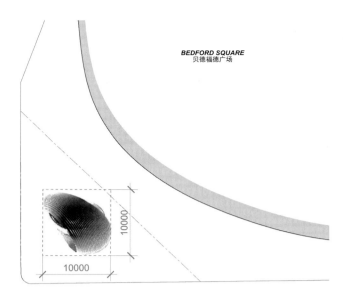

**Site Plan**
总平面图

BEDFORD SQUARE
贝德福德广场

## Profile
The delicate [C]space is the winning entry in the AADRL10 Pavilion competition, which was held to celebrate the tenth anniversary of the AA Design Research Lab.

## Architectural Design
The striking presence of the pavilion invites inspection from a distance. The design is proposed to be entirely constructed from Fiber-C, a thin fiber reinforced cement panel that is normally used as a cladding solution. Upon closer interaction, the ambiguity through the merging of sinuous curves, structural performance, and programmatic functions creates a single continuous form. As you move around, the surface varies from opaque to transparent, producing a stunning three-dimensional moiré.

The surface encloses while also providing a route through for passing pedestrians blurring the distinction between inside and outside.

## Technical Design
The jointing system in the pavilion exploits uses a simple interlocking cross joint which is tightened by a set of locking neoprene gaskets. Close consultation with the Fiber-C technical department in Austria and extensive material testing are required to develop the design.

Over a period of 6 weeks 16 iterations of the design model are analyzed before a structural solution is found. In parallel to the digital modeling, numerous rapid prototypes, scale models and full scale physical mock-ups are built to develop the design of individual elements and test the tolerance and fit of entire assemblies. The final pavilion constructed from 850 individually unique profiles that are nested on standard 13mm flat sheets.

**Top View**
俯视图

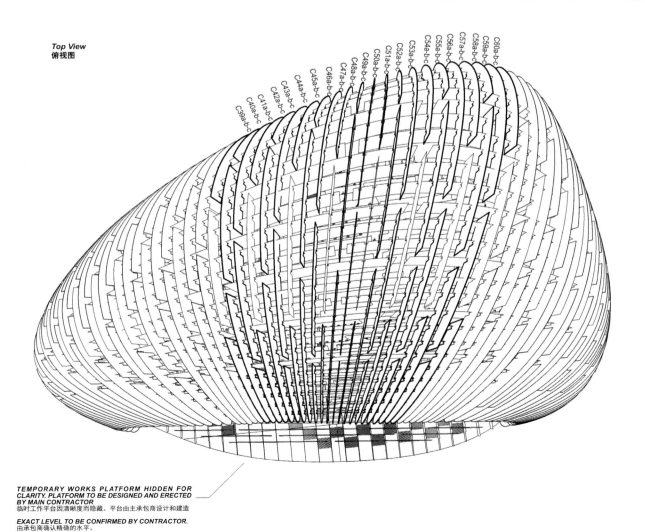

TEMPORARY WORKS PLATFORM HIDDEN FOR CLARITY. PLATFORM TO BE DESIGNED AND ERECTED BY MAIN CONTRACTOR
临时工作平台因清晰度而隐藏。平台由主承包商设计和建造

*EXACT LEVEL TO BE CONFIRMED BY CONTRACTOR.*
由承包商确认精确的水平。

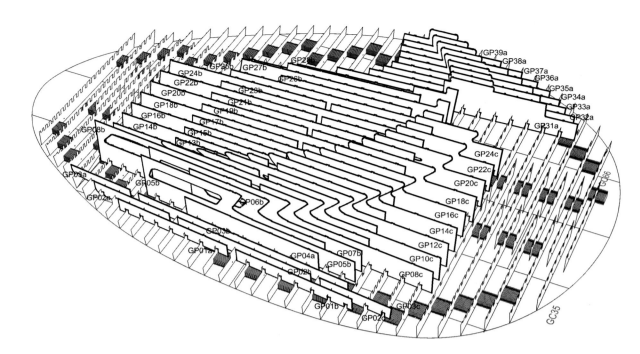

*Rear Layout View*
后部布局视图

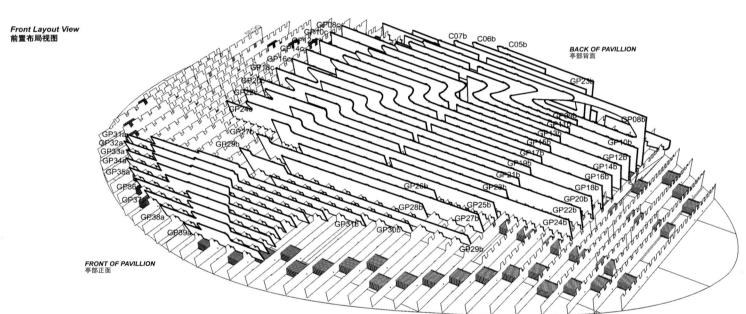

*Front Layout View*
前置布局视图

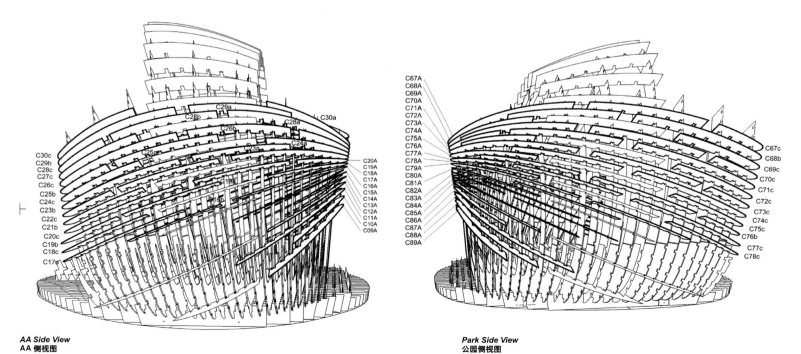

*AA Side View*
AA 侧视图

*Park Side View*
公园侧视图

258　面向未来——建筑趋势2015 FACING TO THE FUTURE–ARCHITECTURE TREND 2015

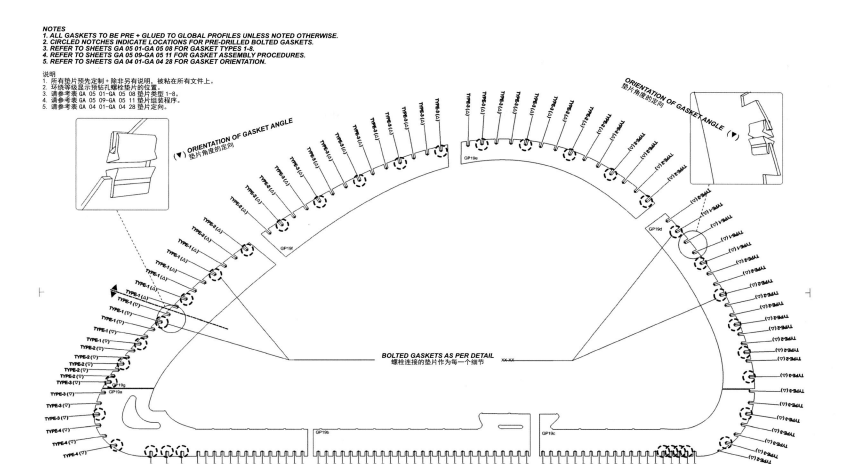
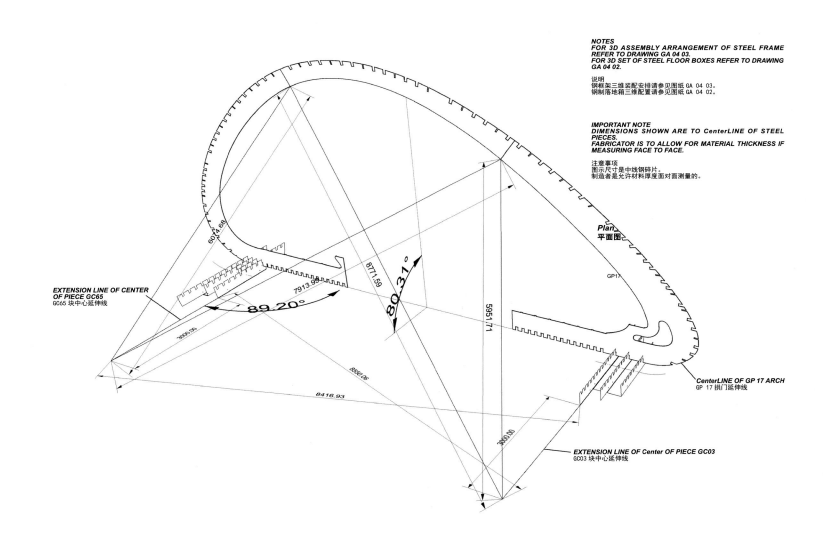

# 上海 JNBY
JNBY

| | |
|---|---|
| 设计单位：HHD_FUN | Designed by: HHD_FUN |
| 开发商：江南布衣 | Client: JNBY |
| 　　　　美国棉花协会 | COTTON USA |
| 设计团队：章 健　王振飞　王鹿鸣 | Project Team: Zhang Jian, Wang Zhenfei, Wang Luming, |
| 　　　　Thomas Clifford Bennett | Thomas Clifford Bennett |
| 　　　　Mehrnoush Rad　Christian Olav | Mehrnoush Rad, Christian Olav |
| 　　　　H & J International | H & J International |

**项目概况**

该项目是HHD_FUN为JNBY和美国棉花协会在上海新天地举办的品牌互动体验活动设计的一个可变的临时建筑，可适用于相关的现场音乐会、时装秀、展览及公司推广等宣传活动。

**建筑设计**

为秉承JNBY和美棉的设计理念，该建筑在设计上充分体现实用性和环保性。建筑师运用最新的参数化设计理念和拓扑理念，创造出一个独特的空间结构。

整个主体由3组不同形态、共6个可互相拼接的组件单元构成，不同形态的组件单元是通过对以三角形为基础的基本单元进行几何拓扑变形后生成。通过组件间不同的对接方式，该建筑可呈现出若干种可能的形态。组件上的拱形部分，可作为建筑空间的入口或与其他组件的内部结合面，而组件间相互对应的尺寸关系则大大增加了整体结构变化的可能性。

在钢结构搭起的主体外侧，覆盖的是一种由美国棉花协会提供的弹力防水材料，这种材料除制作过程非常低碳外，还兼具半透明特性。这使得即使在夜间，即便身处活动场外，依然可看到内部激光及灯光营造出的绝美视觉效果。

建筑易拆装特性及低碳材料的使用，使之无疑成为一个生态环保建筑，也成为JNBY和美国棉花的"自然、环保"理念的延续。

参数化设计
拓扑理念
低碳材料

Parametric Design
Topological Analysis
Low-carbon Material

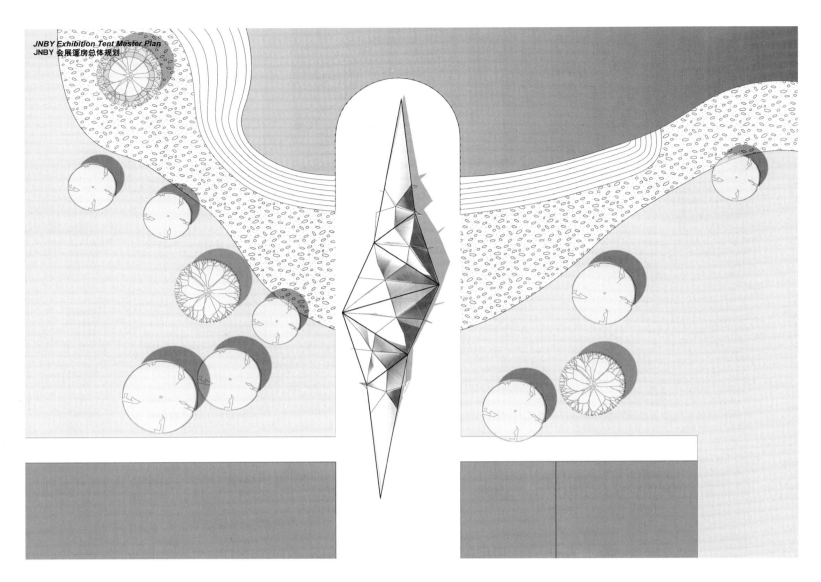

*JNBY Exhibition Tent Master Plan*
JNBY 会展篷房总体规划

## Profile

This project is a transformable temporary structure for brand interactive events of the JNBY and COTTON USA, held in Shanghai Xintiandi, with an ability to take on numerous different promos including live music and corporate entertainment.

## Architectural Design

The design is developed alongside the notion of JNBY and COTTON USA, so a practical and environmentally-friendly approach is taken. With the use of the latest parametric design tools and topological analysis, the unique structural design is formed.

The whole structure consists of six interlocking components, sharing three varied designs. Each design is achieved from a process of continuous deformation and manipulation of one triangular surface. The structure can be assembled into one of many possible combinations with an ability to be easily transformed from one form to the next, or in a more fashion related term, from one pose to another. The archways, acting as an entrance or an interlocking face, have corresponding dimensions to other archways and so increasing the number of possible overall forms.

The tent itself is created from a steel structure with a taut elastic waterproof material with translucent properties. The material, provided by COTTON USA is made from low-carbon footprint products. Even at night, outside of the structure, the splendid effect created by the internal laser and light can be seen.

The biologically friendly installation with a major advantage of being able to be easily re-installed and re-used in the future makes it a sustainable structure and extension of "natural, environmental-friendly" notion of JNBY and COTTON USA.

*JNBY Exhibition Tent Interior Layout Plan*
JNBY 会展篷房内部布局规划

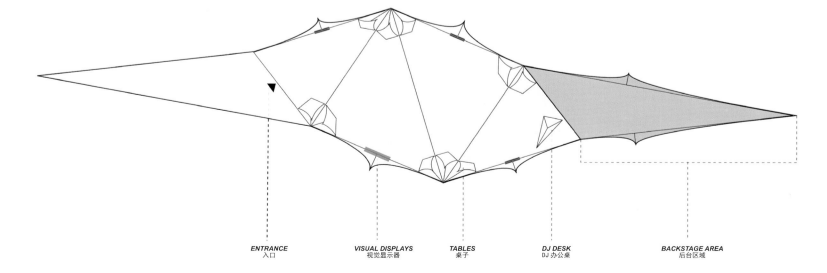

ENTRANCE
入口

VISUAL DISPLAYS
视觉显示器

TABLES
桌子

DJ DESK
DJ 办公桌

BACKSTAGE AREA
后台区域

Section A-A
剖面 A—A

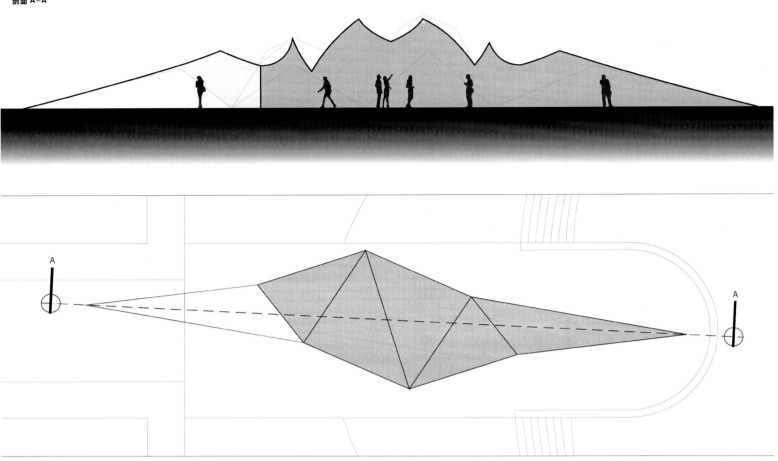

**Formation of The Structure**
结构形成

**FORMATION OF AN ISOSCELES TRIANGLE.**
一个等腰三角形的形式。

**ORIGINAL TRIANGLE DIVIDED BY 4 TRIANGLES INTO 12 SECTIONS, IDENTIFYING 7 CONTROL POINTS.**
原来的三角形被四个三角形划分为 12 个部分，确定 7 个控制点。

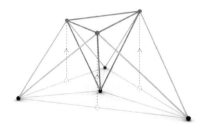

**MID-LENGTH CONTROL POINTS OF ORIGINAL TRIANGLE ARE RAISED, FORMING A NEW ISOSCELES TRIANGLE AT THE TOP.**
原来三角形的边线中长的控制点被提升，在顶端形成一个新的等腰三角形。

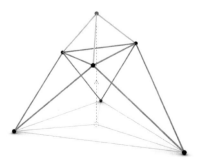

**Center CONTROL POINT OF ORIGINAL TRIANGLE IS RAISED, ABOVE THE NEWLY FORMED TRIANGLE.**
原来三角形的中心控制点被提升，在新形成的三角形上边形成三角形。

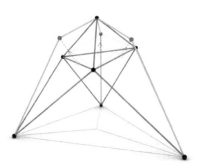

**MID-LENGTH CONTROL POINTS OF THE TOP TRIANGLE ARE CREATED AND PULLED OUT WARDS.**
顶端三角形的边线中长控制点被建立并且被拉伸到区域之外。

**IN LINE WITH THE ORIGINAL MID-LENGTH CONTROL POINTS, A NEW CONTROL POINT IS CREATED, RAISED AND PULLED OUT WARDS.**
在原来的中长控制点线上，一个新的控制点形成，然后被提升，拉伸到区域之外。

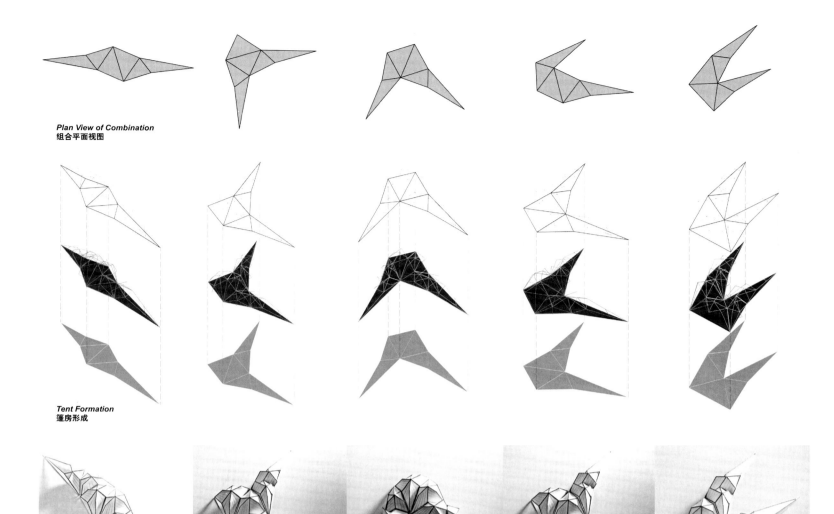

**Plan View of Combination**
组合平面视图

**Tent Formation**
蓬房形成

**Physical Model**
实体模型

## 公司简介 Company Profile

### 70F architecture

70F architecture 是一家正在不断成长壮大的设计事务所，由 Bas ten Brinke 和 Carina Nilsson 成立于 1999 年。

70F 团队对需求结构、室内或物品设计很感兴趣。建筑、室内设计、设计三个部门和公司常任合伙人一起共同努力寻求解决任何问题的最佳方法。公司所接手的项目包括从动物房到人居房屋，从桥梁到天线设计，从办公楼到学校、医疗建筑等各个领域。

丰厚的设计经验、良好的客户关系使公司得到了越来越多的广泛好评。

### 70F architecture

70F architecture is a growing office, founded by Bas ten Brinke and Carina Nilsson in 1999.

The 70F group work with great pleasure on any demanded structure, interior or object. The three departments – architecture, interior design and design, with the fixed partners, 70F architecture tries to find the best solution for any given problem. Structures involved range from animal housing to people housing, bridges to antennas, offices to schools and health care buildings.

Rich design experience and excellent client relations make 70F architecture receive wide appreciation.

### AART architects

AART architects 是一个扎根于北欧建筑传统的位于丹麦的建筑设计公司，以通过创造感性的构架使最平常的生活变得充满意义为使命，以建造最高质量建筑为目标，致力于建筑项目的研究和发展。多年来，公司已经储备了广泛的知识材料、技术和用户需求，形成了概念、建筑、绿色、健康四个研究团队，使公司在概念发展、项目工程、能源优化和促进健康方面有了长足的发展。项目主要包括萨尔普斯堡的 Inspiria 科学中心，埃尔西诺文化院等。

始终坚持以人为本、生态环保的建筑理念使得 AART architects 多次获得包括 Inspiria 科学中心获得的 2012 年度国际 Prime Property 评奖团特别奖、2012 年度挪威 Arnstein Arnberg 大奖，DANVA 总部大楼获得的 2011 年度得斯坎讷堡自治市颁发的建筑大奖、2011 年度 Bæredygtig Beton 大奖等各种重要奖项。

### AART architects

AART architects is a Danish architectural firm which, with roots in the Nordic architectural tradition, working with the community as a value-creating element. The aim of AART is to always achieve the highest architectural quality. AART has engaged in several research and development projects over the years and has established the research department AART+ which consists of four research teams: AART Concept, AART Build, AART Green and AART Health. The team forms a common thread in professional development and give added value to each project by challenging standard approaches and developing new methods and processes in the fields of concept development, project engineering, energy optimization and health-promoting. The main projects carried out by AART include Inspiria Science Center in Sarpsborg, and the Culture Yard in Elsinore.

Always adhere to the concept of "people-oriented and environmental-friendly", AART has achieved many important awards including the jury's special prize at the international Prime Property Awards 2012, Norwegian Arnstein Arnberg Prize 2012, the Architecture Prize 2011 by Skanderborg Municipality and Bæredygtig Beton Prize 2011.

### Antonino Cardillo

Antonino Cardillo 毕业于巴勒莫建筑学院，是一位基于伦敦的巡回建筑师。设计杂志《Wallpaper》在 2009 年将他选为 30 名最有才华的设计师之一，评价其为当下时代最具影响力的建筑师。

其作品遍布世界各地，在包括第四届鹿特丹建筑双年展在内的不同场馆进行展示，注重探索古今语言的界线，忽视建筑空间上世界观、信仰、传统和文化上的差异，使建筑成为缩短差异的一种方式。

### Antonino Cardillo

Born in Sicily and graduated in Architecture at Palermo, Antonino Cardillo is an itinerant architect based in London. He has been selected among the thirty best new young architectural practices from around the World in Wallpaper magazine in 2009. Cardillo is appraised as the most significant architects of our time.

His works were exhibited at different venues, including the 4th International Architecture Biennale of Rotterdam. Through his works he explores the boundaries between ancient and modern languages. His architectonic spaces attempt to get to a new syncretic synthesis reconciling different world views, beliefs, traditions and cultures: he interprets Architecture as a way to bridge differences.

### Architectenbureau Marlies Rohmer

Architectenbureau Marlies Rohmer 位于荷兰阿姆斯特丹，在城市改造、总体规划、住宅项目、学校建筑、医疗中心、室内设计等多个领域具有 25 年的经验。

公司拥有近 15 名员工，是一家基于网络的组织机构，将各个领域融合起来，相互渗透，对多种社会、文化现象的调查推动着事务所的设计进程。公司在项目整个过程，各个阶段都可提供服务，其认为每一项任务都是独一无二的，需要有特定的解决策略，因此相当注重对建筑细节的刻画以创造具有识别性的建筑。

经过多年的发展，公司项目几乎遍布整个欧洲，并曾多次获得重要国际大奖，例如：2002 年和 2008 年的荷兰学校建筑奖，2003 年在布宜诺斯艾利斯国际教育论坛中获得国际学校建筑奖、2009 年获得奈梅亨建筑奖，2012 年获得荷兰设计奖、国际房地产联盟奖等。

### Architectenbureau Marlies Rohmer

Architectenbureau Marlies Rohmer is an Amsterdam (NL) based practice. Its field of operation and its twenty-five years' experience ranges from urban renewal and master planning, housing projects, school buildings and care centers to interior design.

With a staff of approximately fifteen, Marlies Rohmer Office is a network-based organization in which various disciplines are integrated and enrich one another. The design process is always driven by research into a variety of social and cultural phenomena. The firm is well equipped to carry projects through from start to finish, all stages of the process. To Marlies Rohmer Office, every assignment is approached as a unique project, demanding specific solutions. The results are a powerful and recognizable architecture with a strong detailing.

Over the years, the office has been working throughout Europe. The firm got several awards, including the Dutch School Building Award 2002 and 2008, the International School Building Award 2003 of the Buenos Aires International Forum of Educational Architecture, the Golden Pyramid 2009, the Dutch Design Award 2012 and the FIABCI Award 2012.

### Architetto Michele De Lucchi S.r.l

Michele De Lucchi 被认为是天生的设计人才，早在 1992 年到 2002 年他在 Olivetti 担任设计总监，在此期间他获得与惠普、飞利浦、西门子等电子巨头的合作机会。

设计师的创意与专业水平，来自于他对建筑、设计、科技以及工艺的自我探索。其主要作品有：米兰三年展中心、罗马展览博物馆、柏林新博物馆、意大利的 le Gallerie、米兰斯卡拉广场、Radison 酒店、巴统服务大楼、第比利斯和平桥等。

### Architetto Michele De Lucchi S.r.l

Michele De Lucchi was born in 1951 in Ferrara and graduated in architecture in Florence. During the period of radical and experimental architecture he was a prominent figure in movements like Cavart, Alchymia and Memphis. De Lucchi has designed furniture for the most known Italian and European companies. From 1992 to 2002, he has been Director of Design for Olivetti and developed experimental projects for Compaq Computers, Philips, Siemens and Vitra. His professional work has always gone side-by-side with a personal exploration of architecture, design, technology and crafts. He designed buildings for museums including the Triennale di Milano, Palazzo delle Esposizioni di Roma, Neues Museum Berlin and the le Gallerie d'Italia Piazza Scala in Milan. In the last years he developed many architectural projects for private and public client in Georgia, that include the Ministry of Internal Affairs and the bridge of Peace in Tbilisi, the Radison Hotel and Public Service Building in Batumi.

### KWK Promes

KWK Promes 建筑师事务所成立于 1999 年，其创办人 Robert Konieczny 毕业于格利维策的西里西亚工业大学，曾 6 次被密斯·凡德罗基金会提名。

公司在 Robert Konieczny 的带领下，经过多年的不懈发展，赢得了社会的广泛好评。2007 年 KWK Promes 公司名列由"Scalae"发行的世界 44 位最佳青年建筑师排行榜，同年《Wallpaper》杂志将 KWK Promes 列于世界 101 家最出色的建筑工作室之一。

除此之外，公司作品 Aatrial 住宅和"隐藏住宅"被芝加哥国际建筑与设计博物馆评审团列于世界最佳住宅；OUTrial 住宅以其独特的革新设计成为 2009 年度 LEAF 大奖的最终获得者；"安全住宅"于 2009 年成为获得建筑节奖的年度大奖等。

### KWK Promes

KWK Promes was established in 1999. A founder of KWK Promes – Robert Konieczny, was graduated from New Jersey Institute of Technology, six times nominee of the European Award of Mies van der Rohe Foundation.

Over the years, KWK Promes has received more and more appreciation. The office was listed among 44 best young architects of the world published by "Scalae". At the same year, the "Wallpaper" magazine issued the practice as one of the 101 most exciting architecture studios in the world.

Additionally, the International Jury of Museum of Architecture and Design Chicago put the Aatrial House and the Hidden House on the list of the best houses in the world. In 2009 the Safe House was in the World Architecture Festival Awards 2009 finale among the best realizations in the world and the OUTrial House was a finalist of LEAF Awards 2009 in Berlin for its innovatory design.

## Camenzind Evolution

得益于从尼古拉斯·格雷姆肖伦敦工作室和伦佐·皮亚诺巴黎工作室中学习和吸取的多年的国际项目经验，Stefan Camenzind 于 2005 年正式创立了 Camenzind Evolution 建筑设计事务所。从业的短短十余年间，Stefan Camenzind 及其领衔的事务所办公建筑、公共文化设施、商业项目、高档社区和老建筑的改造等诸多领域建树不凡。

Camenzind Evolution 建筑设计事务所秉持在理性中融入浪漫的设计哲学，这使他们能够以简单的和复合逻辑的方式来应对那些即使是最复杂的项目，并提出巧妙、和谐的解决方案。其设计的作品虽看似平淡，仔细品味却回味无穷。

### Camenzind Evolution

The architecture practice Camenzind Evolution was founded in 2004 by Stefan Camenzind and has grown to a team of 15 to 20 architects. Together with his business partner Tanya Ruegg, he recently achieved a number of internationally award-winning projects. These are characterized not only by the architecture, but also by their user-oriented development processes that support innovative approaches and endorse a consistent delivery.

## Tham & Videgård Arkitekter

Tham & Videgård Arkitekter 位于瑞典斯德哥尔摩，由联合创始人、总建筑师 Bolle Tham 和 Martin Videgård 领导，是一家侧重于建筑和设计的先进的现代化公司，包括从大型城市规划到建筑设计、室内设计和物体设计等。

公司的目标是以特定的环境和各个项目的条件为出发点，创造出独特的、合宜的建筑。从早期的绘图到后期场地监管，公司积极地投入到整个项目的全过程。在 Tham & Videgård Arkitekter 中，建筑师进行广泛的研究来结合各类直接的方案。这种工作方法形成创新性思维，以推动项目的进展，反过来促进之后项目的完成。

经过多年的发展，公司已有充足的能力来应对各类挑战，在瑞典以及国外还拥有包括政府、商业、私人等在内的众多客户，获得一系列重要奖项，如：

2008 年花园住宅获得瑞典南泰利耶区域最佳新建筑奖、2011 年 Tellus 托儿所获得斯德哥尔摩城市环境奖、2012 年中庭式住宅获得 ECOLA 二氧化碳最佳新建筑一等奖等。

### Tham & Videgård Arkitekter

Tham & Videgård Arkitekter is based in Stockholm, Sweden, and directed by co-founders and chief architects Bolle Tham and Martin Videgård. It is a progressive and contemporary practice that focuses on architecture and design – from large scale urban planning through to buildings, interiors and objects. The practice objective is to create distinct and relevant architecture with the starting point resting within the unique context and specific conditions of the individual project. Taking an active approach, the office is involved throughout the whole process, from developing the early sketch to the on-site supervision. Within the office architects combine straight forward solutions with extensive research. The method of work encourages innovative thinking to drive the development of the project, which in turn facilitates the subsequent realization within the logics of efficient contemporary production.

Over the years, T&V is experienced in dealing with various challenges. Commissions include public, commercial and private clients in Sweden and abroad. T&V has received a lot of awards including: Garden house is awarded the best new building in the Södertälje region, Sweden; Tellus Nursery School wins the Stockholm Urban Environment Award 2011; Atrium House wins the first prize for CO2-optimized New Buildings at the ECOLA Award 2012.

## J.MAYER H. Architects

J.MAYER H. Architects 1996 年成立于德国柏林，是一个聚焦建筑、新技术运用的多学科交叉的设计工作室。事务所的创立者和主要负责人是 Jürgen Mayer H.，他先后就读于德国斯图加特大学、柯柏联盟学院及普林斯顿大学。事务所的业务范围主要包括建筑设计和城市规划。事务所注重探讨新材料、新技术、人以及自然之间的关系。

### J. MAYER H. Architects

Founded in 1996 in Berlin, Germany, J. MAYER H Architects studio focuses on works at the intersection of architecture, communication and new technology. From urban planning schemes and buildings to installation work and objects with new materials, the relationship between the human body, technology and nature form the background for a new production of space.

Jürgen Mayer H. is the founder and principal of this cross-disciplinary studio. He studied at Stuttgart University, The Cooper Union and Princeton University. His work has been published and exhibited worldwide and is part of numerous collections including MoMA New York and SF MoMA. National and international awards include the Mies-van-der-Rohe-Award Emerging Architect Special Mention 2003, Winner Holcim Award Bronze 2005 and Winner Audi Urban Future Award 2010. Jürgen Mayer H. has taught at Princeton University, University of the Arts Berlin, Harvard University, Kunsthochschule Berlin, the Architectural Association in London, the Columbia University, New York and at the University of Toronto, Canada.

## Sid Lee Architecture

Sid Lee Architecture 是一家成立于 2009 年，与建筑公司 Nomade 合并的建筑及城市创造公司，位于加拿大蒙特利尔，在阿姆斯特丹有一家卫星办公室。

Sid Lee Architecture 拥有建筑师、城市设计师 Jean Pelland 和 Martin Leblanc 的天赋和技能，是一家完善的创造性机构。公司致力于通过将协同设计过程与媒体、科技应用的完美结合探索城市设计和建筑，善于挑战旧俗，产生了一种新型的设计知识与跨学科文化，使公司能进行从室内设计到城市规划的各类城市与建筑项目。

这一新的设计文化在完成各种规模的标志性项目中取得成功，体现了现代客户的需求和期望，表达了对客户的高度忠诚。这也正是 Sid Lee Architecture 不断走向成功的关键。

### Sid Lee Architecture

Sid Lee Architecture is an architectural and urban creativity firm, based in Montréal (Canada), with a satellite office in Amsterdam (Holland).

Founded in 2009 with the integration of the architectural firm Nomade, Sid Lee Architecture is the result of the combined talents and skills of architects and urban designers Jean Pelland and MartinLeblanc. It is a well-established creative agency, exploring urban design and architecture through an innovative combination of collaborative design processes, media and technology applications. Challenging conventionality generates a new kind of design knowledge and cross – disciplinary culture, empowering the firm to work on a large diversity of urban and architectural ventures, from interior design to urban planning.

This new design culture succeeds in delivering landmark projects of diverse sizes that embody the demands and expectations of today's clients. The firm has distinguished itself not only by its design excellence but also by high client fidelity.

## Studio Rodighiero Associati

凭借 20 年在建筑和工程领域的经验，Giovanni Rodighiero 于 1986 年建立了 Studio Rodighiero Associati，在其快速成长的过程中，事务所为公共机构和私人客户打造了众多大规模的项目，比如学校、住宅、商业和生产中心等。

1996 年，Studio Rodighiero Associati 在 Massimo、Francesco 以及 Giovanni 的积极合作下组建为一家咨询公司，在这一新的形态下，事务所为自己打造了一个开放的工作室，进行不同规模和主题的具体设计咨询。事务所汇聚了建筑和设计领域的人才，在制图、多媒体和艺术领域具备开放和创新的审美意识。另外，该事务所还具有一定的灵活性，可以从复杂的建筑和城市规划跨度到 Logo 设计或是客户独特身份打造等更细微的表达。近期，事务所获得了来自美国、西班牙和中国的合作伙伴。

### Studio Rodighiero Associati

Studio Rodighiero Associati was founded in 1986 by Giovanni Rodighiero, following a two decade experience in the fields of architecture and engineering. In a period of heightened growth, the studio produced large-scale projects both for public entities and private clients such as schools, residences, commercial and productive centers.

In 1996 Studio Rodighiero Associati is reborn as a consulting firm thanks to the active collaboration of Massimo, Francesco and Giovanni. In its new form the studio offers itself as an open workshop for specific design consultations spanning inter-disciplinarily in scale and theme of activity. With its fusion of multiple talents in the architectural and design fields, but also with an open and innovative sense of aesthetic in areas of graphics, multimedia and art, Studio Rodighiero Associati is able to add to its solid matrix of architectural and engineering projects a layer of novelty including graphic design, packaging, and website and product design. The firm denotes a supple flexibility in its ability to span from the complexity of an architectural and urbanistic spectrum to that of a yet more delicate expression such as the design of a logo or the creation of a client's unique identity. The studio has recently achieved partnerships in the U.S.A., Spain and China.

## Tonkin Liu

Tonkin Liu 是一家一流的建筑事务所，于 2002 年由 Mike Tonkin 和 Anna Liu 共同创立而成。其业务范围包括建筑设计、艺术设计和景观设计，同时，致力于为前瞻性的客户提供与项目场地、使用群体以及当地文化相协调的设计。该事务所认为，每一个项目都是人类与自然关系的具体体现，对自然的深入观察和了解，往往能够激发出一些开创性的设计和施工技术，因此，在设计过程中，他们通过对自然的关注来展示仿生学的价值以及对自然元素的大胆运用。

### Tonkin Liu

Tonkin Liu, established by Mike Tonkin and Anna Liu in 2002, is an award-winning architectural practice whose work encompasses architecture, art and landscape. It provides forward-thinking clients with designs that are finely tuned to the project sites, the people who will occupy it, and the culture that surrounds it at the time. The company believes that each project embodies the relationship between man and nature. Further research and observation of the nature may inspire pioneering construction techniques. During the design process, the company boldly uses biomimicry and the elements of nature in its projects.

## Office for Metropolitan Architecture

Office for Metropolitan Architecture 是一家顶级的国际化合伙人公司，专注于当代建筑设计、城市规划、文化分析。公司由 Rem Koolhaas、Ellen van Loon、Reinier de Graaf、Shohei Shigematsu、Iyad Alsaka、David Gianotten，6 名合伙人共同建立，并在鹿特丹、纽约、北京、香港、多哈均设有办事处。

公司目前所设计的在建建筑主要包括台北表演艺术中心、北京电视文化中心、深圳证券交易所。另外，已经完成的项目主要有：中国中央电视台总部新大楼、荷兰驻德柏林大使馆、西雅图中央图书馆、芝加哥伊利诺理工学院校园中心等。

一直以来，公司作品就广受社会各界青睐，获得了一系列国际大奖，主要包括：2000 年普利兹克建筑奖、2003 年日本皇家世界文化奖、2004 年英国皇家建筑师协会奖、2005 年欧盟当代建筑密斯·凡德罗奖、2010 年威尼斯双年展终身成就金狮奖等。

### Office for Metropolitan Architecture

Office for Metropolitan Architecture (OMA) is a leading international partnership practicing architecture, urbanism, and cultural analysis. OMA is led by six partners - Rem Koolhaas, Ellen van Loon, Reinier de Graaf, Shohei Shigematsu, Iyad Alsaka, and David Gianotten - and sustains an international practice with offices in Rottrdam, New York, Beijing, Hongkong and Doha, etc.

The buildings under construction by OMA include the Taipei Performing Arts Centre; the Television Cultural Centre in Beijing; Shenzhen Stock Exchange. OMA's recently completed projects include the headquarters for China Central Television, the Netherlands Embassy in Berlin, the Seattle Central Library and the IIT Campus Center in Chicago, etc.

The work of Rem Koolhaas and OMA has won several international awards including the Pritzker Architecture Prize in 2000, the Praemium Imperiale (Japan) in 2003, the RIBA Gold Medal (UK) in 2004, the Mies van der Rohe - European Union Prize for Contemporary Architecture (2005) and the Golden Lion for Lifetime Achievement at the 2010 Venice Biennale.

## Synthesis Design + Architecture

Synthesis Design + Architecture 是由 Alvin Huang 创建的一家新兴的现代设计公司，在建筑设计、基础设施建设、室内设计、装置设计、展览设计、家具设计等领域积累了多年的专业实践经验，其卓越的设计工作已获得国际认可。该团队由多学科的专业设计人员组成，包括注册设计师和建筑设计者，以及在美国、英国、丹麦、葡萄牙、台湾接受了专业性教育与培训的计算机专家。

Synthesis Design + Architecture 也是一家极具前瞻性的国际性设计公司，其设计统筹了性能、技术和工艺之间的关系，平衡了现实与想象之间的差距，以实际、实用的手法实现超凡的设计。

### Synthesis Design + Architecture

Synthesis is an emerging contemporary design practice with collective professional experience in the fields of architecture, infrastructure, interiors, installations, exhibitions, furniture, and product design. The firm's work has already begun to achieve international recognition for its design excellence. Its diverse team of multidisciplinary design professionals includes registered architects, architectural designers and computational specialists educated, trained, and raised in the USA, UK, Denmark, Portugal, and Taiwan. This diverse cultural and disciplinary background has supported our expanding portfolio of international projects in the USA, UK, Russia, Thailand, and China.

## Plajer & Franz Studio

经过十多年的创造性合作，Alexander Plajer 和 Werner Franz 在 1996 年成立了 Plajer & Franz 设计工作室，开展了大量令人印象深刻的设计作品，并拥有国际化的客户群。零售业的品牌构造和企业标识开发、高级酒店及度假村的设计成为该工作室的专业核心。

Plajer & Franz Studio 的独特之处在于其不断地探索和跨领域、跨学科精神。正是因为他们将各个领域所汲取的经验加以运用，才能完成优美、细腻的创新性设计。工作室从项目的概念孕育到设计、管理都是由精选的 50 名建筑师、室内设计师和绘图设计师团队实施完成，对细节的注重、强大的规划技能以及完美的风格刻画，使其在国际上享有盛誉。

### Plajer & Franz Studio

In over a decade of creative and imaginative partnership, Plajer & Franz, founded in 1996 by architects Alexander Plajer and Werner Franz, has built up an impressively broad-ranging portfolio with an international client base. The development of brand architecture and corporate identity in retail as well as the design of premium hotels and resorts form the core of their expertise.

The key to Plajer & Franz's freshness of vision lies in their continuous exploration and cross fertilization between disciplines and areas of experience. Their ability to deliver show innovative design with elegant and meticulous finishing lies in being able to take what they learn in one area and applying it. At Plajer & Franz, all project stages, from concept to design as well as roll-out supervision, are carried out in-house by the hand-picked team of 50 architects, interior designers and graphic designers. Plajer & Franz has an international reputation for innovative excellence, quality down to the smallest detail, great planning skills and a superb sense of style.

## 三磊建筑设计有限公司

三磊建筑设计有限公司是一家拥有建筑工程甲级设计资质、城乡规划乙级资质、国家高新技术企业认证、ISO9001 质量体系认证的综合性设计咨询服务机构。三磊设计注重团队合作，具备扎实的工程技术经验，可在公共建筑、都市与住区、结构设计与咨询、机电设计与咨询等领域为客户提供创造性和专业性的服务。

三磊善于对各个项目的功能、成本、市场、营销、文化、文脉、运营、可持续发展等因素进行整合思考，凭借专业能力，在业主利益、使用者利益与公共利益之间建立沟通桥梁，努力寻求完美的解决方案。为实现这一目标，三磊在所有项目的规划设计中建立了整合资源、以人为本、锐意创新和可持续发展的四项基本原则。

### Sunlay Design

Sunlay Design is a comprehensive design consulting firm possessing Class A design qualification of construction engineering, Class B qualification of town & country planning, national qualification of high and new tech enterprise, ISO-9001 Quality System Certification. Sunlay Design emphasizes on team cooperation. With solid engineering technology, it can offer creative professional services for clients on public buildings, urban & residential area, structural design & consultation, electromechanical design & consultation and etc.

Sunlay Design is good at integrating different factors at each project: cost, market, marketing, culture, context, operation, sustainable development and reaching comprehensive balance. By virtue of professional competence, it aims to build the communication bridge between the interests of the owners, users and public interests, and to seek the perfect solution. To achieve this goal, Sunlay Design establishes the four cardinal principles in the planning and design of all projects. They are integration of resources, to be people-oriented, innovation and sustainable development.

## Tony Owen Partners

2004 年，在商业建筑和住宅项目方面有着丰富经验的 NDM 与有设计天赋的 Tony Owen 合作，建立了 Tony Owen Partners。这是一家新兴的建筑事务所，其设计师将先进的设计与可持续性原则和商业价值结合起来，以构建挑战常规的、可实践的建筑。事务所采用最新的 2D CAD 技术，并致力于推进 3D 建模和可视化软件的发展，在建筑设计、城市规划和室内设计等领域获得发展。

其主要作品有：悉尼多佛海茨区的莫比斯住宅、Eliza 豪华公寓、比尔私人住宅、波士顿大学学生宿舍、哈雷戴维森总部、Paramount 酒店、Fractal 咖啡厅、奥斯陆歌剧院、阿布扎比女士俱乐部、NSW 教师联盟、波浪住宅、坎特伯雷城市中心等。

### Tony Owen Partners

Tony Owen Partners was formed in 2004 when NDM, with 10 years of commercial and residential experience combined with the awarded design talents of Tony Owen. Since that time Tony Owen Partners has grown rapidly to be an emerging mid-sized practice focusing on an idea based approach to commercial projects. This combination combines a genuinely progressive approach to design with a firm with a strong track record in proven documentation, management and deliverability.

Tony Owen Partners offer Architectural, Interiors and Urban Planning services. Tony Owen Partners currently has a full time staff of 20 including 15 architectural staff, 3 interior designers and in house 3-D rendering facilities. Tony Owen Partners utilizes the latest 2D CAD technology and continues to push the boundaries in 3D Modelling and visualization software. Tony Owen Partners's core capabilities include: progressive design that sets projects apart; design based on sound commercial and sustainable principles; a proven track record in deliverability; proven results with authorities.

## de Architekten Cie.

de Architekten Cie. 是一家全球性的建筑设计公司，具有 30 多年的建设设计和规划经验。其主要业务范围包括总体规划、城市规划、建筑设计和室内设计。

公司的主要创始人 Pi de Bruijn 1967 年毕业于代尔夫特理工大学建筑学院，其后分别在伦敦萨瑟克区的伦敦市委员会建筑系和阿姆斯特丹市政房屋署工作，并于 1978 年成为 Oyevaar Van Gool De Bruijn 建筑事务所 BNA 办事处的合伙人。1988 年，Pi de Bruijn 和 Frits van Dongen、Carel Weeber、Jan Dirk Peereboom Voller 共同建立了 Branimir Medić & Pero Puljiz, de Architekten Cie.。

### de Architekten Cie.

de Architekten Cie. is a global architectural design company with over 30 years of architectural design and planning experience. The business scope covers overall planning, urban planning, architectural design and interior design. Pi de Bruijn, the chief founder of the company, completed his studies at the Faculty of Architecture at Delft University of Technology in 1967. He then left to work at the Architects Department of the London City Council in Southwark, London. On his return to Amsterdam, he worked for the Municipal Housing Department, until he established himself as an independent architect in 1978, as a partner in the Oyevaar Van Gool De Bruijn Architecten BNA bureau. In 1988 he founded de Architekten Cie. together with Frits van Dongen, Carel Weeber and Jan Dirk Peereboom Voller, and has been a partner ever since.

## Mark Dziewulski Architect

Mark Dziewulski Architect 建立近 20 年以来，积累了丰富的经验，在业界获得了广泛的好评。事务所在美国、欧洲和亚洲都有项目，其作品获得多项建筑大奖，包括美国建筑师学会优秀建筑设计奖、国际设计与开发奖、太平洋海岸建筑设计奖。其作品也多次在《泰晤士报》《纽约时报》等杂志上发表。

事务所有着独特的设计风格，在公共领域设计了众多标志性建筑。其设计是对戏剧性建筑设计的严谨表达，同时也是对环境、对使用者的关怀。他们采用革新技术和诗意的、令人印象深刻的形态来营造强烈的场所感，赋予建筑独特而个性的魅力。

事务所的主要作品有：北京联合国贸易总部、波兰克拉科夫航空航天博物馆、中国宝安市政大厅、加利福尼亚 TRIANGLE 演艺中心、加利福尼亚克罗克文化艺术中心、伦敦国家美术馆和特拉法加广场等。

### Mark Dziewulski Architect

The office of Mark Dziewulski Architect has grown significantly in experience and reputation since it was established nearly twenty years ago. They carry out projects in the USA, Europe and Asia. They have received many design awards, including three Excellence in Design awards from the American Institute of Architects for "Significant Works of Architecture", as well as the International Design and Development Award and the Pacific Coast Builders Design Award two years in a row. Their work includes important cultural and institutional buildings as well as commercial and residential projects. Increasingly recognized as a design leader, their buildings have been published in over a hundred books and magazines in fourteen countries, including The Times in London and several articles in The New York Times.

The office was set up with the aim to build architecturally significant projects, creating iconic buildings in the public realm. The designs are intended to be a rigorous expression of dramatic architectural design, whilst also being sensitive to their context and users. The projects combine the use of technological innovation with poetic and sculptural forms to create a strong sense of "place". Mark Dziewulski is personally involved with each project from start to completion, so that the work reflects an individual and distinctive design sensibility. Each new commission receives a fresh approach based on its contextual and programmatic influences. The office is committed to environmentally sensitive construction and has created several technologically innovative buildings.

## Neil M. Denari Architects

Neil M. Denari Architects 是一家美国洛杉矶的建筑事务所，致力于建筑设计、城市设计、室内设计、景观设计、制图设计以及全球文化现象的探索等多重领域。

事务所自 1998 年成立以来，已横跨多个大洲工作，为多种客户进行多样化设计。事务所开展的项目旨在寻求创新的方法来应对当今世界所面临的复杂问题，不管是一件家具设计还是城市设计规划，事务所始终坚持以强有力、能产生共鸣的、实用的方式来解决这些问题。

该事务所的工作在过去的 15 年中已成为当代建筑文化中一股强大的影响力。如今，事务所仍旧继续开发新项目，以此展现其对创新设计和卓越建筑物的执着，例如：Alan-Voo 住宅、Endeavor 人才机构、日本三菱联合金融集团、纽约 HL23 公寓塔等。

近年来，Neil M. Denari Architects 已 2 次获得美国建筑师协会国家级大奖、8 次美国建筑师协会洛杉矶分奖，还在 2005 年获得先进建筑奖。其建筑作品多次在洛杉矶、旧金山、纽约现代艺术博物馆以及东京 Mori 博物馆等地展示，还曾在 2008 年被列入 Taschen 当代建筑百科全书。

### Neil M. Denari Architects

Neil M. Denari Architects (NMDA) is a Los Angeles based office dedicated to exploring urban design, building design, interiors, landscape design, graphic design and global cultural phenomenon.

Neil M. Denari Architects has been working across multiple continents since 1988, designing at all scales for a variety of clients and conditions. The office seeks out projects that demand new and innovative solutions to the complex issues facing the world today. Whether in a piece of furniture or in an urban design plan, ambitions of the office always are to materialize these questions in a powerful, evocative, and functional way.

The work of the office has become, over the last 15 years, an influential force in the culture of contemporary architecture. Today, NMDA continues to develop through new projects to demonstrate their commitment to design innovation and construction excellence, such as the Alan-Voo Residence, the Endeavor Talent Agency, built projects in Japan for the Mitsubishi United Financial Group, and for HL23 in New York, a 14 story residential tower.

In the past five years, NMDA has won 2 National AIA Honor awards, 8 LA Chapter AIA Honor awards, and a 2005 Progressive Architecture Award. Their work has been exhibited at the MOCA Los Angeles, San Francisco MOMA, MOMA New York, and the Mori Museum in Tokyo. In 2008, Neil Denari was included in Taschen's Encyclopedia of Modern Architecture.

## SAREA Alain Sarfati Architecture

SAREA Alain Sarfati Architecture 是一家专业从事项目设计到完工阶段全程领导的建筑工程公司，并致力于城市规划，由 Alain Sarfati 在 1983 年创立于巴黎。

Alain Sarfat 既是建筑师，还是城市规划师，从 SAREA 成立以来，就一直担任公司负责人。法国国家骑士勋章、法国国家军官勋章得主，法国国家住宅大奖得主，法兰西学院奖得主的 Alain Sarfat 的设计原则是对未来使用者的考虑，用多样的方法来应对不同的建筑环境，在其每项设计中，都用最具创新性、最合意的方式来确保建筑所起到的社会、文化、艺术和科技的作用。

经过多年的发展，公司的作品广受各界好评，并获得众多国际奖项。

### SAREA Alain Sarfati Architecture

SAREA is a building engineering company specialized in leading projects from the designing stage through to the completion of the project, and in urban planning, established by Alain Sarfati in 1983.

Alain Sarfati is both an architect and a town planner. He is Officer of the National Order of Arts and Letters, Knight of the National Order of Merit, and Knight of the National Order of the Legion of Honor. He has been awarded prizes by the Institute of France and is one of the top prizewinners in the National Housing Awards. Alain Sarfati's guiding principle is always to think of the future users. He responds to the diversity of situations with a diversity of approaches and shapes, and he seeks untiringly to ensure that architecture plays a social, cultural, artistic and technical role in his search for an innovative, well-adapted solution for each of the design he creates.

Over the years, the work of SAREA Alain Sarfati Architecture has received wide appreciation and international awards.

## Moore Ruble Yudell Architects & Planners

30 多年以前，Charles Moore、John Ruble、Buzz Yudell 创立了 Moore Ruble Yudell Architects & Planners。他们三人热于探索基于场所与当地居民之间的对话而衍生的新建筑，关注人文环境，鼓励社区群体间的交流和联系，这些价值观也成为了该公司在建筑规划与设计时的核心理念。

作为一家拥有近 60 名成员的建筑事务所，该公司强调成员间的合作与交流，同时不断引入各类型的资深设计师和工程顾问，以灵活、高效地为不同难度的各类型项目提供解决方案。

### Moore Ruble Yudell Architects & Planners

The founding partners – Charles Moore, John Ruble, and Buzz Yudell established Moore Ruble Yudell Architects & Planners over thirty years ago. They share a passion for an original architecture that grows out of an intense dialogue with places and people, celebrates human activity, enhances and nurtures community. These values continue to guide their process, providing the core principles for a wide-ranging exploration of planning and architecture.

With an office of some sixty people, Moore Ruble Yudell has been able to meet the challenges of complex programs and contemporary project delivery, while maintaining a close involvement by the partners. As the practice has expanded – both programmatically and geographically – it has grown in its technical capability and its skill in leveraging the multiple talents of the firm, often bringing in increasingly specialized design and engineering consultants. Moore Ruble Yudell has formed successful national and international alliances, gaining flexibility to effectively address the needs of large and small-scale projects with a global reach.

## UN Studio

UN Studio 由本·范·伯克尔和卡洛琳·博斯于 1988 年组建，是一家专门从事建筑设计、城市开发和基础工程建设的建筑设计事务所。公司名 UN Studio 代表的是 United Network Studio，强调团体的协作与配合。事务所致力于在设计、技术、专业知识和管理等方面不断提升自身质量，以在建筑领域做出应有的贡献。

2009 年亚洲 UN Studio 建立，其第一个办事处设在中国上海，该办事处由最开始致力于杭州来福士广场项目的设计，逐渐扩展成为一个全方位服务的设计公司，有着全面、专业的跨国建筑师团队。

UN Studio 设计的作品是环境可持续发展、市场需求与客户意愿的完美结合，其主要作品有：杭州来福士广场、Karbouw、Remu 发电站、Villa Wilbrink、伊拉斯谟斯大桥、摩比斯宅邸、Het Valkhof 博物馆、Prince Claus 大桥、梅赛德斯奔驰博物馆、Arnhem 中央火车站等。

### UNStudio

UNStudio, founded in 1988 by Ben van Berkel and Caroline Bos, is a Dutch architectural design studio specializing in architecture, urban development and infrastructural projects. The name, UNStudio, stands for United Network Studio, referring to the collaborative nature of the practice. In 2009 UNStudio Asia was established, with its first office located in Shanghai, China. UNStudio Asia is a full daughter of UNStudio and is intricately connected to UNStudio Amsterdam. Initially serving to facilitate the design process for the Raffles City project in Hangzhou, UNStudio Asia has expanded into a full-service design office with a multinational team of all-round and specialist architects.

# 公司简介 Company Profile

## NL Architects

NL Architects 总部位于阿姆斯特丹，由 Pieter Bannenberg、Walter van Dijk、Kamiel Klaasse 和 Mark Linnemann 于 1997 年建立。在 NL 看来，当今的建筑与郊区存在的问题及发展策略是密不可分的，这也是他们构建创新、前瞻性项目的关键。NL 认为建筑是以一种实验性形式展现的活动，融合了经济、文化、技术和环境等因素，城市则是一个平衡了经济发展与环境保护的生态系统。

事务所的精选项目包括格罗宁根广场、阿姆斯特丹菲英 K 街区、韩国环形住宅、赞斯塔德 A8ernA、乌特勒支篮球吧、乌特勒支 WOS8。其获得的奖项包括：2006 年城市公共空间欧洲地区大奖；2005 年密斯·凡·德罗大奖；2004 年 N.A.I 大奖；2003 年里特维德奖以及 2001 年鹿特丹设计奖。

## NL Architects

Amsterdam-based NL Architects was founded in 1997 by Pieter Bannenberg (1959), Walter van Dijk (1962), Kamiel Klaasse (1967) and Mark Linnemann (1963).

In the view of NL, architecture today cannot be separated from suburban issues and strategies, which is their key to develop innovative and forward-looking devices and arrangements. Architecture is presented as a field of experimental activity, at the crossroads of economic, programmatic, technical and environmental thinking. The city is perceived as an ecosystem, an environment where logical systems of urban growth and natural factors, consumption and production, flux and stasis are balanced; and where recycling is set up as a method of stable and sustainable functioning, whether it has to do with waste, energy, materials or even architecture.

Its selected works include Groninger Forum (Groningen, 2006–2011), Funen Blok K (Amsterdam, 2010), Loop House (Korea, 2006), A8ernA (Zaanstad, 2006), Basket Bar (Utrecht, 2003), WOS 8 (Utrecht, 1997).

NL Architects participated in many exhibitions around the world: Out There, Architecture Biennale Venice (Venice, 2008), New Trends of Architecture in Europe and Asia-Pacific (Shanghai, 2007), Sign as Surface (traveling Exhibition USA, 2003), Fresh Facts, Biennale Venice (Venice, 2002), NL Lounge, Dutch Pavilion Biennale Venice (Venice, 2000).

Among the numerous awards there are the European Price for Urban Public Space (2006), Mies van der Rohe award (2005), N.A.I award (2004), Rietveld Price (2003), and the Rotterdam design price (2001).

## Zecc Architecten

Zecc Architecten 以"规划、建筑技术、美观"为三大主题，是一家拥有成熟专业团队的建筑事务所。

公司注重建筑细节的刻画，客户群涉及企业、小型企业主、开发商、社会住宅以及政府组织等，非常广泛。一直以来，公司都善于倾听客户的想法，将客户的问题作为任务的基本目标，旨在通过与客户及建造团队的交流创造出整体式设计，同时又尊重项目所设定的条件。

除此之外，公司还开展改造项目，包括将学校建筑、教堂、水塔等改造成独特的住宅建筑。将建筑的功能性、可持续性、美观性结合起来是公司的力量核心，由此产生最精彩的设计理念——根越深，视野越开阔。

## Zecc Architecten

With the three topics "program", "construction technique" and "aesthetics", Zecc Architecten owns a highly developed expertise.

Zecc Architecten attaches great importance to fine details of architecture, serving a range of professional clients: corporations, small business owners, developers, organizations for social housing and the government. They listen to the clients and translate the questions of clients in a grounded vision on the tasks. Zecc aims to create an integral design through a dialogue with client or a construction team, while honoring the boundary conditions of the assignment.

Zecc also transforms former school buildings, churches and a water tower into unique residential homes. Zecc works from three ground principles: what they build is functional, sustainable and engages all senses at the same time. This is the core of Zecc's strength. Since these bring forth the most wonderful ideas: "The deeper the roots, the higher the vision."

## Za Bor Architects

Za Bor Architects 由 Arseniy Borisenko 和 Peter Zaytsev 于 2003 年创立。事务所将现代美学思想融入到建筑设计及室内设计中，其丰富的设计手法以及复杂动感的建筑形态使设计作品脱颖而出，并成为事务所的标志性风格。事务所的室内设计尤其突出地展示出事务所的设计风格，其室内的内嵌式和独立式家具皆出自建筑师之手。

Za Bor Architects 参与过六十多个项目的设计，其中包括二十多个住宅项目、一个办公大楼、一个木屋定居项目还有二十多个办公室项目。其主要作品有：蚂蚁——Pushkinskiy 电影院、莫斯科寄生办公室、莫斯科夏季露台、俄罗斯"Mr.R"住宅、Yandex 现代办公空间等。事务所的客户涉及媒体、商业、政府、公司等多个领域，其中包括 Forward Media Group、Yandex、Inter RAO UES 以及莫斯科工商总局等机构。

## Za Bor Architects

Za Bor Architects is a Moscow based architectural office founded in 2003 by principals Arseniy Borisenko and Peter Zaytsev.

The bureau's objects are created mainly in contemporary aesthetics. What distinguishes them is an abundance of architectural methods used both in the architectural and interior design, as well as a complex dynamical shape which is a visiting card of Za Bor projects. The interiors demonstrate this feature especially brightly, since for all their objects architects create built-in and free standing furniture themselves. Many conceptual and implemented design-projects of Za Bor Architects were awarded at international and Russian exhibitions and competitions. At the moment Za Bor Architects is involved in variety of projects at several countries.

Za Bor Architects have been involved in more than 60 projects including residential houses, an office building, a cottage settlement, and many offices. Among the clients of Za Bor Architects there are media, business and government companies such as Forward Media Group, Yandex, Inter RAO UES, Moscow Chamber of Commerce and Industry and others.

## 加拿大 C.P.C. 建筑设计顾问有限公司

加拿大 C.P.C. 建筑设计顾问有限公司于 1994 年在加拿大温哥华成立，1995 年公司开始进入中国市场，为政府及房地产开发商提供城市规划、高端住宅设计、商业项目策划及设计、公共建筑设计、景观设计、室内设计等全方位的设计服务。随着工程项目的日益增加，2001 年 CPC 公司将中国的办事处置于上海，以便更好地为业主提供即时的服务。

其主要作品有：深圳国际会展中心、杭州文化广场、苏州金鸡湖酒店、大连大学城、北京万柳购物中心、外滩 15#、上海嘉定新城、上海浦江智谷工业园区规划及建筑设计、上海宝山顾村规划等。

其获得的主要奖项和荣誉包括：2012 年梅陇镇西地块配套商品房荣获"房型设计奖"；2010 年都江堰友爱学校重建项目荣获年度工程勘察设计"四优"一等奖；2005 年"古北瑞仕花园"项目荣获全国人居经典建筑规划设计方案竞赛"规划、建筑双金奖"。

## The C.P.C. Group

The Coast Palisade Consulting Group (C.P.C. Group) is an architectural design firm operating in Vancouver and Shanghai. The firm specializes in architectural design, planning, urban design and interior design. Most members of the firm have more than one university degree in architecture or related fields such as planning and urban design.

The C.P.C. Group offers clients creative design solutions that are market and user responsive and feasible to construct from both a technical and budget perspective. The company consists of a core group of long-term staff and a large pool of designers, planners, computer technologists, model makers, etc.

## C.F.Møller Architects

C.F.Møller Architects 是斯堪的纳维亚半岛历史最悠久、规模最大的建筑机构之一，业务范围涵盖方案分析、城市规划、总体规划、景观设计、建筑工程设计及其他诸多领域。事务所于 1924 年创立，以简约、明快、朴素的理念指导各项实践。并根据每个项目的基址特点，结合国际发展趋势和地域差异对理念进行重新解读和诠释。

事务所以改革和创新为发展理念，力图打造独具吸引力和发展前景的工作环境，使每位员工都能接受高要求设计项目的挑战。多年来，事务所屡获国内外设计大奖，其作品多次在国内外的展会上展出。其主要作品有：法尔斯特岛新封闭式州立监狱、奥尔胡斯 Incuba 科学公园、奥尔胡斯大学礼堂、奥尔胡斯艺术大楼扩建项目、奥尔胡斯低能耗办公大楼、国家海事博物馆扩建、巴里考古博物馆等。

## C.F.Møller Architects

C.F.Møller Architects is one of Scandinavia's oldest and largest architectural practices. Its award-winning work involves a wide range of expertise that covers all architectural services, landscape architecture, product design, healthcare planning and management advice on user consultation, change management, space planning, logistics, client consultancy and organizational development. Simplicity, clarity and unpretentiousness, the ideals that have guided its work since the practice was established in 1924, are continually re-interpreted to suit individual projects, always site-specific and based on international trends and regional characteristics.

C.F. Møller regards environmental concerns, resource-consciousness, healthy project finances, social responsibility and good craftsmanship as essential elements in its work, and this holistic view is fundamental to all its projects, all the way from master plans to the design.

Today C.F. Møller has about 320 employees. The head office is in Aarhus, Denmark and it has branches in Copenhagen, Aalborg, Oslo, Stockholm and London.

## Philippe SAMYN and PARTNERS, architects & engineers

Philippe SAMYN and PARTNERS, architects & engineers 是一家由 Ir Philippe SAMYN 博士领导的私人公司。随着其附属公司 Ingenieursbureau Jan MEIJER、FTI、DAE、AirSR 的相继建立，该公司也在建筑设计和建筑工程各领域表现得极为活跃。其业务范围涵盖了规划设计、城市规划、景观设计、建筑设计、室内设计、建筑物理、MEP 和工程结构、工程建设管理、成本规划与控制、工程造价管理等多方面。

Philippe SAMYN and PARTNERS, architects & engineers 的设计方案建立在"质疑"的基础上,可用"为什么"理念来概括。该公司尝试着接手各种类型的项目,并悉心听取客户的意见和需求。

**Philippe SAMYN and PARTNERS, Architects & Engineers**

Philippe SAMYN and PARTNERS, Architects & Engineers is a private company owned by its partners and lead by its design partner Dr Ir Philippe Samyn. With the establishment of its affiliated companies Ingenieursbureau Jan MEIJER, FTI, DAE, and AirSR, it is active in all fields of architecture and building engineering. The firm's client services include Planning and Programming, Urban Planning, Landscaping and Architectural Design, Interior Design, Building Physics, MEP and Structural Engineering, Project and Construction Management, Cost and Planning Control, Quantity Surveying, Safety and Health Coordination.

Philippe Samyn's architectural and engineering design approach is based on questioning, which can be summarized as a "why" methodology. The firm approaches projects openly to all sorts of possibilities whilst listening closely to its clients' demands.

## HHD_FUN

HHD_FUN 是一个由年轻建筑师,设计师,程序设计师等组成的致力于设计和研究的事务所。参数化设计及可持续发展是他们的主要研究方向。他们致力于将不同领域的知识创造性地带入建筑设计,希望基由不同的设计方法和设计过程创造出有别于常规的"意料之外"的设计。数学、几何学、算法技术、建筑信息模型、电子学、人工智能学等都是他们涉猎的领域。事务所同时与包括艺术家、服装设计师、数学家、工程师等在内的各界人士进行跨界合作,并希望通过这样的合作探索新设计的可能性。

**HHD_FUN**

As the Beijing Branch of HHDesign, HHD_FUN is a design and research studio with interests in bringing knowledge from various fields outside of architecture and integrating these means into the design of architecture. Parametric design and sustainability is the main research direction of HHD_FUN. The mathematics, geometric principles, algorithms, BIM, Artificial Intelligence and etc. are one portion of their approaches as the means in architecture generation. They collaborate with artist, fashion designer, mathematician, engineer, etc. and seeing these as opportunities of exploring new possibilities of design.

## Forte, Gimenes & Marcondes Ferraz Arquitetos

Forte, Gimenes & Marcondes Ferraz Arquitetos (FGMF) 由 Forte, Gimenes & Marcondes Ferraz 成立于 1999 年,旨在创造无材料种类使用限制以及无建造技术限制的现代建筑,以探索建筑和人类的联系。

公司具有丰富的职业及学术经验,在每个项目中都追求创造性,为客户提供独具特色的设计方案。在 FGME,没有死板、定式的设计方程式,面对每一项挑战,都从零开始,利用绘画作为研究工具,创造出城市中新的建筑物蓝图。

**Forte, Gimenes & Marcondes Ferraz Arquitetos**

Founded in 1999 by fellow students from FAU-USP, Forte, Gimenes & Marcondes Ferraz (FGMF) was conceived with the cause of making a contemporary architecture, without any restraints regarding the use of varied material and building techniques, exploring the connection between architecture and Man.

Based on the professional and academic experience of its partners, FGMF pursuits an innovative and distinct approach at every project proposed. There are no rigid or preconceived formulas: at each challenge, the firm starts from scratch, using the drawing as out research tool for the elaboration of a new conceit for a building, an object, in a city.